Dynamics – Formulas and Problems

Dietmar Gross · Wolfgang Ehlers
Peter Wriggers · Jörg Schröder
Ralf Müller

Dynamics – Formulas and Problems

Engineering Mechanics 3

Springer

Dietmar Gross
Division of Solid Mechanics
TU Darmstadt
Darmstadt
Germany

Jörg Schröder
Institute of Mechanics
Universität Duisburg-Essen
Essen
Germany

Wolfgang Ehlers
Institute of Applied Mechanics
Universität Stuttgart
Stuttgart
Germany

Ralf Müller
Engineering Mechanics
TU Kaiserslautern
Kaiserslautern
Germany

Peter Wriggers
Institute of Continuum Mechanics
Leibniz Universität Hannover
Hannover
Germany

ISBN 978-3-662-53436-6 ISBN 978-3-662-53437-3 (eBook)
DOI 10.1007/978-3-662-53437-3

Library of Congress Control Number: 2016951667

Printed on acid-free paper

This Springer imprint is published by Springer Nature
The registered company is Springer-Verlag GmbH Germany
The registered company address is: Heidelberger Platz 3, 14197 Berlin, Germany

Preface

This 3rd volume of the *Formulas and Problems* concludes the series to the basic courses in Engineering Mechanics.

Experience shows that the field of Dynamics is particularly difficult for students, because besides the concept of force now additional kinematic quantities occur, which must be brought into relation with each other and with the forces. Therefore, with numerous purely kinematical problems, we tried to deepen the understanding of the relevant geometric quantities and their description in different coordinate systems. Likewise, only by exercises, i.e. by an independent treatment of problems, one can gain experience, which basic principle leads to the solution in the simplest way. Often there are several approaches possible. Therefore we demonstrate this frequently so that the reader can realize the advantages and disadvantages of the alternatives.

As in the 1st and 2nd volume, we deliberately placed the emphasis on the principal way how to to apply the theory and not in numerical results. The correct formulation of the relevant basic equations and their solution is in the beginning much more important than numerical calculations without a deeper understanding of the background.

Experience also shows that it is an illusion to believe that simply reading and trying to comprehend the presented solutions leads to an understanding of the theory. Neither does it improve the problem solving skills. Therefore, we strongly recommend that the reader first tries to solve the problems independently, possibly by using other approaches. Let us emphasize that a collection of formulas and examples is only an additional aid when studying mechanics and it cannot replace a textbook. When the reader is not familiar with one or the other formula or concept, it is necessary to brush up the theory with the help of a textbook; a number of titles can be found in the list of references.

Darmstadt, Stuttgart, Hannover,
Essen and Kaiserslautern, Summer 2016

D. Gross
W. Ehlers
P. Wriggers
J. Schröder
R. Müller

Table of Contents

Literature

Textbooks

Gross, D., Hauger, W., Schröder, J., Wall, W., Govindjee, S., Engineering Mechanics 3, Dynamics, 2nd edition, Springer 2014

Gross, D., Hauger, W., Wriggers, P., Technische Mechanik, vol 4: Hydromechanics, Elements of Avanced Mechanics, Numerical Methods (in German), 9th edition, Springer 2014

Beatty, M.F., Principles of Engineering Mechanics, vol 2: Dynamics, Springer 2005

Beer, F., Johnston, E.R., Cornwell, P., Vector Mechanics for Engineers: Dynamics, 10th edition, McGraw-Hill Education 2012

Hibbeler, R.C., Engineering Mechanics: Dynamics, 14th edition. Pearson 2016

Meriam, J.L., Kraige, L.G., Bolton, J.N., Engineering Mechanics: Dynamics, 8th edition, Wiley 2016

Plesha, M., Costanzo, F., Gray, G., Engineering Mechanics: Dynamics, 2nd edition, McGraw-Hill 2012

Pytel, A., Kiusalaas, J., Engineering Mechanics: Dynamics, 4th edition, Cengage Learning 2016

Shames, I.H., Engineering Mechanics: Dynamics, 4th edition, Pearson 1996

Collection of Problems

Beer, F., Johnston, E.R., Cornwell, P., Vector Mechanics for Engineers: Dynamics, 10th edition, Solution Manual, McGraw-Hill 2012

Nelson, E.W., et al. Engineering Mechanics - Dynamics, 765 fully solved problems, Schaum's Outlines, McGraw-Hill Education 2010

Gray, G.L., Costanzo, F., Plesha, M.E., Solutions Manual, Engineering Mechanics: Dynamics, 1st edition, McGraw-Hill 2009

Hibbeler, R.C., Practice Problems Workbook, Engineering Mechanics: Dynamics, Pearson 2015

Notation

In the problem solutions the following symbols are used:

$\uparrow$: abbreviation for *equation of motion (impulse law) in direction of arrow.*

$\overset{\curvearrowleft}{A}$: abbreviation for *angular momentum theorem relative to point A with given positive rotation direction.*

$\leadsto$ abbreviation for *it follows.*

Chapter 1

Kinematics of a Point

1

The position of a point P in space is described by the **position vector**

$$\boldsymbol{r}(t)\,.$$

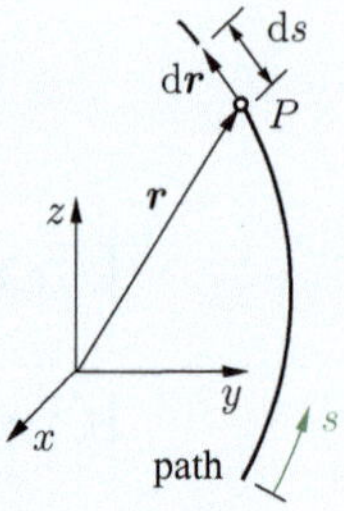

As P moves, its path is given by $\boldsymbol{r}(t)$.

From the displacement $\mathrm{d}\boldsymbol{r}$ of point P in a neighboring position during time $\mathrm{d}t$ follows its **velocity**

$$\boldsymbol{v} = \frac{\mathrm{d}\boldsymbol{r}}{\mathrm{d}t} = \dot{\boldsymbol{r}}\,.$$

The velocity is always tangent to the path (trajectory). With the arc–length s and $|\mathrm{d}\boldsymbol{r}| = \mathrm{d}s$ the speed of P is given by

$$v = \frac{\mathrm{d}s}{\mathrm{d}t} = \dot{s}\,.$$

The change of the velocity vector $\mathrm{d}\boldsymbol{v}(t)$ during time $\mathrm{d}t$ is called **acceleration**

$$\boldsymbol{a} = \frac{\mathrm{d}\boldsymbol{v}}{\mathrm{d}t} = \dot{\boldsymbol{v}} = \ddot{\boldsymbol{r}}\,.$$

The acceleration generally is *not* directed tangent to the path (trajectory)!

The vectors $\boldsymbol{r}$, $\boldsymbol{v}$ and $\boldsymbol{a}$ can be represented in different coordinate systems as follows:

a) Cartesian Coordinates with the unit vectors $\boldsymbol{e}_x$, $\boldsymbol{e}_y$, $\boldsymbol{e}_z$:

$$\boldsymbol{r} = x\,\boldsymbol{e}_x + y\,\boldsymbol{e}_y + z\,\boldsymbol{e}_z\,,$$

$$\boldsymbol{v} = \dot{x}\,\boldsymbol{e}_x + \dot{y}\,\boldsymbol{e}_y + \dot{z}\,\boldsymbol{e}_z\,,$$

$$\boldsymbol{a} = \ddot{x}\,\boldsymbol{e}_x + \ddot{y}\,\boldsymbol{e}_y + \ddot{z}\,\boldsymbol{e}_z\,.$$

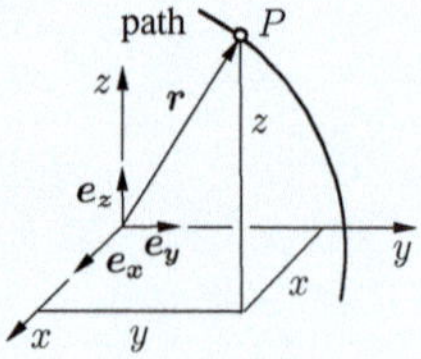

b) Cylindrical Coordinates with the unit vectors $\boldsymbol{e}_r$, $\boldsymbol{e}_\varphi$, $\boldsymbol{e}_z$:

$$\boldsymbol{r} = r\,\boldsymbol{e}_r + z\,\boldsymbol{e}_z\,,$$

$$\boldsymbol{v} = \dot{r}\,\boldsymbol{e}_r + r\dot{\varphi}\,\boldsymbol{e}_\varphi + \dot{z}\,\boldsymbol{e}_z\,,$$

$$\boldsymbol{a} = (\ddot{r} - r\dot{\varphi}^2)\,\boldsymbol{e}_r + (r\ddot{\varphi} + 2\dot{r}\dot{\varphi})\,\boldsymbol{e}_\varphi + \ddot{z}\,\boldsymbol{e}_z\,.$$

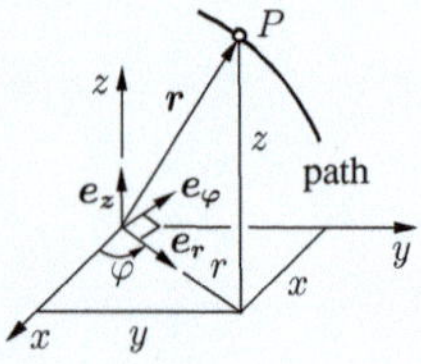

c) Serret-Frenet Frame with the unit vectors $\boldsymbol{e}_t$, $\boldsymbol{e}_n$, $\boldsymbol{e}_b$ in tangential, principal normal and binormal direction.

$$\boldsymbol{v} = v\,\boldsymbol{e}_t\,,$$

$$\boldsymbol{a} = \dot{v}\,\boldsymbol{e}_t + \frac{v^2}{\rho}\,\boldsymbol{e}_n\,.$$

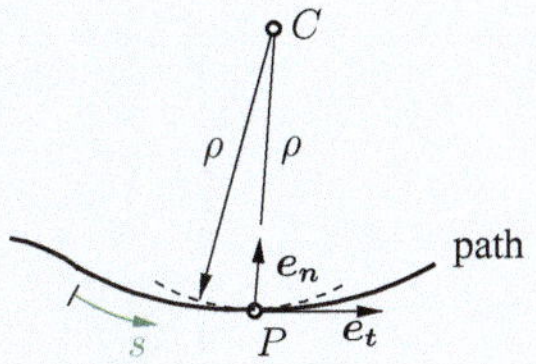

Here are:

ρ	=	radius of curvature (distance between P and center of curvature C),
$v = \dot{s} = \dfrac{\mathrm{d}s}{\mathrm{d}t}$	=	speed,
$a_t = \dot{v} = \dfrac{\mathrm{d}v}{\mathrm{d}t}$	=	tangential acceleration,
$a_n = \dfrac{v^2}{\rho}$	=	normal acceleration (centripetal acceleration).

Remarks: The two acceleration components a_t, a_n are located in the so-called *osculating plane*. The acceleration vector points always to the 'interior' of the path.

Rectilinear motion

Position $\quad x(t)\,,$

Velocity $\quad v = \dfrac{\mathrm{d}x}{\mathrm{d}t} = \dot{x}\,,$

Acceleration $\quad a = \dfrac{\mathrm{d}v}{\mathrm{d}t} = \dot{v} = \ddot{x}\,.$

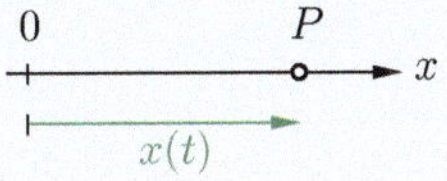

Circular motion ($r = \text{const}$)

Position $\quad s = r\varphi(t)\,,$

Velocity $\quad v = r\dot{\varphi} = r\omega\,,$

Tangential acceleration $\quad a_t = r\ddot{\varphi} = r\dot{\omega}\,,$

Centripetal acceleration $\quad a_n = \dfrac{v^2}{r} = r\omega^2$

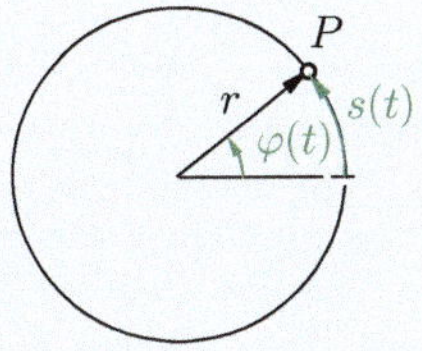

with $\omega = \dot{\varphi} =$ angular velocity.

Planar motion in polar coordinates

From the relations for cylindrical coordinates follow for $z = 0$, $\dot{\varphi} = \omega$

$$\boldsymbol{v} = v_r \boldsymbol{e}_r + v_\varphi \boldsymbol{e}_\varphi \,, \qquad \boldsymbol{a} = a_r \boldsymbol{e}_r + a_\varphi \boldsymbol{e}_\varphi$$

with

radial velocity	$v_r = \dot{r}\,,$
circular velocity	$v_\varphi = r\omega\,,$
radial acceleration	$a_r = \ddot{r} - r\omega^2\,,$
circular acceleration	$a_\varphi = r\dot{\omega} + 2\dot{r}\omega\,.$

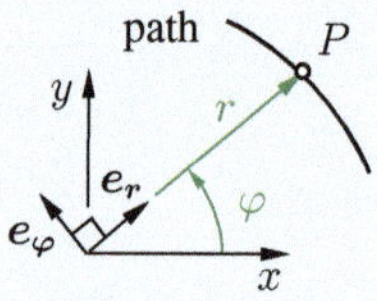

Remark: In case of a *central motion* the circular acceleration vanishes. From $a_\varphi = r\dot{\omega} + 2\dot{r}\omega = (r^2\omega)^{\cdot}/r = 0$ then follows the 'Law of Equal Areas' (KEPLER'S 2nd Law) $r^2\omega = \text{const}$.

Kinematic basic problems for a rectilinear motion

At initial time t_0 the initial position x_0 and initial velocity v_0 are assumed to be given.

Given	Sought
$a = 0$	$v = v_0 = \text{const}\,, \qquad x = x_0 + v_0 t$ *uniform motion*
$a = a_0 = \text{const}$	$v = v_0 + a_0 t\,, \qquad x = x_0 + v_0 t + \frac{1}{2} a_0 t^2$ *uniform acceleration*
$a = a(t)$	$v = v_0 + \int_{t_0}^{t} a(\bar{t})\mathrm{d}\bar{t}\,, \quad x = x_0 + \int_{t_0}^{t} v(\bar{t})\mathrm{d}\bar{t}$
$a = a(v)$	$t = t_0 + \int_{v_0}^{v} \frac{\mathrm{d}\bar{v}}{a(\bar{v})} = f(v)\,, \quad x = x_0 + \int_{t_0}^{t} F(\bar{t})\mathrm{d}\bar{t}$ with the inverse function $v = F(t)$
$a = a(x)$	$v^2 = v_0^2 + 2\int_{x_0}^{x} a(\bar{x})\mathrm{d}\bar{x}\,, \quad t = t_0 + \int_{x_0}^{x} \frac{\mathrm{d}\bar{x}}{v(\bar{x})} = g(x)$ the inverse function of $t = g(x)$ gives $x = G(t)$

Remarks:

- The relations above can also be used for a general motion by replacing x through s and a through the tangential acceleration a_t. The normal acceleration then follows from $a_n = v^2/\rho$.
- If the velocity is given as a function of the position, the acceleration is found from

$$a = v\frac{\mathrm{d}v}{\mathrm{d}x} = \frac{\mathrm{d}}{\mathrm{d}x}\left(\frac{v^2}{2}\right) .$$

Problem 1.1 The minimum distance b between two vehicles shall be as big as the distance which the rear vehicle covers within $t_S = 2\,\mathrm{s}$ at its constant velocity. **P1.1**

a) Determine the minimum distance x_p required for passing.

b) Determine the minimum time t_p a car (length $l_1 = 5\,\mathrm{m}$, constant speed $v_1 = 120\,\mathrm{km/h}$) needs staying on the fast lane for passing a truck (length $l_2 = 15\,\mathrm{m}$, speed $v_2 = 80\,\mathrm{km/h}$) correctly? Disregard the time for changing the lanes.

Solution

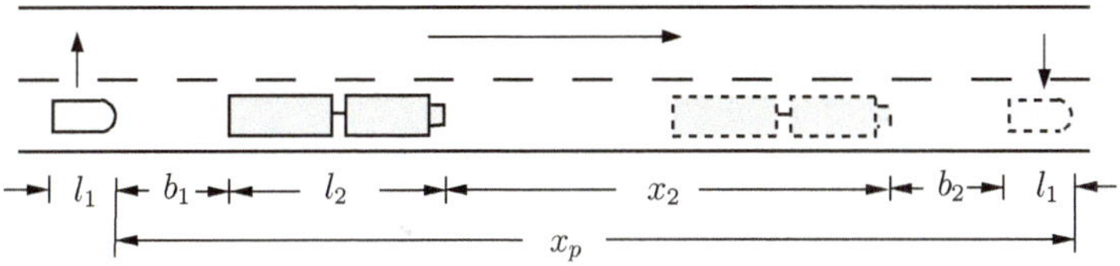

a) For uniform motion the minimum distances follow with $1\,\mathrm{km/h} = 1000\,\mathrm{m}/3600\,\mathrm{s}$ as

$$b_1 = v_1\, t_S = \frac{120}{3.6} \cdot 2 = \frac{200}{3}\,\mathrm{m}\,, \qquad b_2 = v_2\, t_S = \frac{80}{3.6} \cdot 2 = \frac{400}{9}\,\mathrm{m}\,.$$

Thus, the required distance for passing is given by

$$x_p = b_1 + l_2 + x_2 + b_2 + l_1\,.$$

Furthermore, the relations

$$x_2 = v_2\, t_p\,, \qquad x_p = v_1 t_p$$

hold. Elimination of t_p yields

$$\underline{\underline{x_p}} = \frac{b_1 + b_2 + l_1 + l_2}{1 - \frac{v_2}{v_1}} = \frac{\frac{200}{3} + \frac{400}{9} + 5 + 15}{1 - \frac{80}{120}} = \frac{1180}{3} = \underline{\underline{393,33\,\mathrm{m}}}\,.$$

b) Thus, the minimum time for passing is

$$\underline{\underline{t_p}} = \frac{x_p}{v_1} = \frac{1180 \cdot 3,6}{3 \cdot 120} = \underline{\underline{11,8\,\mathrm{s}}}\,.$$

P1.2 **Problem 1.2** To simulate absence of gravity, vacuum drop-shafts are used. Given is a shaft with a depth of $l = 200$ m.

Determine the maximum available test time t_1 and test distance x_1 during free fall, when the sample after passing the test distance is decelerated with $a_{II} = -50\,g$ to $v = 0$?

Solution Because the sample is released from rest ($x_0 = v_0 = 0$), during free fall with $a_I = \text{const} = g$, the velocity and position are

$$v_I = gt\,, \qquad x_I = \frac{g}{2}t^2\,.$$

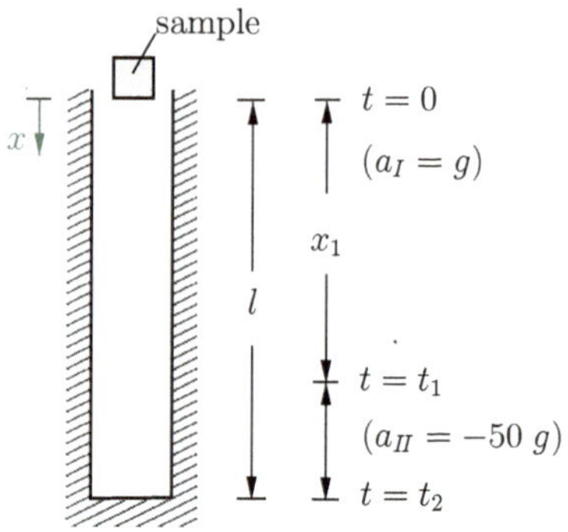

During the deceleration phase with $a_{II} = -50\,g$, velocity and position are given by

$$v_{II} = v_{II_0} - 50\,gt\,,$$
$$x_{II} = x_{II_0} + v_{II_0}\,t - 50\,gt^2/2\,.$$

It shall be noted that the integration constants x_{II_0} and v_{II_0} have no direct physical meaning.

For $t = t_2$ the following conditions must hold:

$$v_{II}(t_2) = 0 \quad \rightsquigarrow \quad v_{II_0} = 50\,gt_2\,,$$
$$x_{II}(t_2) = l \quad \rightsquigarrow \quad x_{II_0} = l - v_{II_0}\,t_2 + \frac{50}{2}\,gt_2^2 = l - 25\,gt_2^2\,.$$

From the transition conditions

$$v_I(t_1) = v_{II}(t_1) \quad \rightsquigarrow \quad gt_1 = 50\,g(t_2 - t_1)\,,$$
$$x_I(t_1) = x_{II}(t_1) \quad \rightsquigarrow \quad \frac{g}{2}\,t_1^2 = l - \frac{50}{2}\,gt_2^2 + 50\,gt_1t_2 - \frac{50}{2}\,gt_1^2$$

the test time t_1 and subsequently x_1 can be determined:

$$\underline{\underline{t_1}} = \sqrt{\frac{100\,l}{51\,g}} = \sqrt{\frac{100\cdot 200}{51\cdot 9.81}} = \underline{\underline{6.32\text{ s}}}\,,$$

$$\underline{\underline{x_1}} = x_I(t_1) = \frac{g}{2}\,t_1^2 = \frac{g}{2}\,\frac{100\,l}{51\,g} = \frac{50}{51}\,l = \underline{\underline{196\text{ m}}}\,.$$

Problem 1.3 Between 2 stations an underground covers a distance of 3 km. Given are the starting acceleration $a_a = 0.2\,\mathrm{m/s^2}$, the braking deceleration $a_d = -0.6\,\mathrm{m/s^2}$ and the maximum speed $v^* = 90\,\mathrm{km/h}$. P1.3

Determine the acceleration distance, the deceleration distance, the distance during uniform motion and the travel time.

Solution From the constant acceleration a_a within the starting phase the velocity follows as

$$v_a = a_a t\ .$$

With the given maximum speed we obtain the starting time

$$t_a = \frac{v^*}{a_a} = \frac{90 \cdot 1000}{3600 \cdot 0.2} = 125\,\mathrm{s}$$

and the acceleration distance

$$\underline{\underline{s_a}} = \frac{1}{2}\,a_a t_a^2 = \frac{1}{2} \cdot 0.2 \cdot 125^2 = \underline{\underline{1563\,\mathrm{m}}}\ .$$

During braking with constant deceleration a_d the velocity is given by

$$v_d = v^* + a_d t\ .$$

Thus, the time t_d until stop ($v_d = 0$) is

$$t_d = -\frac{v^*}{a_d} = -\frac{90 \cdot 1000}{3600 \cdot (-0.6)} = 41.67\,\mathrm{s}\ ,$$

and for the associated braking distance follows

$$\begin{aligned}\underline{\underline{s_d}} &= v^* t_d + \frac{1}{2}\,a_d t_d^2 = \frac{90 \cdot 1000}{3600} \cdot 41,67 - \frac{1}{2} \cdot 0.6 \cdot 41.67^2 \\ &= 1041.75 - 520.92 = \underline{\underline{521\,\mathrm{m}}}\ .\end{aligned}$$

For the phase with constant velocity v^* remains a distance of

$$\underline{\underline{s^*}} = 3000 - s_a - s_d = \underline{\underline{916\,\mathrm{m}}}$$

and an associated time

$$t^* = \frac{s^*}{v^*} = \frac{916 \cdot 3600}{90 \cdot 1000} = 36.64\,\mathrm{s}\ .$$

Thus, the total travel time is

$$\underline{\underline{T}} = t_a + t^* + t_d = 203.31\,\mathrm{s} = \underline{\underline{3.39\,\mathrm{min}}}\ .$$

P1.4 **Problem 1.4** A car driver approaches a traffic light with the speed of $v_0 = 50\,\text{km/h}$. At a distance of $l = 100\,\text{m}$ the lights turn to 'Red'. The 'Red' and 'Yellow' phase takes $t^* = 10\,\text{s}$. The driver wants passing the traffic lights just when the lights turn back to 'Green'.

a) Determine the necessary constant deceleration a_0, when the driver is braking along the entire distance?

b) Determine the velocity v_1 of the car when arriving at the lights?

c) Draw the diagrams $a(t)$, $v(t)$ and $x(t)$.

Solution For constant acceleration a_0 we have with $x(t=0)=0$

$$v = v_0 + a_0 t\,,$$

$$x = v_0 t + a_0 \frac{t^2}{2}\,.$$

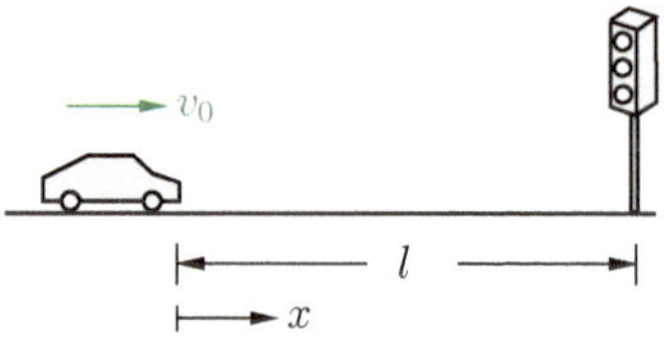

a) The 2nd equation leads with the condition $x(t^*) = l$ to

$$\underline{\underline{a_0}} = \frac{2}{t^{*2}}(l - v_0 t^*) = \frac{2}{10^2}\left(100 - \frac{50 \cdot 1000}{3600}\cdot 10\right) = \underline{\underline{-0.78\,\frac{\text{m}}{\text{s}^2}}}\,.$$

The negative sign indicates that the car decelerates.

b) With the now known deceleration during braking, the 1st equation yields

$$\underline{\underline{v_1}} = v(t^*) = 50\cdot\frac{1000}{3600} - 0.78\cdot 10$$

$$= 6.09\,\frac{\text{m}}{\text{s}} = \underline{\underline{21.9\,\frac{\text{km}}{\text{h}}}}\,.$$

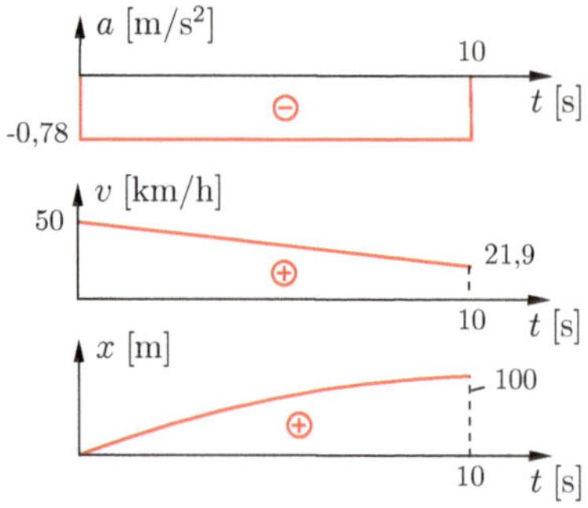

c) Integration of the *constant* acceleration yields a *linear* velocity plot, a second integration a *parabolic* path-time diagram.

Problem 1.5 A vehicle moves according to the given speed-time diagram. **P1.5**

Determine the occuring accelerations, the covered distance and draw the diagrams $x(t)$, $a(t)$, $v(x)$ and $a(x)$.

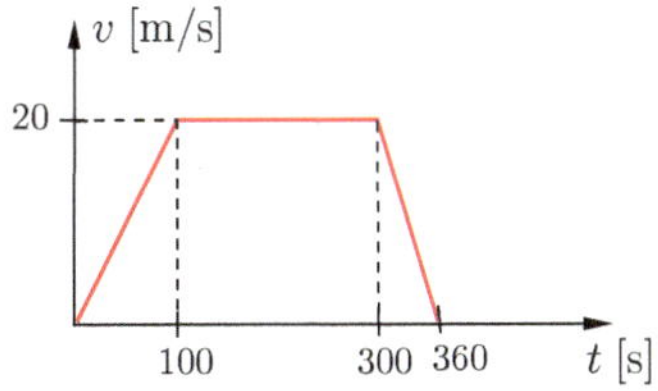

Solution It is advantageous to devide the motion into 3 time sections:

1. Section $0 \le t_1 \le 100\,\text{s}$ (starting with constant acceleration a_1): From $v_1 = a_1 t_1$ follows

$$\underline{\underline{a_1}} = \frac{v_1(100)}{100} = \frac{20}{100} = \underline{\underline{\tfrac{1}{5}\,\text{m/s}^2}}\,, \qquad x_1 = \tfrac{1}{2}\,a_1 t_1^2\,,$$

$$\underline{\underline{s_1}} = x_1(100) = \tfrac{1}{2}\cdot\tfrac{1}{5}(100)^2 = \underline{\underline{1000\,\text{m}}}\,, \quad v_1(x_1) = \sqrt{2a_1 x_1}\,.$$

2. Section $0 \le t_2 \le 200\,\text{s}$ (uniform motion): From $v_2 = 20\,\text{m/s} = \text{const}$ results

$$\underline{\underline{a_2 = 0}}\,, \quad x_2 = v_2 t_2\,, \quad \underline{\underline{s_2}} = x_2(200) = 20\cdot 200 = \underline{\underline{4000\,\text{m}}}\,.$$

3. Section $0 \le t_3 \le 60\,\text{s}$ (braking with constant deceleration a_3): With $v_3 = 20\,\text{m/s} + a_3 t_3$ we obtain

$$\underline{\underline{a_3}} = -\frac{20}{60} = \underline{\underline{-\tfrac{1}{3}\,\text{m/s}^2}}\,, \qquad x_3 = 20\,t_3 + \tfrac{1}{2}\,a_3 t_3^2\,,$$

$$\underline{\underline{s_3}} = x_3(60) = 20\cdot 60 - \tfrac{1}{2}\cdot\tfrac{1}{3}(60)^2 = \underline{\underline{600\,\text{m}}}\,, \quad v_3 = \sqrt{400 + 2a_3 x_3}\,.$$

In total, the vehicle covers the distance

$$\underline{\underline{s}} = s_1 + s_2 + s_3 = 1000 + 4000 + 600 = \underline{\underline{5600\,\text{m}}}\,.$$

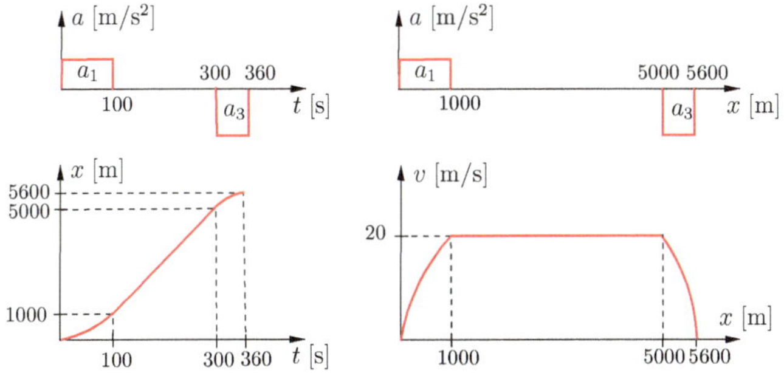

P1.6

Problem 1.6 Taking air resistance into account, the acceleration of a free falling body can be described approximately by $a(v) = g - \alpha v^2$. Here g is the gravity acceleration and α a constant.

Determine the velocity $v(t)$ of the body that is released from rest.

Solution According to the table on page 4 we have for a given $a(v)$

$$t = t_0 + \int_{v_0}^{v} \frac{\mathrm{d}\bar{v}}{g - \alpha \bar{v}^2} .$$

If the motion starts at $t_0 = 0$ we obtain with the initial condition $v(t_0) = v_0 = 0$

$$t = \int_0^v \frac{\mathrm{d}\bar{v}}{g - \alpha\bar{v}^2} = \frac{1}{\alpha}\int_0^v \frac{\mathrm{d}\bar{v}}{(\sqrt{\frac{g}{\alpha}} - \bar{v})(\sqrt{\frac{g}{\alpha}} + \bar{v})}$$

and after partial fraction decomposition

$$t = \frac{1}{\alpha}\,\frac{1}{2\sqrt{\frac{g}{\alpha}}}\int_0^v \left(\frac{1}{\sqrt{\frac{g}{\alpha}} - \bar{v}} + \frac{1}{\sqrt{\frac{g}{\alpha}} + \bar{v}}\right)\mathrm{d}\bar{v}$$

$$= \frac{1}{2\sqrt{g\alpha}}\left[-\ln(\sqrt{\frac{g}{\alpha}} - \bar{v}) + \ln(\sqrt{\frac{g}{\alpha}} + \bar{v})\right]_0^v = \frac{1}{2\sqrt{g\alpha}} \ln\frac{\sqrt{\frac{g}{\alpha}} + v}{\sqrt{\frac{g}{\alpha}} - v} .$$

Solving for v yields

$$e^{2\sqrt{g\alpha}\,t} = \frac{\sqrt{\frac{g}{\alpha}} + v}{\sqrt{\frac{g}{\alpha}} - v} \quad \rightsquigarrow \quad \underline{\underline{v = \sqrt{\frac{g}{\alpha}}\,\frac{e^{2\sqrt{g\alpha}\,t} - 1}{e^{2\sqrt{g\alpha}\,t} + 1}}} .$$

With the hyperbolic function $\tanh\varphi = \frac{e^{\varphi} - e^{-\varphi}}{e^{\varphi} + e^{-\varphi}} = \frac{e^{2\varphi} - 1}{e^{2\varphi} + 1}$ the result also can be written as

$$\underline{\underline{v = \sqrt{\frac{g}{\alpha}}\tanh\sqrt{g\alpha}\,t}} .$$

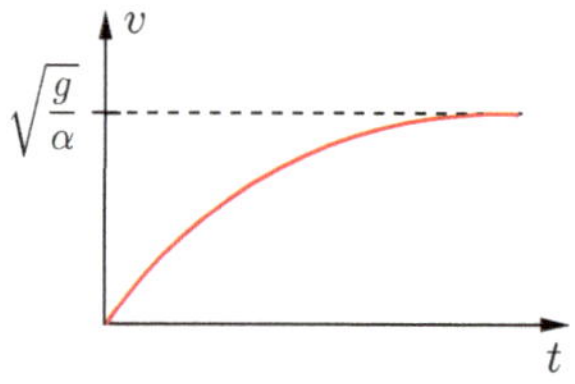

From this last representation the limit value

$$\lim_{t\to\infty} v(t) = \sqrt{g/\alpha} ,$$

can be recognized, i.e. after sufficiently large time the body practically falls with constant speed ($a = 0 \rightsquigarrow v = \sqrt{g/\alpha}$).

Problem 1.7 On account of the gas expansion, a piston (diameter d) moves in a cylinder. Here, the acceleration a of the piston is proportional to the current gas pressure p, i.e. $a = c_0\, p$, where for the gas pressure Boyle-Mariotte's gas law pV = const is valid. The initial state is given by the pressure p_0, the piston location l_0 and the initial velocity $v_0 = 0$. P1.7

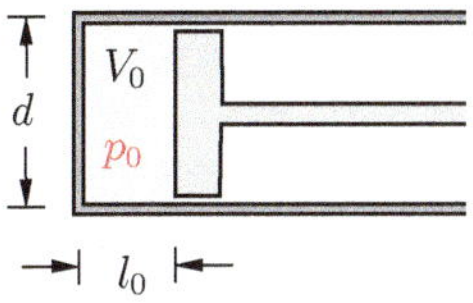

Determine the velocity v of the piston as a function of its position.

Solution For a piston displacement x we obtain, according to the gas law $p_0 V_0 = pV$, the relation

$$p_0\,\frac{\pi d^2}{4}\,l_0 = p\,\frac{\pi d^2}{4}\,(l_0 + x)\,,$$

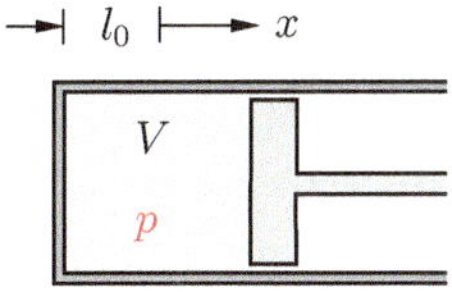

where the pressure is given by

$$p = p_0\,\frac{l_0}{l_0 + x}\,.$$

This leads to the acceleration

$$a(x) = c_0\, p = c_0\, p_0\,\frac{1}{1 + x/l_0} = a_0\,\frac{1}{1 + x/l_0}\,.$$

Here $a_0 = c_0\, p_0$ is the initial acceleration at $x = 0$. Using the table on page 4 for a given $a(x)$, the velocity is determined by

$$v^2 = v_0^2 + 2\int\limits_{x_0}^{x} a(\bar{x})\mathrm{d}\bar{x} = 2\int\limits_{0}^{x}\frac{a_0}{1 + \bar{x}/l_0}\mathrm{d}\bar{x} = 2 l_0\, a_0 \ln\left(1 + \frac{x}{l_0}\right)$$

$$\leadsto \quad \underline{\underline{v = \sqrt{2\, l_0\, a_0 \ln\left(1 + \frac{x}{l_0}\right)}}}\,.$$

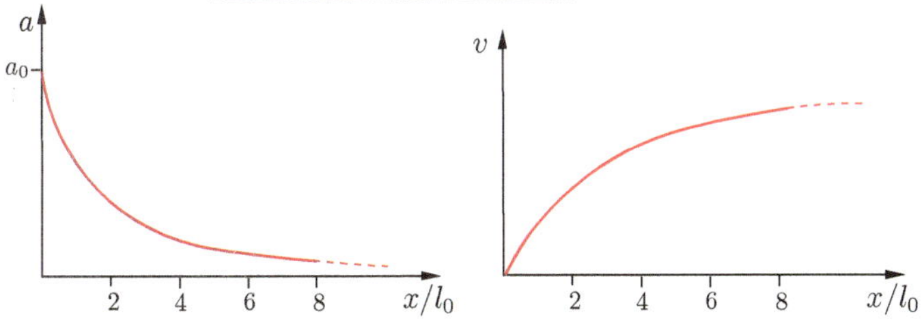

Remark: Because the acceleration decreases with increasing x, the velocity increase drops continuously.

P1.8

Problem 1.8 The acceleration of a point P, moving along a straight line, is directed to the point Z and its magnitude is inverse proportional to the distance x.

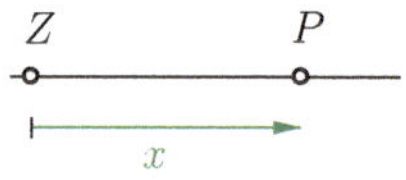

For $t = 0$ the point P has the distance $x_0 = 2$ m, the velocity $v_0 = 4$ m/s and the acceleration $a_0 = -3\ \text{m/s}^2$.

a) Determine the velocity v_1 for the distance $x_1 = 3$ m.

b) At what distance x_2 the velocity is zero?

Solution According to the problem description, the acceleration is $a = -c/x$, where c can be determined from the given initial conditions:

$$c = -a_0 x_0 = -(-3) \cdot 2 = 6\ (\text{m/s})^2 .$$

Knowing $a(x)$, the velocity is obtained from (see table on page 4)

$$v^2 = v_0^2 + 2 \int_{x_0}^{x} a(\bar{x}) \mathrm{d}\bar{x} = v_0^2 + 2 \int_{x_0}^{x} \left(-\frac{c}{\bar{x}} \mathrm{d}\bar{x}\right) = v_0^2 - 2c \ln \frac{x}{x_0}$$

$$\rightsquigarrow \quad v(x) = \pm\sqrt{v_0^2 - 2c \ln x/x_0} .$$

a) Hence, the velocity for $x_1 = 3$ m results as

$$\underline{\underline{v_1}} = \binom{+}{-}\sqrt{16 - 12 \ln \frac{3}{2}} = \underline{\underline{3,34\ \frac{\text{m}}{\text{s}}}} .$$

b) The velocity is zero for:

$$v = 0 \quad \rightsquigarrow \quad v_0^2 - 2c \ln \frac{x}{x_0} = 0 \quad \rightsquigarrow \quad \underline{\underline{x_2 = x_0\, \mathrm{e}^{v_0^2/2c}}} = 2\,\mathrm{e}^{4/3} = \underline{\underline{7,59\ \text{m}}} .$$

Remarks:

- The velocity-position diagram is symmetric with respect to the x-axis.
- The motion can also take place in the domain of negative x. Because of the discontinuity at $x = 0$, the equations then must be formulated new, considering the direction change of a.

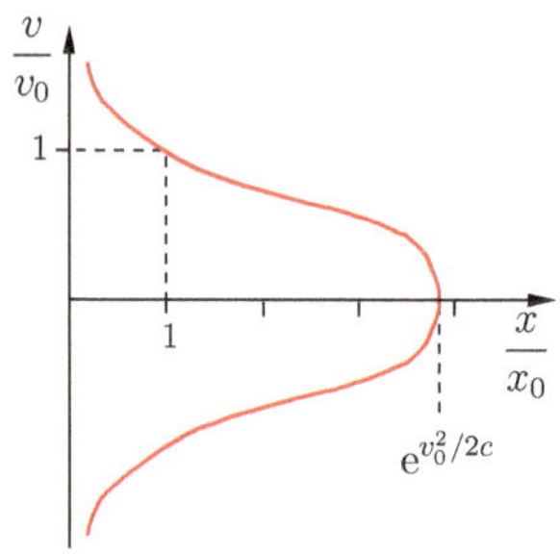

Problem 1.9 A ball is thrown vertically upwards with an initial velocity $v_{01} = 20$ m/s. Two seconds later, a second ball likewise is thrown vertically upwards with $v_{02} = 18$ m/s. **P1.9**

Determine the height H where the two balls meet.

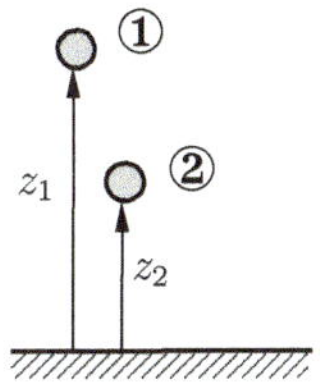

Solution We start counting time t when launching the 1st ball. Considering the given initial values, the direction of gravity g and the time difference $\Delta t = 2$ s, we have

$$z_1 = v_{01} t - \frac{g}{2} t^2 ,$$

$$z_2 = v_{02}(t - \Delta t) - \frac{g}{2}(t - \Delta t)^2 .$$

From the condition $z_1 = z_2$ for meeting, the meeting time t^* is obtained:

$$v_{01} t^* - \frac{g}{2} t^{*2} = v_{02}(t^* - \Delta t) - \frac{g}{2}(t^{*2} - 2t^* \Delta t + \Delta t^2)$$

$$\rightsquigarrow \quad t^* = \frac{\Delta t \left(v_{02} + \frac{1}{2} g \Delta t \right)}{v_{02} - v_{01} + g \Delta t} = 3.16 \text{ s} .$$

Introducing t^* into the equation for z_1 (or z_2) yields the height:

$$\underline{\underline{H}} = z_1(t^*) = 20 \cdot 3.16 - 4.9 \cdot 3.16^2 = \underline{\underline{14.27 \text{ m}}} .$$

The solution can be illustrated by the position-time diagram:

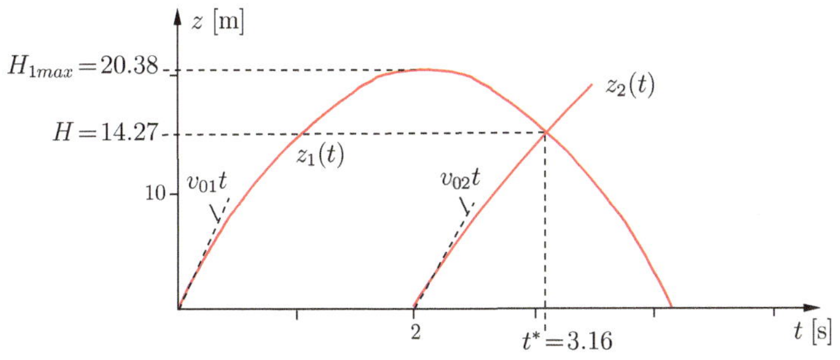

P1.10 **Problem 1.10** From the rim of a cliff, 50 m above the sea level, a ball is thrown vertically upwards with an initial velocity of 10 m/s.

a) Determine the maximum height the ball reaches above sea level.

b) When the ball impinges on sea surface?

c) Determine the velocity of the ball when impinging on sea surface.

Solution With $a = -g$ and the initial velovity v_0 we have

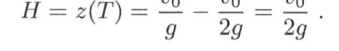

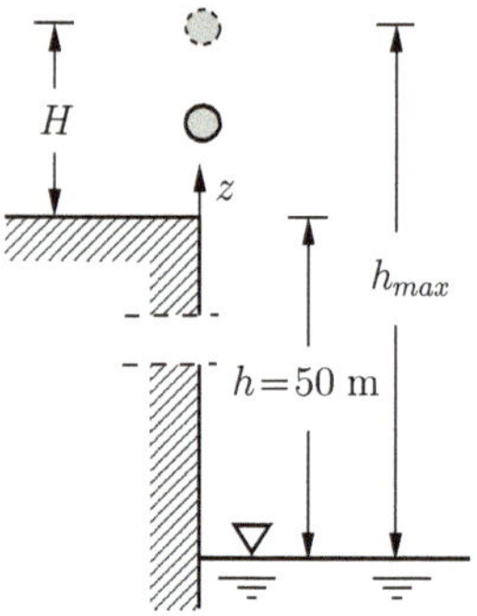

$$v = \dot{z} = v_0 - gt\,,$$

$$z = v_0 t - \frac{g}{2}\,t^2\,.$$

a) The *rise time* T follows from the condition $v(T) = 0$:

$$v_0 - gT = 0 \quad \rightsquigarrow \quad T = \frac{v_0}{g}\,.$$

Therefore, the *throw height* H is given by

$$H = z(T) = \frac{v_0^2}{g} - \frac{v_0^2}{2g} = \frac{v_0^2}{2g}\,.$$

With $v_0 = 10$ m/s the maximum height h_{max} is obtained as

$$\underline{\underline{h_{max}}} = h + H = h + \frac{v_0^2}{2g} = 50 + \frac{10^2}{2 \cdot 9.81} = \underline{\underline{55.1\text{ m}}}\,.$$

b) The time t_i until the ball impinges on sea surface is obtained from the condition $z(t_i) = -h$:

$$-h = v_0 t_i - \frac{g}{2}\,t_i^2 \quad \rightsquigarrow \quad \underline{\underline{t_i}} = \frac{1}{g}\left\{v_0 \underset{(-)}{+} \sqrt{v_0^2 + 2gh}\right\} = \underline{\underline{4.37\text{ s}}}\,.$$

Note: The formally possible minus sign in front of the root is inapplicable. It would lead to a negative time t_i.

c) With $t = t_i$ the impact velocity follows as

$$\underline{\underline{v(t_i)}} = v_0 - g t_i = 10 - 9.81 \cdot 4.37 = \underline{\underline{-32.87\,\frac{\text{m}}{\text{s}}}}\,.$$

The minus sign indicates that the velocity is downwards directed, i.e. opposite to the chosen coordinate z.

Problem 1.11 The crew of a balloon, that is moving upwards in a cloud with constant speed v_0, wants to determine the current hight h_0 above ground. For this purpose a gauging member is released from the gondola that falls down and explodes when hitting the ground. After time t_1 the crew hears the detonation. **P1.11**

Determine the height h_0 for the following data: $v_0 = 5$ m/s, $g = 9.81$ m/s^2, $t_1 = 10$ s, $c = 330$ m/s (speed of sound).

Solution Introducing x downwards from the position where the member is launched ($t = 0$), the falling time t_m until hitting the ground is obtained from

$$x(t_m) = \frac{1}{2}\, g\, t_m^2 - v_0 t_m = h_0$$

as

$$t_m = \frac{v_0}{g}\left\{1 \underset{(-)}{+} \sqrt{1 + \frac{2gh_0}{v_0^2}}\right\} .$$

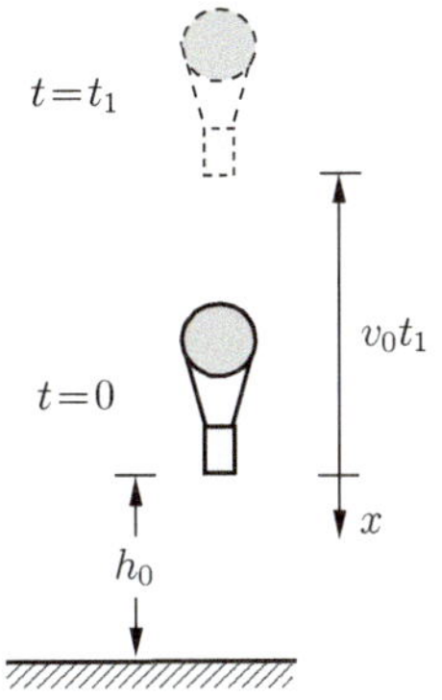

Only the positive root is meaningful since t_m must be positive.

The sound covers the distance $h_0 + v_0 t_1$ because during time t_1 the baloon is rising the distance $v_0 t_1$. Therefore the sound requires the time

$$t_s = \frac{\text{sound distance}}{\text{sound speed}} = \frac{h_0 + v_0 t_1}{c} .$$

The total time is given by the sum of falling time and the time of sound:

$$t_1 = t_m + t_s = \frac{v_0}{g}\left\{1 + \sqrt{1 + \frac{2gh_0}{v_0^2}}\right\} + \frac{h_0 + v_0 t_1}{c} .$$

After rearranging and squaring we obtain

$$\underline{\underline{h_0}} = \frac{c - v_0}{g}\left\{g t_1 + c\left[1 \overset{(+)}{-} \sqrt{1 + 2\,\frac{g t_1}{c - v_0}}\right]\right\}$$

$$= \frac{325}{9{,}81}\left\{98{,}1 + 330\left[1 - \sqrt{1 + 2\,\frac{98{,}1}{325}}\right]\right\} = \underline{\underline{338\ \text{m}}} .$$

The solution with the positive root leads to a mechanically unreasonable result.

P1.12 **Problem 1.12** The two points P_A and P_B start their motion along a circular path at the same time $t = 0$ at A and B, respectively. Point P_A moves with the given constant speed v_0 and point P_B starts from rest with a constant angular acceleration $\dot{\omega}_0$.

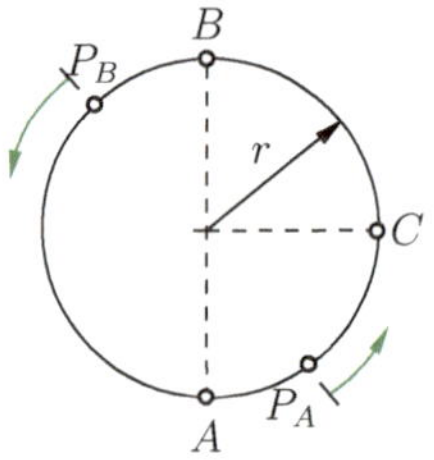

a) Determine $\dot{\omega}_0$ such that both points arrive at C at the same time.

b) Calculate the angular velocity of P_B when passing A?

c) Determine the normal acceleration of both points at C.

Solution For P_A follows from $v_A = \dot{s}_A = \text{const} = v_0$ and by considering the initial condition $s_A(0) = 0$,

$$s_A = v_0 t \,.$$

For P_B we obtain from $\dot{\omega}_B = \ddot{\varphi}_B = \text{const} = \dot{\omega}_0$ with $\varphi_B(0) = 0$, $\dot{\varphi}_B(0) = 0$

$$\varphi_B = \frac{1}{2}\dot{\omega}_0 t^2, \quad \dot{\varphi}_B = \omega_B = \dot{\omega}_0 t \quad \rightsquigarrow \quad s_B = \frac{1}{2}\dot{\omega}_0 r\, t^2, \quad v_B = \dot{\omega}_0 r\, t\,.$$

a) Because both points shall arrive at the same time t_C at C, it follows with the different distances $s_A(t_C) = \pi r/2$ and $s_B(t_C) = 3\pi r/2$

$$\frac{1}{2}\,\pi r = v_0 t_C\,, \quad \frac{3}{2}\,\pi r = \frac{1}{2}\dot{\omega}_0 r\, t_C^2 \quad \rightsquigarrow \quad t_C = \frac{\pi r}{2v_0}\,, \quad \underline{\underline{\dot{\omega}_0 = \frac{12 v_0^2}{\pi r^2}}}\,.$$

b) With the angle π between B and A and the known $\dot{\omega}_0$ we can calculate the time t_A when P_B arrives at A and the respective angular velocity:

$$t_A^2 = \frac{2\pi}{\dot{\omega}_0} = \frac{\pi^2 r^2}{6 v_0^2} \quad \rightsquigarrow \quad t_A = \frac{\pi r}{\sqrt{6}\, v_0} \quad \rightsquigarrow \quad \underline{\underline{\omega_B(t_A)}} = \dot{\omega}_0 t_A = \underline{\underline{2\sqrt{6}\,\frac{v_0}{r}}}\,.$$

c) The normal accelerations at C are found by using the relation $a_n = r\omega^2 = v^2/r$:

$$\underline{\underline{a_{nA} = \frac{v_0^2}{r}}}\,, \qquad \underline{\underline{a_{nB}}} = \omega_B^2(t_C)\, r = \dot{\omega}_0^2 t_C^2\, r = \underline{\underline{36\,\frac{v_0^2}{r}}}\,.$$

Problem 1.13 At the fixed point A a point P starts moving along a circular path with radius r. Its motion is described by the relation $s = ct^2$.

P1.13

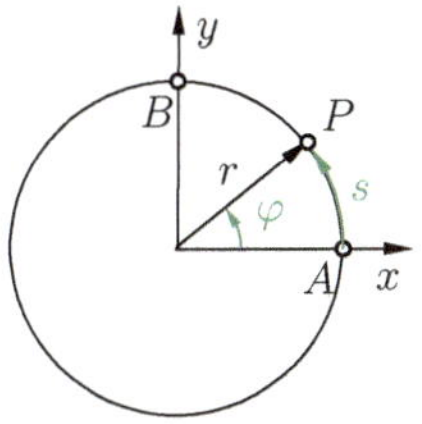

Determine:
a) the velocity components $v_x(t)$ and $v_y(t)$,
b) the velocity at point B,
c) the tangential acceleration $a_t(s)$ and the normal acceleration $a_n(s)$.

Solution From $s = ct^2$ the speed follows as

$$v = \dot{s} = 2ct\,.$$

a) Because the velocity is always tangential to the path, its cartesian components at an arbitrary point are

$$v_x = -v \sin\varphi\,, \qquad v_y = v\cos\varphi\,.$$

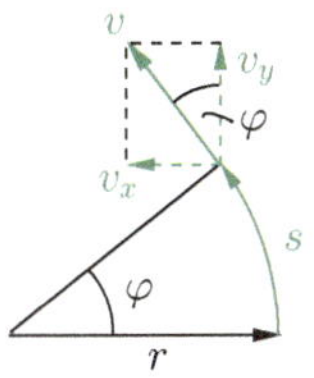

Hence, with

$$\varphi = \frac{s}{r} = \frac{ct^2}{r}$$

we obtain

$$\underline{\underline{v_x = -2ct\sin\frac{ct^2}{r}}}\,, \qquad \underline{\underline{v_y = 2ct\cos\frac{ct^2}{r}}}\,.$$

b) In point B we have

$$s(t_B) = \frac{\pi r}{2} = ct_B^2 \quad \rightsquigarrow \quad t_B = \sqrt{\frac{\pi r}{2c}}\,,$$

$$\underline{\underline{v(t_B)}} = 2ct_B = 2c\sqrt{\frac{\pi r}{2c}} = \underline{\underline{\sqrt{2\pi rc}}}\,.$$

c) From

$$a_t = \dot{v}\,, \qquad a_n = \frac{v^2}{r} = \frac{4c^2t^2}{r}\,,$$

with $\dot{v} = 2c$ and $ct^2 = s$, follow the results

$$\underline{\underline{a_t = 2c}}\,, \qquad \underline{\underline{a_n = \frac{4cs}{r}}}\,.$$

Remark: While the tangential acceleration remains constant, the normal acceleration increases linearly with s.

P1.14 **Problem 1.14** A point M moves along a half circle. The projection of its motion on the diameter AB is an uniform motion with the speed c.

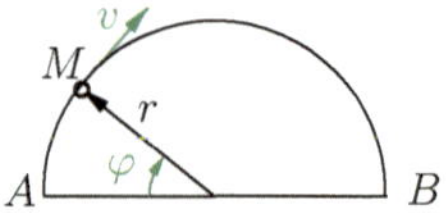

Determine

a) the speed $v(\varphi)$ and the magnitude $a(\varphi)$ of the acceleration,

b) the angle between the acceleration vector and the diameter AB.

Solution a) From the condition $v \sin\varphi = c$ follows

$$\underline{\underline{v(\varphi) = \frac{c}{\sin\varphi}}} .$$

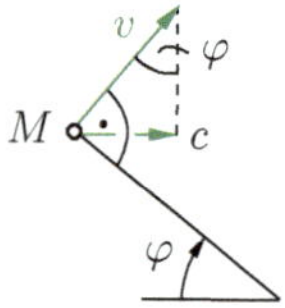

The acceleraton components a_t and a_n are determined with $r\dot{\varphi} = v$ as

$$a_t = \frac{\mathrm{d}v}{\mathrm{d}t} = \frac{\mathrm{d}v}{\mathrm{d}\varphi}\dot{\varphi} = -\frac{c}{\sin^2\varphi}\cos\varphi\,\frac{v}{r} = -\frac{c^2}{r}\frac{\cos\varphi}{\sin^3\varphi} ,$$

$$a_n = \frac{v^2}{r} = \frac{c^2}{r\sin^2\varphi} .$$

This finally leads to

$$\underline{\underline{a}} = |\boldsymbol{a}| = \sqrt{a_t^2 + a_n^2} = \frac{c^2}{r}\sqrt{\frac{\cos^2\varphi}{\sin^6\varphi} + \frac{1}{\sin^4\varphi}}$$

$$= \frac{c^2}{r\sin^3\varphi}\sqrt{\cos^2\varphi + \sin^2\varphi} = \underline{\underline{\frac{c^2}{r\sin^3\varphi}}} .$$

b) From the figure can be seen:

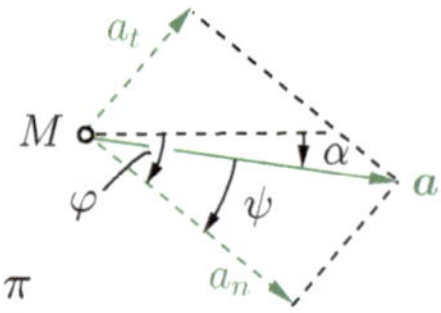

$$\tan\psi = \tan(\varphi - \alpha) = \frac{a_t}{a_n} = \frac{-\frac{c^2}{r}\frac{\cos\varphi}{\sin^3\varphi}}{\frac{c^2}{r\sin^2\varphi}}$$

$$= -\cot\varphi = +\tan(\varphi - \pi/2) \quad \rightsquigarrow \quad \underline{\underline{\alpha = \frac{\pi}{2}}} ,$$

i.e. the acceleration vector is perpendicular to AB.

Remark: the latter result can also be found without any calculation: if a component of v is constant then there is no acceleration in this direction, i.e. $\boldsymbol{a}$ is perpendicular to the direction of this component.

Problem 1.15 A bike B starts moving at A from rest along a circular path (radius r). It accelerates according to the law $a_t(v) = a_0(1 - \kappa v)$ until the speed v_C is reached at C. Here a_0 is the initial acceleration and κ is a constant. **P1.15**

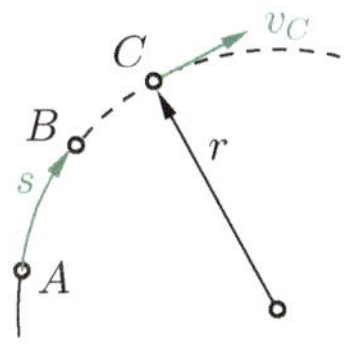

a) Determine the acceleration time t_C and
b) the acceleration distance s_C.
c) Calculate the normal acceleration a_n at $t = t_C/2$ for $\kappa = 1/(2v_C)$.

Solution a) From the given acceleration law $a_t(v) = \dot{v} = a_0(1 - \kappa v)$ follows with the initial velocity $v_0 = 0$ and $t_0 = 0$ (see page 4)

$$t(v) = \frac{1}{a_0}\int_0^v \frac{d\bar{v}}{1-\kappa\bar{v}} = -\frac{1}{a_0\kappa}\ln(1-\kappa v)\,.$$

The acceleration time is determined from the condition $v(t_C) = v_C$:

$$\underline{\underline{t_C}} = t(v_C) = \underline{\underline{-\frac{1}{a_0\kappa}\ln(1-\kappa v_C)}}\,.$$

b) The inverse function of $t(v)$ is given by

$$e^{-a_0\kappa t} = 1 - \kappa v \quad \rightsquigarrow \quad v(t) = \frac{1}{\kappa}\left(1 - e^{-a_0\kappa t}\right).$$

This leads by integration and by considering the initial condition $s_0 = s(0) = 0$ to

$$s(t) = \int_0^t v(\bar{t})d\bar{t} = \frac{1}{a_0\kappa^2}\left(a_0\kappa t + e^{-a_0\kappa t}\right)$$

and thus to the acceleration distance

$$\underline{\underline{s_C}} = s(t_C) = \underline{\underline{\frac{1}{a_0\kappa^2}\left(-\ln(1-\kappa v_C) + 1 - \kappa v_C\right)}}\,.$$

c) The normal acceleration within the time span $0 \le t \le t_a$ is given by

$$a_n = \frac{v^2}{r} = \frac{1}{r\kappa^2}\left(1 - e^{-a_0\kappa t}\right)^2.$$

Inserting $t = t_C/2$ and $\kappa = 1/2v_C$, we obtain

$$\underline{\underline{a_n}} = \frac{4v_C^2}{r}\left(1 - \left[1 - \frac{v_C}{2v_C}\right]^{1/2}\right)^2 = \underline{\underline{\frac{4v_C^2}{r}\left(\frac{3}{2} - \sqrt{2}\right) = 0.343\,\frac{v_C^2}{r}}}\,.$$

P1.16

Problem 1.16 A point mass P is released from rest at A on a circular path in a vertical plane. Due to gravity it experiences a tangential acceleration $g\cos\varphi$.

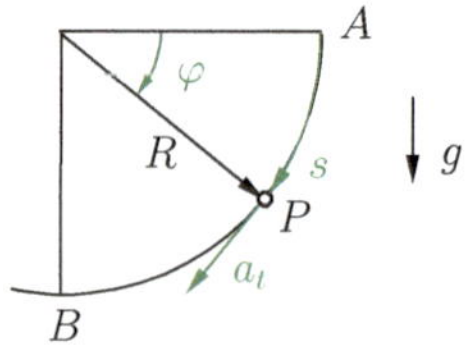

Determine the velocity and the magnitude of acceleration in dependence on the angle φ.

Solution From the tangential acceleration

$$a_t = g\cos\varphi = \dot{v} = a_t(\varphi)$$

follows with $s = R\varphi$ and the initial velocity $v_0 = 0$ (see page 4)

$$v^2 = 2\int_0^{\varphi} g\cos\bar{\varphi}\; R\,\mathrm{d}\bar{\varphi} = 2gR\sin\varphi$$

$$\rightsquigarrow \quad \underline{\underline{v = \sqrt{2gR\sin\varphi}}}\ .$$

Hence, the acceleration components are

$$a_t = g\cos\varphi\,, \qquad a_n = \frac{v^2}{R} = 2g\sin\varphi\,.$$

Therewith, the magnitude of acceleration is given by

$$a = |\boldsymbol{a}| = \sqrt{a_t^2 + a_n^2} = g\sqrt{\cos^2\varphi + 4\sin^2\varphi}$$

$$\rightsquigarrow \quad \underline{\underline{a = g\sqrt{1 + 3\sin^2\varphi}}}\ .$$

Remarks:

- At A the acceleration components are $a_t = g$ and $a_n = 0$. (Point P has solely a tangential acceleration.)
- At B we have $a_t = 0$ and $a_n = 2g$. (Pure normal acceleration upwards directed.)
- With the height difference $h = R\sin\varphi$ the velocity can be represented by $v = \sqrt{2gh}$.

Problem 1.17 A satellite S is moving along a circular path around the earth if its normal acceleration just equals the gravitational acceleration $g\,R^2/r^2$ (gravity on earth surface $g = 9.81$ m/s^2, earth radius $R = 6370$ km). **P1.17**

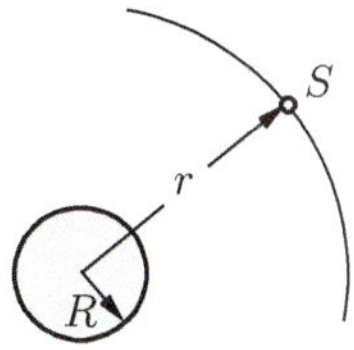

a) In which distance H above earth surface a satellite is circling that has the speed of 25000 km/h?

b) Determine the required speed of a satellite whose orbit lies 1000 km above earth surface.

c) What time requires a satellite for 1 circle (orbital period) in a height of $H = 400$ km?

d) Determine the height of a geostationary satellite.

Solution a) From the necessary normal acceleration

$$a_n = v^2/r = g\,R^2/r^2$$

follows with the given speed

$$r = g\,\frac{R^2}{v^2} \quad \leadsto \quad \underline{\underline{H}} = r - R = R\Big(g\,\frac{R}{v^2} - 1\Big) = \underline{\underline{1884\text{ km}}}\,.$$

b) From the same equation with a given distance yields

$$\underline{\underline{v}} = R\,\sqrt{\frac{g}{r}} = R\,\sqrt{\frac{g}{R+H}} = \underline{\underline{26457\ km/h}}\,.$$

c) With the given speed $v = R\sqrt{g/r}$ and the arc length $L = 2\pi r$ of the orbit, the orbital period is obtained:

$$\underline{\underline{T}} = \frac{L}{v} = 2\pi\frac{r^{3/2}}{R\sqrt{g}} = 5547\text{ s} = \underline{\underline{1.54\text{ h}}}\,.$$

d) A geostationary satellite has the same angular velocity as the earth:

$$\omega_E = 2\pi/(24\text{ h})\,.$$

Thus, we have

$$a_n = r\omega_E^2 = g\Big(\frac{R}{r}\Big)^2 \quad \leadsto \quad r = \Big(g\,\frac{R^2}{\omega_E^2}\Big)^{1/3}$$

or numerically evaluted

$$r = \Big(9.81\cdot 10^{-3}\cdot(3600)^2\,\frac{(6370)^2}{(2\pi)^2}\,24^2\Big)^{1/3} = 4.22\cdot 10^4\text{ km}\,,$$

$$\leadsto \quad \underline{\underline{H}} = r - R \approx \underline{\underline{36000\text{ km}}}\,.$$

P1.18

Problem 1.18 A point is moving with constant speed v_0 in a plane along the given path $r(\varphi) = b\,e^{\varphi}$ (logarithmic spiral). For $t = 0$ the angle is $\varphi = 0$.

Determine the angular velocity $\dot{\varphi}$ in dependence on φ and t as well as the radial velocity $\dot{r}$.

Solution In polar coordinates the speed is represented by

$$v = \sqrt{\dot{r}^2 + r^2\dot{\varphi}^2}\,.$$

Introducing $v = v_0$ and

$$\dot{r} = \frac{\mathrm{d}r}{\mathrm{d}\varphi}\frac{\mathrm{d}\varphi}{\mathrm{d}t} = \dot{\varphi}\frac{\mathrm{d}r}{\mathrm{d}\varphi} = \dot{\varphi} b\,e^{\varphi}$$

we obtain

$$v_0 = \sqrt{\dot{\varphi}^2 b^2 e^{2\varphi} + b^2 e^{2\varphi}\dot{\varphi}^2} = \sqrt{2}\,b\,e^{\varphi}\dot{\varphi}\,.$$

Solving for $\dot{\varphi}$ yields

$$\underline{\underline{\dot{\varphi} = \frac{v_0}{\sqrt{2}\,b}\,e^{-\varphi}}}$$

and thus

$$\underline{\underline{\dot{r}}} = \dot{\varphi} b\,e^{\varphi} = \underline{\underline{v_0/\sqrt{2} = \mathrm{const}}}\,.$$

To determine the dependence on time we find from

$$\dot{\varphi} = \frac{\mathrm{d}\varphi}{\mathrm{d}t} = \frac{v_0}{\sqrt{2}\,b}\,e^{-\varphi}$$

by separation of variables

$$e^{\varphi}\mathrm{d}\varphi = \frac{v_0}{\sqrt{2}\,b}\,\mathrm{d}t$$

and integration, taking into account the initial condition $\varphi(t = 0) = 0$,

$$e^{\varphi} - 1 = \frac{v_0}{\sqrt{2}\,b}\,t \qquad \rightsquigarrow \qquad e^{\varphi} = 1 + \frac{v_0}{\sqrt{2}\,b}\,t\,.$$

Introduction into $\dot{\varphi}(\varphi)$ finally yields

$$\underline{\underline{\dot{\varphi} = \frac{v_0}{\sqrt{2}\,b}\,\frac{1}{1 + \frac{v_0}{\sqrt{2}\,b}\,t}}}\,.$$

P1.19

Problem 1.19 A bar rotates about A according to the law $\varphi = \kappa t$. Along the bar a knuckle K slides with prescibed speed $\dot{r} = v_0 - at$ and initial conditions $r(0) = 0$, $\varphi(0) = 0$.

a) Calculate κ and a such that $r_{max} = l$ is reached at $\varphi^* = 2\pi$.

b) Determine the path $r(\varphi)$ of K and

c) the magnitude of velocity and acceleration at $\varphi^{**} = \pi$.

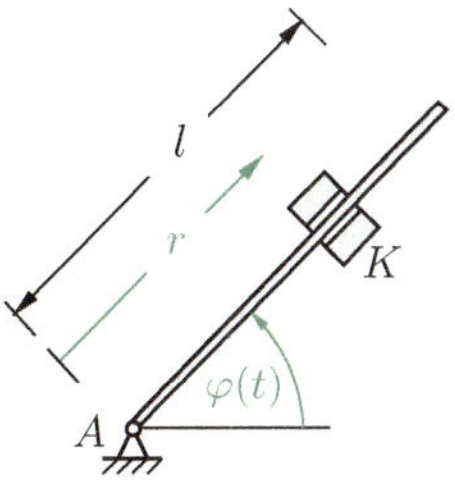

Solution From the given functions and initial conditions we first determine the components of position, velocity and acceleration in polar coordinates:

$$\dot{r} = v_0 - at \quad \rightsquigarrow \quad r = v_0 t - at^2/2\,, \quad \ddot{r} = -a\,,$$

$$\varphi = \kappa t \quad \rightsquigarrow \quad \dot{\varphi} = \kappa\,, \quad \ddot{\varphi} = 0\,.$$

a) r_{max} is reached when $\dot{r} = 0$, i.e. at time $t^* = v_0/a$. Thus, the prescribed conditions lead to

$$r(t^*) = l \quad \rightsquigarrow \quad \frac{v_0^2}{a} - \frac{v_0^2}{2a} = l \quad \rightsquigarrow \quad \underline{\underline{a = \frac{v_0^2}{2l}}}\,,$$

$$\varphi(t^*) = \varphi^* = 2\pi \quad \rightsquigarrow \quad \kappa\,\frac{v_0}{a} = 2\pi \quad \rightsquigarrow \quad \underline{\underline{\kappa = \pi\,\frac{v_0}{l}}}\,.$$

b) By eliminating t from $r(t)$ and introducing a and κ we find

$$\underline{\underline{r(\varphi)}} = v_0\frac{\varphi}{\kappa} - a\,\frac{\varphi^2}{2\kappa^2} = \underline{\underline{l\Big(\frac{\varphi}{\pi} - \frac{\varphi^2}{4\pi^2}\Big)}}\,.$$

c) The magnitude of velocity and acceleration is obtained from the respective components at $\varphi = \varphi^{**} = \pi$, i.e. at time $t^{**} = \pi/\kappa = l/v_0$:

$$v_r = \dot{r} = v_0 - \frac{v_0^2}{2l}\frac{l}{v_0} = \frac{v_0}{2}\,, \quad v_\varphi = r\dot{\varphi} = \Big(v_0\frac{l}{v_0} - \frac{1}{2}\frac{v_0^2}{2l}\frac{l^2}{v_0^2}\Big)\pi\frac{v_0}{l} = \frac{3\pi}{4}\,v_0\,,$$

$$\underline{\underline{v}} = \sqrt{v_r^2 + v_\varphi^2} = \underline{\underline{\sqrt{1/4 + 9\pi^2/16}\;v_0 = 2.41\,v_0}}\,,$$

$$a_r = \ddot{r} - r\dot{\varphi}^2 = -\frac{v_0^2}{2l} - \frac{3l}{4}\,\pi^2\frac{v_0^2}{l^2} = -\frac{v_0}{l}\Big(\frac{1}{2} + \frac{3}{4}\,\pi^2\Big) = -7.90\,\frac{v_0^2}{l}\,,$$

$$a_\varphi = r\ddot{\varphi} + 2\dot{r}\dot{\varphi} = 2\frac{v_0}{2}\,\pi\frac{v_0}{l} = \pi\,\frac{v_0^2}{l}\,,$$

$$\underline{\underline{a}} = \sqrt{a_r^2 + a_\varphi^2} = \underline{\underline{\sqrt{7.90^2 + \pi^2}\;v_0^2/l = 8.50\,v_0^2/l}}\,.$$

P1.20 **Problem 1.20** From the planar motion of a point we know the radial velocity $v_r = c_0 = \text{const}$ and the radial acceleration $a_r = -a_0 = \text{const}$.

Determine for the initial conditions $r(t{=}0) = 0$ and $\varphi(t{=}0) = 0$:

a) the angular velocity $\omega(t)$,

b) the path (trajectory) $r(\varphi)$,

c) the circular acceleration $a_\varphi(t)$.

Solution **a)** From $v_r = \dot{r} = c_0$ follows with $r(t{=}0) = 0$

$$\ddot{r} = 0 \qquad \text{and} \qquad r = c_0 t\,.$$

Therewith, from $a_r = \ddot{r} - r\omega^2$ we obtain

$$-a_0 = -c_0 t\omega^2 \quad \rightsquigarrow \quad \underline{\underline{\omega = \sqrt{\frac{a_0}{c_0 t}}}}\,.$$

b) By integrating $\omega = \dot{\varphi}$ we find with $\varphi(t{=}0) = 0$

$$\varphi = \int\limits_0^t \omega\,\mathrm{d}\bar{t} = \sqrt{\frac{a_0}{c_0}} \int\limits_0^t \frac{\mathrm{d}\bar{t}}{\sqrt{\bar{t}}} = \sqrt{\frac{a_0}{c_0}}\, 2\sqrt{t} \quad \rightsquigarrow \quad t = \frac{\varphi^2}{4}\,\frac{c_0}{a_0}\,.$$

Introduction into $r = c_0 t$ yields the equation for the path trajectory

$$\underline{\underline{r(\varphi) = \frac{c_0^2}{4a_0}\varphi^2}}\,.$$

c) The circular acceleration can be determined from

$$a_\varphi = r\ddot{\varphi} + 2\dot{r}\dot{\varphi}$$

with

$$\dot{\varphi} = \omega = \sqrt{\frac{a_0}{c_0 t}} \qquad \text{and} \qquad \ddot{\varphi} = -\frac{1}{2}\sqrt{\frac{a_0}{c_0 t^3}}$$

as

$$\underline{\underline{a_\varphi}} = c_0 t\Big(-\frac{1}{2}\sqrt{\frac{a_0}{c_0 t^3}}\Big) + 2c_0\sqrt{\frac{a_0}{c_0 t}} = \underline{\underline{\frac{3}{2}\sqrt{\frac{a_0 c_0}{t}}}}\,.$$

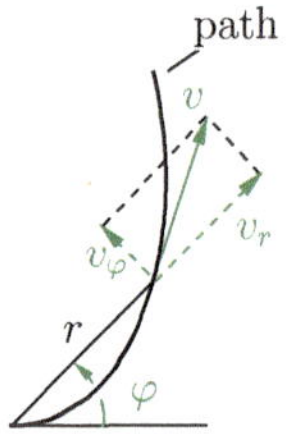

Remark: The circular velocity is given by $v_\varphi = r\omega = \sqrt{a_0 c_0 t} = c_0\varphi/2\,.$

Problem 1.21 An obeserver watches the flight of an low flyer P, flying with constant speed v_0 in a flight altitude h. **P1.21**

Calcuate the angular acceleration $\ddot{\varphi}(\varphi)$ of his head and the radial acceleration $\ddot{r}$. Sketch both diagrams.

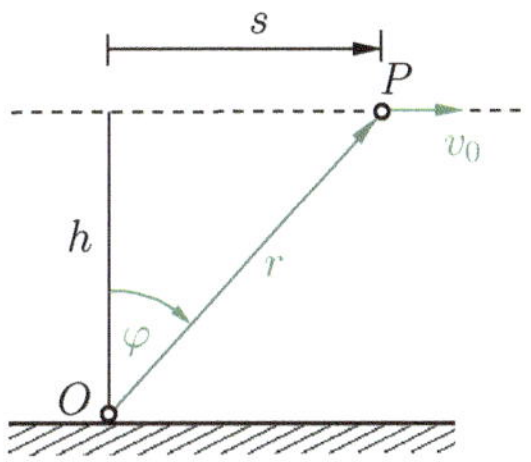

Solution The position of P is given by

$$s = h\tan\varphi \quad \rightsquigarrow \quad \varphi = \arctan\frac{s}{h} \qquad r = \sqrt{h^2+s^2}\,.$$

Differentiation, considering $\dot{s} = v = v_0 = \text{const}$ and $\ddot{s} = 0$, leads to

$$\dot{\varphi} = \frac{1}{1+\left(\frac{s}{h}\right)^2}\frac{\dot{s}}{h} = \frac{v_0 h}{h^2+s^2} = \frac{v_0}{h(1+\tan^2\varphi)} = \frac{v_0}{h}\cos^2\varphi\,,$$

$$\underline{\underline{\ddot{\varphi}}} = \frac{v_0}{h}2\cos\varphi(-\sin\varphi)\dot{\varphi} = \underline{\underline{-2\left(\frac{v_0}{h}\right)^2\sin\varphi\cos^3\varphi}}\,,$$

$$\dot{r} = \frac{2s\dot{s}}{1\sqrt{h^2+s^2}} = \frac{v_0\tan\varphi}{\sqrt{1+\tan^2\varphi}} = v_0\sin\varphi\,,$$

$$\underline{\underline{\ddot{r}}} = v_0\cos\varphi\,\dot{\varphi} = \underline{\underline{\frac{v_0^2}{h}\cos^3\varphi}}\,.$$

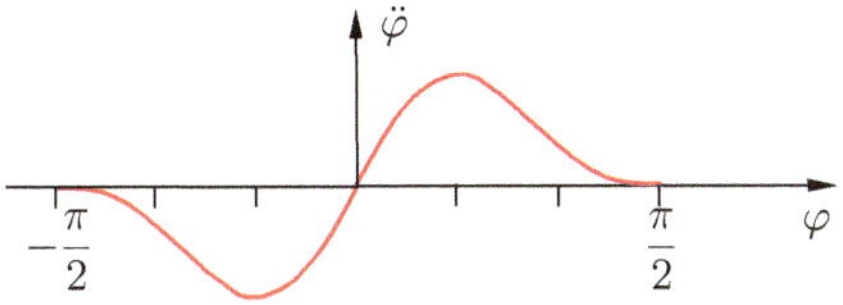

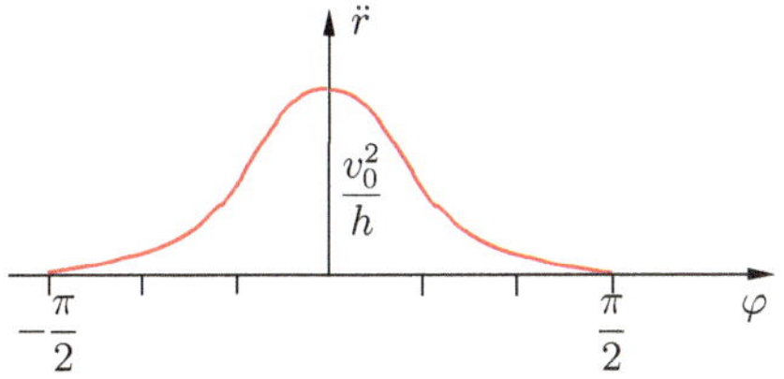

Remark: The maximum angular acceleration occurs at $\varphi = \pm 30^o$; its magnitude is $|\ddot{\varphi}_{\max}| = \frac{3}{8}\sqrt{3}\left(\frac{v_0}{h}\right)^2$. The maximum radial acceleration occurs at $\varphi = 0$. Note: the total acceleration of P is zero!

P1.22

Problem 1.22 A fox F and a rabbit R spot each other and start running at the same time. The rabbit runs with constant speed v_R straight to the save warren W, and the fox, to catch the rabbit, with constant speed v_F along the curve $r = 4\sqrt{2}\, a\, \varphi/\pi$.

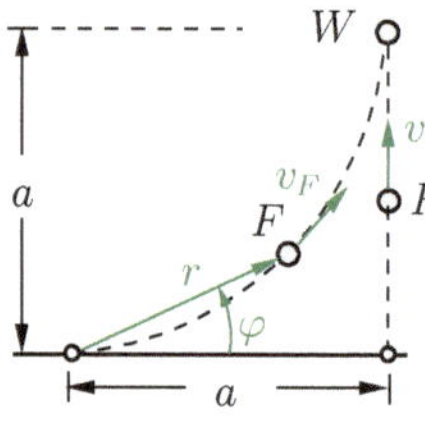

Determine the necessary speed v_R of the rabbit such that it will not be caught by the fox and at what time t_R it should arrive at W?

Hint: $\int \sqrt{1+x^2}\,\mathrm{d}x = \frac{1}{2}\left[x\sqrt{1+x^2} + \ln\left(x + \sqrt{1+x^2}\right)\right]$

Solution We first determine the time t_F the fox needs to reach W. From the path $r(\varphi)$ follow the velocity components

$$v_r = \dot{r} = \frac{\mathrm{d}r}{\mathrm{d}\varphi}\frac{\mathrm{d}\varphi}{\mathrm{d}t} = \frac{4\sqrt{2}\,a}{\pi}\dot{\varphi}\,, \qquad v_\varphi = r\dot{\varphi} = \frac{4\sqrt{2}\,a}{\pi}\varphi\dot{\varphi}\,.$$

Since v_F is constant, we obtain

$$v_F = \sqrt{v_r^2 + v_\varphi^2} = \frac{4\sqrt{2}\,a}{\pi}\,\dot{\varphi}\sqrt{1+\varphi^2} = \frac{4\sqrt{2}\,a}{\pi}\frac{\mathrm{d}\varphi}{\mathrm{d}t}\sqrt{1+\varphi^2}\,.$$

Now we separate the variables, integrate over the whole path from start ($t=0$, $\varphi = 0$) until W ($t = t_F$, $\varphi = \pi/4$) and find in this way:

$$v_F \int_0^{t_F} \mathrm{d}\bar{t} = \frac{4\sqrt{2}\,a}{\pi}\int_0^{\pi/4}\sqrt{1+\bar{\varphi}^2}\,\mathrm{d}\bar{\varphi} \quad \leadsto$$

$$v_F t_F = \frac{4\sqrt{2}\,a}{2\pi}\left[\frac{\pi}{4}\sqrt{1+\frac{\pi^2}{16}} + \ln\left(\frac{\pi}{4} + \sqrt{1+\frac{\pi^2}{16}}\right)\right] = 1.548\,a \quad \leadsto$$

$$t_F = 1.548\,\frac{a}{v_F}\,.$$

The time t_R the rabbit needs to arrive at W is calculated from its speed and the distance: $t_R = a/v_R$. To be not caught by the fox, the rabbit must be earlier at W than the fox, i.e. $t_R < t_F$. This condition leads to

$$\underline{\underline{v_R}} > \frac{1}{1.548}v_F = \underline{\underline{0.646\,v_F}}\,,$$

$$\underline{\underline{t_R < 1.548\,\frac{a}{v_F}}}\,.$$

Problem 1.23 The point A moves in a plane with constant speed v_0 along the x-axis. In constant distance a the point A is followed by a point B, moving such that its velocity vector always points to A. At time $t = 0$ point A is located at the origin of the coordinate system and B is on the y-axis. **P1.23**

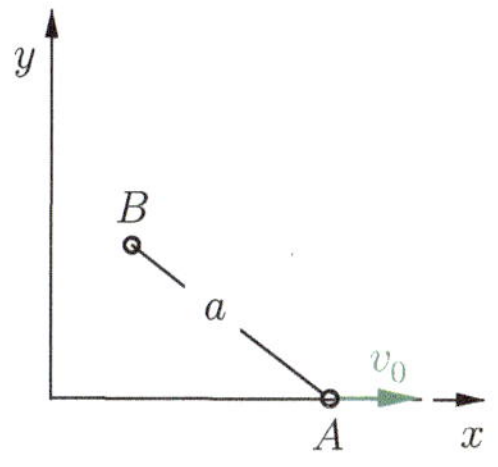

a) Determine the path of B and

b) the speed $v_B(x_A)$?

Hint:

$$\int \frac{\mathrm{d}\alpha}{\sin\alpha} = \ln\Big(\tan\frac{\alpha}{2}\Big)\,, \qquad \sin\alpha = \frac{2\tan\frac{\alpha}{2}}{1+\tan^2\frac{\alpha}{2}}\,, \qquad \cos\alpha = \frac{1-\tan^2\frac{\alpha}{2}}{1+\tan^2\frac{\alpha}{2}}\,.$$

Solution **a)** With the known motion of point A

$$\dot{x}_A = v_0\ , \qquad x_A = v_0\,t$$

and by using the angle $\varphi(t)$ it firstly follow the position coordinates and the velocity components of B:

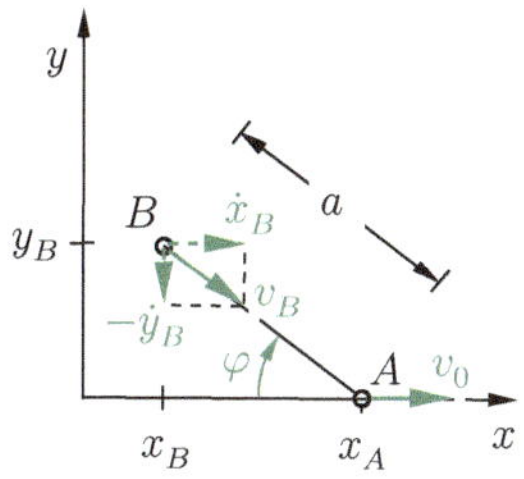

$$\begin{aligned} x_B &= x_A - a\cos\varphi \qquad & y_B &= a\sin\varphi\,,\\ &= v_0\,t - a\cos\varphi\,, & & \\ \dot{x}_B &= v_0\ + a\dot{\varphi}\sin\varphi\,, & \dot{y}_B &= a\dot{\varphi}\cos\varphi\,. \end{aligned}$$

The condition that the velocity vector of B always points to A is expressed by

$$\frac{-\dot{y}_B}{\dot{x}_B} = \frac{y_B}{x_A - x_B} \quad\rightsquigarrow\quad \frac{-a\dot{\varphi}\cos\varphi}{v_0\ + a\dot{\varphi}\sin\varphi} = \frac{a\sin\varphi}{a\cos\varphi}\,.$$

Solving for $\dot{\varphi}$ yields step by step

$$a\dot{\varphi}(\sin^2\varphi + \cos^2\varphi) = -v_0\ \sin\varphi \quad\rightsquigarrow\quad \frac{\mathrm{d}\varphi}{\mathrm{d}t} = -\frac{v_0}{a}\ \sin\varphi\,.$$

Separation of variables and integration lead for $\pi/2 \geq \varphi \geq 0$ to (see hint)

$$\int \frac{\mathrm{d}\varphi}{\sin\varphi} = -\frac{v_0}{a}\int \mathrm{d}t \quad\rightsquigarrow\quad \ln\Big(\tan\frac{\varphi}{2}\Big) = -\frac{v_0}{a}\,t + C\,.$$

The integration constant C is determined from the initial condition:

$$\varphi(0) = \frac{\pi}{2}: \quad \ln 1 = 0 + C \quad \leadsto \quad C = 0\,.$$

Therewith, we obtain the dependence of the angle φ on time t and on position x_A, respectively (remind: $x_A = v_0 t$):

$$\tan\frac{\varphi}{2} = \mathrm{e}^{-v_0\,t/a} \qquad \text{or} \qquad \underline{\underline{\tan\frac{\varphi}{2} = \mathrm{e}^{-x_A/a}}}\,.$$

By the latter equation, the path of B is uniquely described through x_A and the accompanied angle φ.

The parameter representation of the path in cartesian coordinates is obtained by using the formulas of the hint:

$$\underline{\underline{x_B}} = x_A - a\cos\varphi = x_A - a\,\frac{1-\tan^2\frac{\varphi}{2}}{1+\tan^2\frac{\varphi}{2}} = \underline{\underline{x_A - a\,\frac{1-\mathrm{e}^{-2x_A/a}}{1+\mathrm{e}^{-2x_A/a}}}}\,,$$

$$\underline{\underline{y_B}} = a\sin\varphi = a\,\frac{2\tan^2\frac{\varphi}{2}}{1+\tan^2\frac{\varphi}{2}} = \underline{\underline{a\,\frac{2\mathrm{e}^{-2x_A/a}}{1+\mathrm{e}^{-2x_A/a}}}}\,.$$

b) For the velocity we first have with $\dot\varphi = -\frac{v_0}{a}\sin\varphi$

$$\begin{aligned} v_B^2 &= \dot x_B^2 + \dot y_B^2 = v_0^2 + 2v_0\,a\dot\varphi\sin\varphi + a^2\dot\varphi^2\sin^2\varphi + a^2\dot\varphi^2\cos^2\varphi \\ &= v_0^2 + 2v_0\,a\dot\varphi\sin\varphi + a^2\dot\varphi^2 = v_0^2 - 2v_0^2\sin^2\varphi + v_0^2\sin^2\varphi \\ &= v_0^2(1-\sin^2\varphi) = v_0^2\cos^2\varphi \end{aligned}$$

$$\leadsto \quad v_B = v_0\cos\varphi\,.$$

Introducing the result of a) we finally obtain

$$\underline{\underline{v_B(x_A)}} = v_0\,\frac{1-\tan^2\frac{\varphi}{2}}{1+\tan^2\frac{\varphi}{2}} = \underline{\underline{v_0\,\frac{1-\mathrm{e}^{-2x_A/a}}{1+\mathrm{e}^{-2x_A/a}}}}\,.$$

Remarks:

- For the limit case $x_A/a \to \infty$ we obtain $\varphi \to 0$, $y_B \to 0$ and $x_B \to x_A - a$. The velocity of B then is given by $v_B \to v_0$.
- The representation with the angle φ is shorter and more practical than that by cartesian coordinates.
- The results can also be represented by using hyperbolic functions. For example, recasting leads to $v_B(x_A) = v_0\tanh x_A/a$.

Chapter 2

Kinetics of a Point Mass

2

NEWTON's 2nd Law (law of motion): The motion of a point mass under the action of forces is described by

$$\frac{\mathrm{d}(m\,\boldsymbol{v})}{\mathrm{d}t} = \dot{\boldsymbol{p}} = \boldsymbol{F}$$

with $\boldsymbol{F} = \sum \boldsymbol{F}_i$ and the *momentum*

$$\boldsymbol{p} = m\,\boldsymbol{v} \ .$$

Since the mass is constant, Newton's law can also be expressed as

$$m\,\boldsymbol{a} = \boldsymbol{F} \qquad \textit{mass} \times \textit{acceleration} = \textit{force} \ .$$

As an example, this leads for cartesian coordinates to

$$ma_x = \sum F_x \ , \qquad ma_y = \sum F_y \ , \qquad ma_z = \sum F_z \ .$$

Remarks:

- Newton's law is valid in this form only in an inertial reference frame (= reference system that is absolutely at rest or in uniform, rectilinear motion, see also chapter 8),
- Bodies with finite dimensions can be regarded as point masses if their dimensions have no influence on the motion.

Impulse Law: Time integration of the law of motion leads to

$$m\boldsymbol{v} - m\boldsymbol{v}_0 = \int_{t_0}^{t} \boldsymbol{F}\,\mathrm{d}\bar{t} \qquad \text{bzw.} \qquad \boldsymbol{p} - \boldsymbol{p}_0 = \widehat{\boldsymbol{F}}$$

where $\widehat{\boldsymbol{F}} = \int_{t_0}^{t} \boldsymbol{F}\mathrm{d}\bar{t}$ is the *linear impulse*. When no forces are acting ($\boldsymbol{F} = 0$), the linear momentum is conserved:

$$\boldsymbol{p} = m\,\boldsymbol{v} = \text{const} \ .$$

Angular Momentum Theorem: The vector product of Newton's law with the position vector $\boldsymbol{r}$ yields

$$\frac{\mathrm{d}\boldsymbol{L}^{(0)}}{\mathrm{d}t} = \boldsymbol{M}^{(0)} \ ,$$

where

$\boldsymbol{L}^{(0)} = \boldsymbol{r} \times \boldsymbol{p}$ = angular momentum with respect to the fixed point 0,

$\boldsymbol{M}^{(0)} = \boldsymbol{r} \times \boldsymbol{F}$ = moment with respect to the fixed point 0.

If the moment vanishes ($\boldsymbol{M}^{(0)} = 0$), the angular momentum is conserved:

$$\boldsymbol{L}^{(0)} = \boldsymbol{r} \times m\,\boldsymbol{v} = \text{const} \ .$$

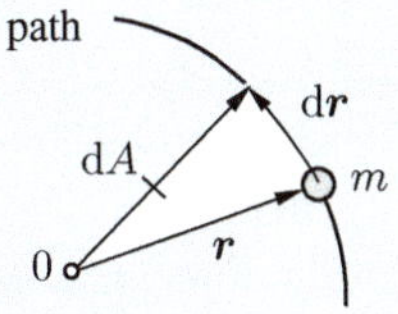

In this case, with

$$\mathrm{d}\boldsymbol{A} = \frac{1}{2}\,\boldsymbol{r} \times \mathrm{d}\boldsymbol{r} \quad \rightsquigarrow \quad \frac{\mathrm{d}\boldsymbol{A}}{\mathrm{d}t} = \frac{1}{2}\,\boldsymbol{r} \times \boldsymbol{v}$$

the *law of areas (Kepler's 2nd law)* is obtained (see page 4):

$$\dot{\boldsymbol{A}} = \text{const} \ .$$

Work–Energy Theorem: Path integration of the law of motion yields

$$\frac{mv_1^2}{2} - \frac{mv_0^2}{2} = \int_{\boldsymbol{r}_0}^{\boldsymbol{r}_1} \boldsymbol{F} \cdot \mathrm{d}\boldsymbol{r} \qquad \text{or} \qquad T_1 - T_0 = U \ ,$$

Kinetic Energy : $\quad T = \frac{1}{2}\,m\,v^2,$

Work of Force $\boldsymbol{F}$: $\quad U = \int \mathrm{d}U = \int \boldsymbol{F} \cdot \mathrm{d}\boldsymbol{r}\,,$

$$\mathrm{d}U = \boldsymbol{F} \cdot \mathrm{d}\boldsymbol{r} = |\boldsymbol{F}|\;|\mathrm{d}\boldsymbol{r}|\cos\alpha.$$

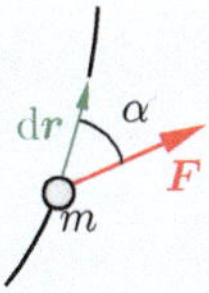

Remarks:
- Forces orthogonal to the path ($\alpha = \pi/2$), do not execute work.
- For a rotation holds $\mathrm{d}U = \boldsymbol{M} \cdot \mathrm{d}\boldsymbol{\varphi}$.

Conservation-of-Energy Law: If the forces according to

$$\boldsymbol{F} = -\operatorname{grad} V = -\Big(\frac{\partial V}{\partial x}\,\boldsymbol{e}_x + \frac{\partial V}{\partial y}\,\boldsymbol{e}_y + \frac{\partial V}{\partial z}\,\boldsymbol{e}_z\Big)$$

can be derived from a potential V ($\widehat{=}$ *conservative forces*), the work is path independent, i.e. given by the potential difference:

$$U = \int_{\boldsymbol{r}_0}^{\boldsymbol{r}_1} \boldsymbol{F} \cdot \mathrm{d}\boldsymbol{r} = V_0 - V_1 \ .$$

From the Work-Energy Theorem then follows

$$T_1 + V_1 = T_0 + V_0 = \text{const} \ .$$

In words: *When the applied forces possess a potential, then the sum of potential energy V and kinetic energy T remains constant during the motion.*

Several Potentials

Gravitational Potential (near earth's surface) $V = mg\,z$

Gravitational Potential (general) $V = -G\,\dfrac{Mm}{r}$

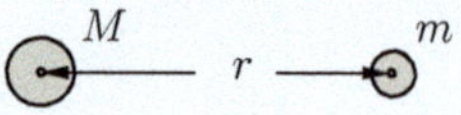

Gravitational constant $G = 6,673 \cdot 10^{-11}\ \mathrm{m^3/kg\,s^2}$

Potential of a spring $V = \dfrac{1}{2}\,kx^2$

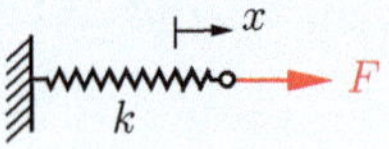

Power

$$P = \frac{\mathrm{d}U}{\mathrm{d}t} = \boldsymbol{F} \cdot \boldsymbol{v} \qquad = \text{Power of a force,}$$

$$P = \boldsymbol{M} \cdot \frac{\mathrm{d}\varphi}{\mathrm{d}t} = \boldsymbol{M} \cdot \boldsymbol{\omega} = \text{Power of a moment.}$$

Projectile Motion

Parabolic trajectory of motion:

$$z = z_0 - \frac{g}{2}\left(\frac{x - x_0}{v_0 \cos\alpha}\right)^2 + (x - x_0)\tan\alpha\ ,$$

Maximum height:

$$h = \frac{1}{2g}(v_0 \sin\alpha)^2\ ,$$

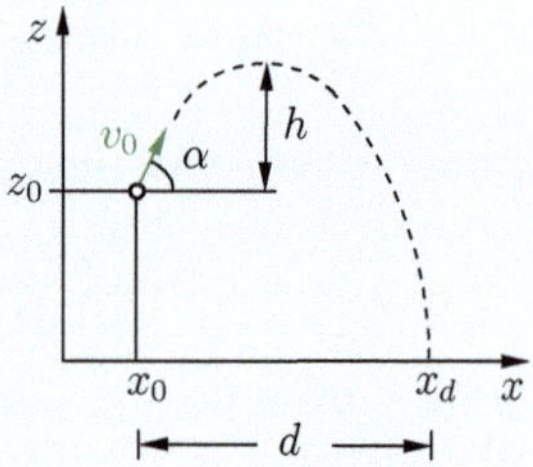

Flight time:

$$t_d = \frac{v_0 \sin\alpha}{g}\left[1 + \sqrt{1 + \frac{2gz_0}{v_0^2 \sin^2\alpha}}\right]\ ,$$

Flight distance:

$$d = v_0^2\,\frac{\sin\alpha\cos\alpha}{g}\left[1 + \sqrt{1 + \frac{2gz_0}{v_0^2 \sin^2\alpha}}\right]\ .$$

Special case $z_0 = 0$:

$$t_d = \frac{2}{g}\,v_0 \sin\alpha\ , \qquad d = \frac{1}{g}\,v_0^2 \sin 2\alpha\ .$$

Problem 2.1 A box of weight W is pushed downwards a rough inclined plane (kinetic friction coefficient μ) with an initial velocity v_0. P2.1

a) Determine the velocity in dependence on the distance.

b) At what distance x_E the box comes to rest? Under what circumstances is this possible?

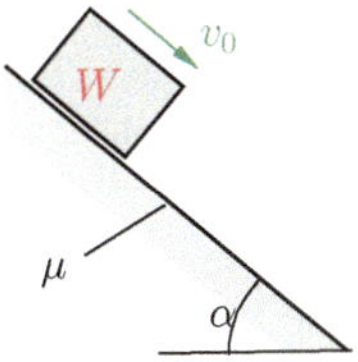

Solution **a)** The law of motion yields in x- and in y-direction

$$\searrow: \quad m\ddot{x} = G\sin\alpha - R\,,$$

$$\nearrow: \quad 0 = N - G\cos\alpha\,.$$

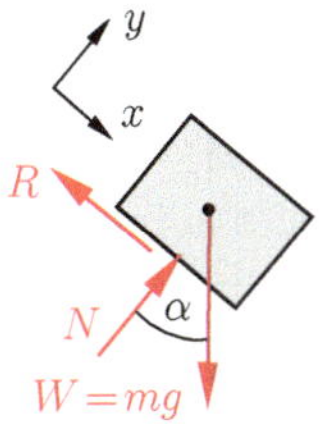

In conjunction with the friction law $R = \mu N$, the acceleration follows as

$$\ddot{x} = g(\sin\alpha - \mu\cos\alpha) = a_0\,.$$

Twice integration, taking into account the initial conditions $x(0) = 0$, $v(0) = v_0$, leads to:

$$v(t) = \dot{x} = v_0 + a_0 t\,, \qquad x(t) = v_0 t + \frac{1}{2}a_0 t^2\,.$$

Therewith, by eleminating the time, we obtain

$$t = \frac{v - v_0}{a_0} \quad \rightsquigarrow \quad x = v_0\frac{v - v_0}{a_0} + \frac{a_0}{2}\,\frac{v^2 - 2vv_0 + v_0^2}{a_0^2} = \frac{v^2 - v_0^2}{2a_0}$$

$$\rightsquigarrow \quad \underline{\underline{v(x) = \sqrt{v_0^2 + 2a_0 x}}}\,.$$

b) From the condition $v(x_E) = 0$ (rest), the covered distance x_E is determined:

$$0 = v_0^2 + 2a_0 x_E \qquad \rightsquigarrow \qquad \underline{\underline{x_E = -\frac{v_0^2}{2a_0}}}\,.$$

From the condition $x_E > 0$ follows $a_0 < 0$, i.e. $\mu > \tan\alpha$.

The same results can be found easier by applying the work-energy theorem $T_1 - T_0 = U$. It leads with

$$U = (mg\sin\alpha)x - Rx, \quad T_0 = \frac{1}{2}mv_0^2, \quad T_1 = \frac{1}{2}mv^2, \quad R = \mu mg\cos\alpha$$

and by solving for v directly to

$$\underline{\underline{v(x) = \sqrt{v_0^2 + 2g(\sin\alpha - \mu\cos\alpha)x}}}\,.$$

P2.2 **Problem 2.2** Two cars, one with and one without ABS, are stopping from the speed $v_0 = 100$ km/h by full breaking. The first without ABS by blocking the wheels, i.e. sliding ('kinetic friction'), the second with ABS with still rolling wheels (ideal 'limiting static friction' assumed).

Determine for both cars the time t_B and the distance s_B for stopping if the coefficients of static and kinetic friction between pavement and tire are $\mu_0 = 0.7$ and $\mu = 0.45$, respectively.

Solution From the equation of motion of the first car (sliding)

$$\rightarrow: \quad m\dot{v} = m\ddot{s} = -R\,, \qquad \uparrow: \quad 0 = N - mg$$

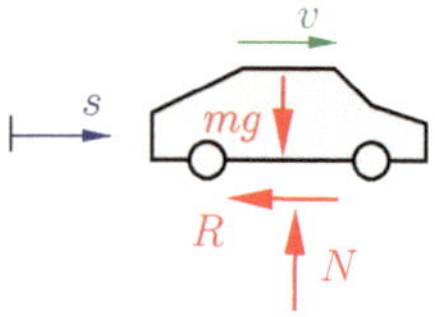

and the friction law

$$R = \mu N\,,$$

it follows

$$\dot{v} = -\mu g\,.$$

Integration yields with $v(t{=}0) = v_0$ and $s(t{=}0) = 0$

$$v(t) = v_0 - \mu g t\,, \qquad s(t) = v_0 t - \frac{1}{2}\,\mu g t^2\,.$$

The stopping time and distance are calculated from the condition $v = 0$:

$$t_B = \frac{v_0}{\mu g}\,, \qquad s_B = s(t_B) = \frac{v_0^2}{\mu g} - \frac{v_0^2}{2\mu g} = \frac{v_0^2}{2\mu g}\,.$$

With the given coefficient of kinetic friction, we obtain

$$\underline{\underline{t_B}} = \frac{100}{3.6 \cdot 0.45 \cdot 9.81} = \underline{\underline{6.3\,\text{s}}}\,, \qquad \underline{\underline{s_B}} = \frac{100^2}{3.6^2 \cdot 2 \cdot 0.45 \cdot 9.81} = \underline{\underline{87\,\text{m}}}\,.$$

For the second car the wheels are still rolling under the limit condition of static friction (ABS), i.e. the friction force is now given by

$$H = H_0 = \mu_0 N\,.$$

This means that in the calculation above only R must replaced by H_0 and μ by μ_0, respectively. Thus, for the car with ABS, we obtain

$$\underline{\underline{t_B}} = 6.3 \cdot \frac{0.45}{0.7} = \underline{\underline{4.05\text{ s}}}\,, \qquad \underline{\underline{s_B}} = 87 \cdot \frac{0.45}{0.7} = \underline{\underline{56\text{ m}}}\,.$$

Remark: Note that stopping time and distance are inverse proportional to the friction coefficient. Note also that, because of the neglected reaction time, the numbers for t_B and s_B in reality might be higher!

Problem 2.3 A parachutist (weight W including parachute) has the initial velocity v_0 imediately after the parachute opens. **P2.3**

a) Determine the velocity v in dependence on t if the air drag is assumed to obey the law $F_d = kv^2$.

b) What limit speed v_l reaches the parachutist?

Solution The law of motion yields

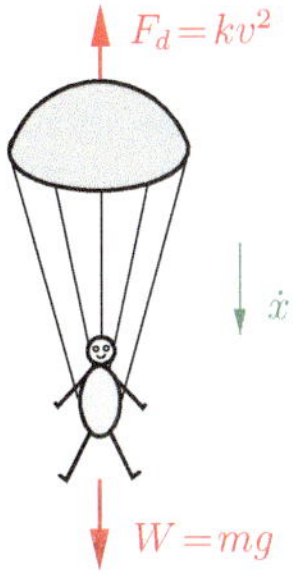

$$\downarrow: \quad ma = m\ddot{x} = mg - kv^2$$

or

$$\ddot{x} = \frac{\mathrm{d}v}{\mathrm{d}t} = g - k_1 v^2 \quad \text{with} \quad k_1 = \frac{k}{m} \ .$$

a) Separation of variables and integration leads to

$$\int_{v_0}^{v} \frac{\mathrm{d}\bar{v}}{g - k_1 \bar{v}^2} = \int_0^t \mathrm{d}\bar{t} \ ,$$

where the time t is counted from the opening of the parachute. With the basic integral

$$\int \frac{\mathrm{d}z}{A - Bz^2} = \frac{1}{\sqrt{AB}} \operatorname{artanh}(\sqrt{B/A}\, z)$$

we obtain

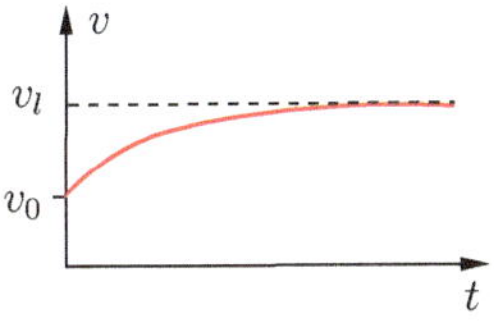

$$\left[\frac{1}{\sqrt{gk_1}} \operatorname{artanh}\sqrt{k_1/g}\, \bar{v}\right]_{v_0}^{v} = t$$

or by solving for $v(t)$

$$\underline{\underline{v(t) = \sqrt{\frac{g}{k_1}} \tanh\left(\sqrt{gk_1}\, t + \operatorname{artanh}\sqrt{\frac{k_1}{g}}\, v_0\right)}} \ .$$

b) For $t \to \infty$, it follows ($\tanh z \to 1$ for $z \to \infty$)

$$\underline{\underline{v_l}} = \sqrt{\frac{g}{k_1}} = \underline{\underline{\sqrt{\frac{W}{k}}}} \ .$$

The same result can be found from the consideration that in the limit case, the acceleration is zero:

$$a = g - k_1 v_l^2 = 0 \quad \rightsquigarrow \quad \underline{\underline{v_l}} = \sqrt{\frac{g}{k_1}} = \underline{\underline{\sqrt{\frac{W}{k}}}} \ .$$

P2.4

Problem 2.4 A computer (weight W=100 N) in a packing case is protected against impact by foam plastics (spring stiffness $k = 100$ N/cm).

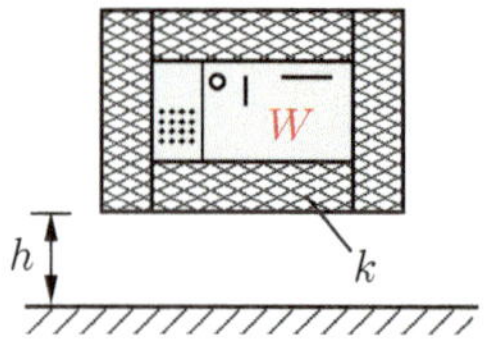

From what height h the case may impinge a hard surface, if the acceleration of the computer shall not be bigger than four times the gravity?

Solution During free fall, the case experiences the acceleration g. After impinging the surface, the foam plastics (≙ linear spring) will be compressed and the computer will be accelerated upwards. Then the motion is described by

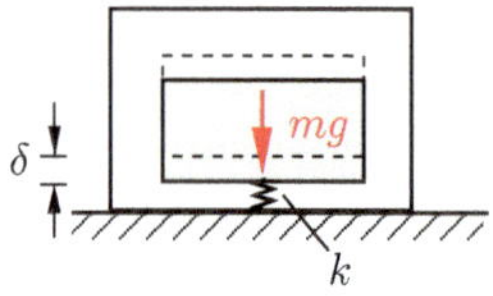

$$\uparrow: \quad m\,a = -mg + k\,\delta\,.$$

From the condition $a_{\max} = 4g$ follows the maximum spring compression

$$\delta_{\max} = \frac{5mg}{k} = 5\,\text{cm}\,.$$

Knowing this limit compression, the allowable height of fall can be determined from the conservation of energy law

$$T_1 + V_1 = T_2 + V_2$$

as follows:

state 1

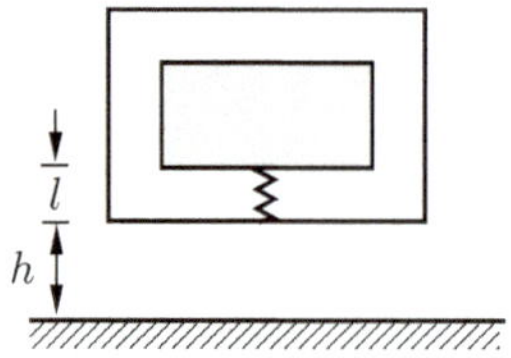

state 2

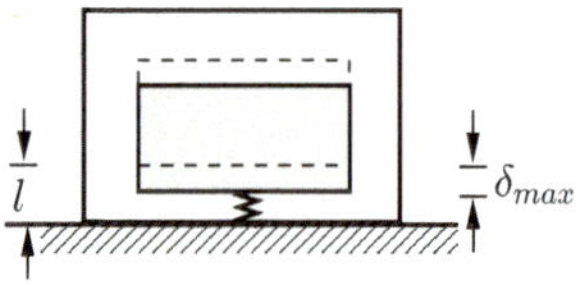

$$T_1 = 0\,, \quad V_1 = mg(l + h)\,, \quad T_2 = 0\,, \quad V_2 = mg(l - \delta_{\max}) + \frac{1}{2}k\delta_{\max}^2\,.$$

Introducing these quantities yields

$$\underline{\underline{h}} = \frac{1}{2}\,\frac{k}{mg}\,\delta_{\max}^2 - \delta_{\max} = \frac{15}{2}\,\frac{mg}{k} = \underline{\underline{7.5\,\text{cm}}}\,.$$

Problem 2.5 On a rough inclined plane (inclination angle α, kinetic friction constant μ), a block (mass m_1) is moving which is connected by a rope with a body of mass m_2. Pulley and rope are regarded as massless. **P2.5**

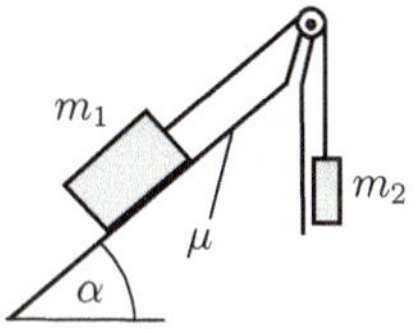

a) Determine the accelerations when m_1 slides upwards and downwards, respectively.

b) What force acts in the rope?

Solution **a)** We cut the rope and formulate for the 3 parts the basic equations, where we first assume upward sliding:

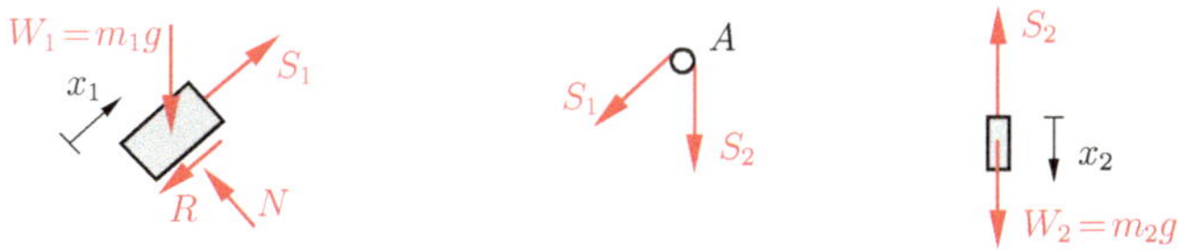

$$\nearrow: \ m_1 a_1 = S_1 - R - W_1 \sin\alpha, \quad \overset{\curvearrowleft}{A}: \ S_1 = S_2, \quad \downarrow: \ m_2 a_2 = W_2 - S_2,$$

$$\nwarrow: \ N = W_1 \cos\alpha, \qquad R = \mu N\,.$$

With the kinematic condition (unextensible rope) $v_1 = v_2$ and consequently $a_1 = a_2 = a$, we obtain

$$\underline{\underline{a^{(u)} = a = g\,\frac{m_2 - m_1(\sin\alpha + \mu\cos\alpha)}{m_1 + m_2}}}\,.$$

For upward sliding, the acceleration must be positive, $a > 0$, and therefore $m_2 > m_1(\sin\alpha + \mu\cos\alpha)$!

For downward sliding, only the direction of R must be changed. Then it follows

$$\underline{\underline{a^{(d)} = a = -g\,\frac{m_1(\sin\alpha - \mu\cos\alpha) - m_2}{m_1 + m_2}}}\,.$$

This case occurs for $a < 0$, i.e. for $m_1(\sin\alpha - \mu\cos\alpha) > m_2$.

b) Independent on the sliding direction, the force in the rope is

$$S = S_2 = S_1 = W_2 - m_2 a_2 = m_2(g - a)\,.$$

Introducing the respective accelerations yields

$$\underline{\underline{S^{(u)} = \frac{m_1 m_2 g(1 + \sin\alpha + \mu\cos\alpha)}{m_1 + m_2}}}, \qquad \underline{\underline{S^{(d)} = \frac{m_1 m_2 g(1 + \sin\alpha - \mu\cos\alpha)}{m_1 + m_2}}}.$$

P2.6

Problem 2.6 A chain (mass m, length l) lying on an inclined frictionless support starts sliding from the sketched position.

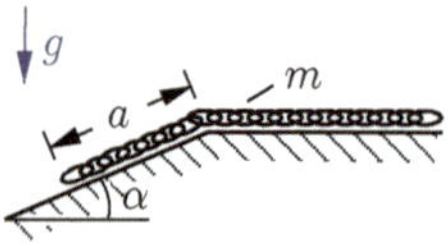

a) Determine the position and velocity as functions of time.

b) Determine the velocity as function of the position by applying the conservation-of-energy law.

Solution **a)** Each chain link experiences the same velocity and acceleration. We therefore consider the chain as a single body of mass m which is driven by a force which depends on the length x of the overhanging part. Thus, the equation of motion reads

$$m\,\ddot{x} = m\,\frac{x}{l}\,g\sin\alpha \quad \rightsquigarrow \quad \ddot{x} - \kappa^2 x = 0\,.$$

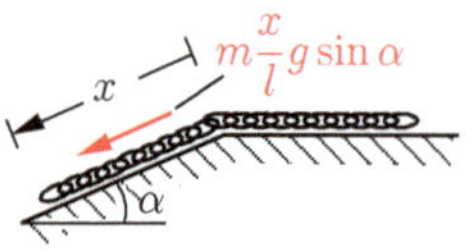

where $\kappa^2 = g\sin\alpha/l$. This differential equation has the solution

$$x(t) = A\cosh\kappa t + B\sinh\kappa t$$
$$\dot{x} = A\kappa\sinh\kappa t + B\kappa\cosh\kappa t\,.$$

The integration constants follow from the initial conditions

$$\dot{x}(0) = 0 \quad \rightsquigarrow \quad B = 0\,, \qquad x(0) = a \quad \rightsquigarrow \quad A = a$$

what finally leads to

$$\underline{\underline{x(t) = a\cosh\kappa t}}\,, \qquad \underline{\underline{\dot{x}(t) = a\kappa\sinh\kappa t}}\,, \qquad \kappa^2 = g\sin\alpha/l.$$

Note that this solution is only valid for $a \le x \le l$. When the complete chain is on the inclined plane, the driving force remains constant!

b) If we use as reference position for zero potential energy the upper horizontal plane, the energy terms in the initial and in the displaced position are given by

$$V_0 = -\frac{a}{2}\sin\alpha\, m\frac{a}{l}g\,, \qquad T_0 = 0\,,$$
$$V_1 = -\frac{x}{2}\sin\alpha\, m\frac{x}{l}g\,, \qquad T_1 = \frac{1}{2}\dot{x}^2 m\,.$$

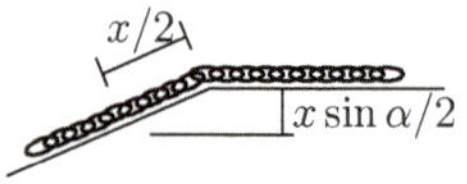

Introducing into $V_0 + T_0 = V_1 + T_1$ and solving for $\dot{x}$ leads to

$$\underline{\underline{\dot{x}(x) = \sqrt{(x^2 - a^2)g\sin\alpha}}}\,.$$

Again, this solution is only valid for $a \le x \le l$.

Problem 2.7 Determine the geometric locus of all points P_1 that is given by the position of all point masses at time $t = t_1$, that are thrown at time $t = 0$ with the *same* initial velocity v_0 from a point P under *different* angles α with respect to the horizontal. Assume that all trajectories are located in the same vertical plane and that there is no air drag. P2.7

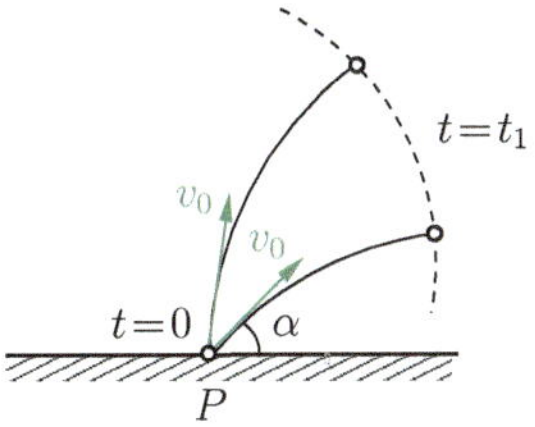

Solution For convenience, the origin of the coordinates is chosen at P. Then it follows from

$$\uparrow: \quad m\ddot{z} = -mg\,, \qquad \rightarrow: \quad m\ddot{x} = 0$$

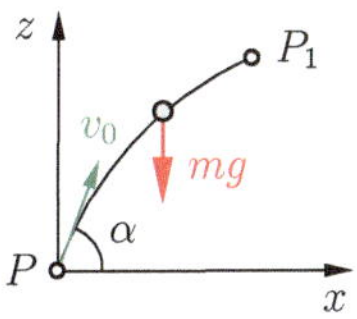

with the initial conditions

$$x(0) = z(0) = 0\,,$$

$$\dot{x}(0) = v_0 \cos\alpha\,,$$

$$\dot{z}(0) = v_0 \sin\alpha$$

by integration

$$x = v_0 t \cos\alpha\,, \qquad z = -\frac{1}{2}gt^2 + v_0 t \sin\alpha\,.$$

Since the solution is sought at time t_1 for arbitrary angles, the angle α must be eliminated. Squaring and adding yields

$$\left.\begin{aligned} x_1^2 &= (v_0 t_1)^2 \cos^2\alpha\,, \\ \left(z_1 + \frac{g}{2}t_1^2\right)^2 &= (v_0 t_1)^2 \sin^2\alpha \end{aligned}\right\}$$

$$\rightsquigarrow \quad \underline{\underline{x_1^2 + \left(z_1 + \frac{g}{2}t_1^2\right)^2 = (v_0 t_1)^2}}\,.$$

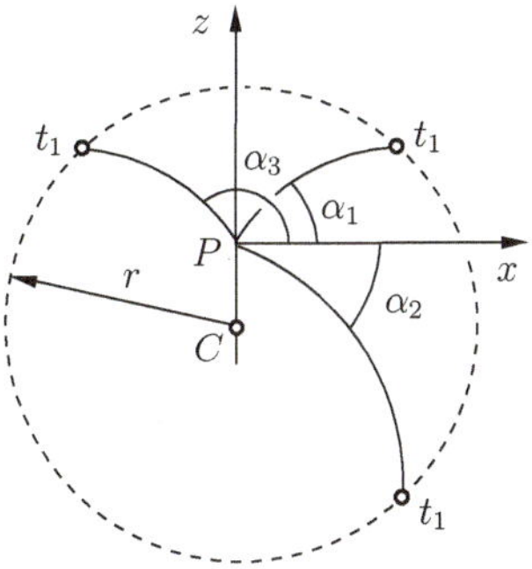

Accordingly, all points P_1 are located on a circle with the radius $r = v_0 t_1$ and the center C at

$$z = -\frac{g}{2}t_1^2\,.$$

P2.8 **Problem 2.8** From the top of a tower, two point masses are thrown with the same initial velocity v_0 under two different angles α_1 and α_2. It is recognized that both masses impinge the surface at the same location.

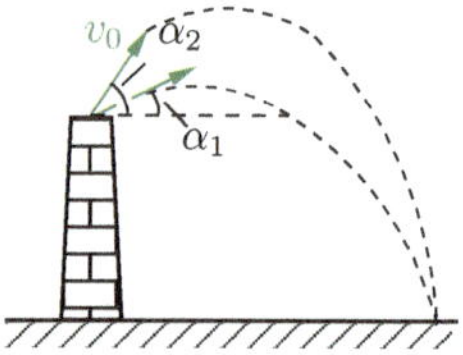

Determine the height of the tower.

Solution The parabolic trajectory of motion is given by (see page 20)

$$z - z_0 = -\frac{g}{2}\left(\frac{x - x_0}{v_0 \cos\alpha}\right)^2 + (x - x_0)\tan\alpha \, .$$

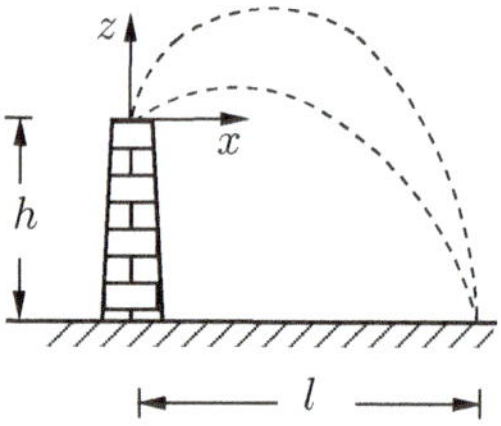

We chose the origin of the coordinates at the top of the tower. Then we have $x_0 = z_0 = 0$, and the point where the masses impinge the surface has the unknown coordinates $x = l$, $z = -h$. For both throws holds:

$$-h = -\frac{g}{2}\,\frac{l^2}{v_0^2 \cos^2\alpha_1} + l\tan\alpha_1 \, ,$$

$$-h = -\frac{g}{2}\,\frac{l^2}{v_0^2 \cos^2\alpha_2} + l\tan\alpha_2 \, .$$

Equating both expressions leads for the horizontal distance l to

$$l = \frac{2v_0^2}{g}\,\frac{1}{\tan\alpha_1 + \tan\alpha_2} \, .$$

Herewith, from the 1st equation, the height is determined as

$$\underline{\underline{h}} = +\frac{g}{2v_0^2}\left(\frac{2v_0^2}{g}\right)^2 \frac{1}{\cos^2\alpha_1}\left(\frac{1}{\tan\alpha_1 + \tan\alpha_2}\right)^2 - \frac{2v_0^2}{g}\,\frac{\tan\alpha_1}{\tan\alpha_1 + \tan\alpha_2}$$

$$= \underline{\underline{\frac{2v_0^2}{g}\,\frac{1}{(\tan\alpha_1 + \tan\alpha_2)\tan(\alpha_1 + \alpha_2)}}} \, .$$

Remark: For the solution, the following formulas are used:

$$\frac{1}{\cos^2\alpha_2} - \frac{1}{\cos^2\alpha_1} = \tan^2\alpha_2 - \tan^2\alpha_1$$

$$= (\tan\alpha_2 - \tan\alpha_1)(\tan\alpha_2 + \tan\alpha_1) \, ,$$

$$\frac{\tan\alpha_1 + \tan\alpha_2}{1 - \tan\alpha_1 \tan\alpha_2} = \tan(\alpha_1 + \alpha_2) \, .$$

Problem 2.9 To rescue shipwrecked persons, a rescue package of mass m is dropped from an airplane, flying with speed $v_0 = 200\,\mathrm{km/h}$ in a height of $h = 150\,\mathrm{m}$. **P2.9**

a) Determine the distance s_B from launching the package until it impacts on sea surface.

b) Calculate the impact velocity v_B

Assuming a high horizontal velocity component v_h, the air drag shall be taken into account by a horizontal drag force $D = \kappa m v_h^2$ with $\kappa = 0.003\,\mathrm{m}^{-1}$.

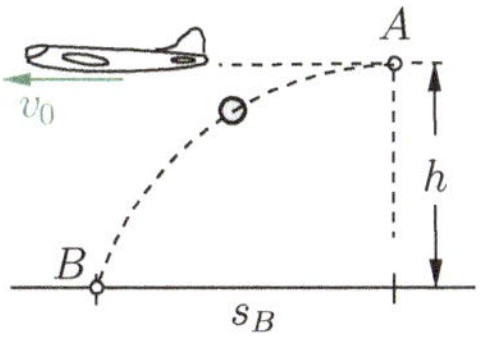

Solution **a)** We introduce an appropriate coordinate system and sketch the free body diagram with the acting drag force D and the weight mg. With $v_h = \dot{x}$, the equations of motion in x- and in z direction read

$$\leftarrow: \quad m\ddot{x} = -m\kappa\,\dot{x}^2\,, \qquad \uparrow: \quad m\ddot{z} = -mg\,.$$

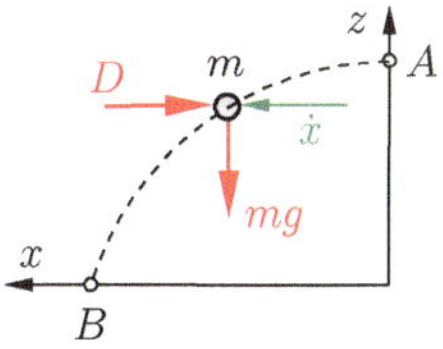

Integration yields with the initial conditions $x(0) = 0$, $z(0) = h$, $\dot{x}(0) = v_0$, $\dot{z}(0) = 0$

$$\int_{v_0}^{\dot{x}} \frac{\mathrm{d}\bar{\dot{x}}}{\bar{\dot{x}}^2} = -\kappa \int_0^t \mathrm{d}\bar{t} \quad \rightsquigarrow \quad \dot{x} = \frac{1}{\frac{1}{v_0} + \kappa t}\,,$$

$$\int_0^x \mathrm{d}\bar{x} = \int_0^t \frac{1}{\frac{1}{v_0} + \kappa\bar{t}}\,\mathrm{d}\bar{t} \quad \rightsquigarrow \quad x = \frac{1}{c_0}\ln(1 + \kappa v_0 t)$$

$$\dot{z} = -gt \quad , \qquad z = -\frac{g}{2}t^2 + h\,.$$

The impact time t_B follows from $z_B = 0$ as

$$t_B = \sqrt{2h/g}\,,$$

and thus, we obtain for the distance

$$\underline{\underline{s_B = x(t_B) = \frac{1}{\kappa}\ln\left(1 + \kappa v_0\sqrt{2h/g}\,\right) = 218\,\mathrm{m}}}\,.$$

b) From the velocity components at impact,

$$\dot{x}(t_B) = \frac{v_0}{1 + \kappa v_0 t_B} = 104\,\mathrm{km/h}\,,$$

$$\dot{z}(t_B) = -g t_B = -54.2\,\mathrm{m/s} = -195\,\mathrm{km/h}\,,$$

results the velocity as

$$\underline{\underline{v_B = \sqrt{\dot{x}^2(t_B) + \dot{z}^2(t_B)} = 221\,\mathrm{km/h}}}\,.$$

P2.10 **Problem 2.10** A rocket without an own propulsion is catapulted vertically upwards from earth's surface with an initial velocity v_0.

a) Determine the maximum flight hight H by considering the change of gravitation and neglecting drag forces.

b) What magnitude of v_0 is required when the rocket shall escape from the gravitation field of earth? (Earth's radius $R = 6370$ km)

Solution **a)** Since only conservative forces are acting, the conservation-of-energy law

$$T_1 + V_1 = T_0 + V_0$$

is appropriate as solution method. The gravitational potential $V = -GMm/r$ according to (force on earth's surface = weight mg)

$$mg = -\frac{\mathrm{d}V}{\mathrm{d}r}\bigg|_{r=R} = G\frac{Mm}{R^2} \quad \rightsquigarrow \quad GM = gR^2$$

can be written as

$$V = -mg\frac{R^2}{r}\,.$$

Thus, the different energies on earth's surface ($r = R$) and final flight height ($r = R + H$) are

$$T_0 = \frac{1}{2}mv_0^2\,, \qquad V_0 = -mgR\,,$$

$$T_1 = 0\,, \qquad V_1 = -mg\frac{R^2}{R+H}\,.$$

Introduction into the energy conservation law yields

$$-mg\frac{R^2}{R+H} = \frac{1}{2}mv_0^2 - mgR \quad \rightsquigarrow \quad \underline{\underline{H = R\,\frac{v_0^2}{2gR - v_0^2}}}\,.$$

b) The 'escape velocity' v_0^* is found from

$$H \to \infty \quad \rightsquigarrow \quad \underline{\underline{v_0^* = \sqrt{2gR}}} = 11180\,\frac{\mathrm{m}}{\mathrm{s}} \approx \underline{\underline{40000\,\frac{\mathrm{km}}{\mathrm{h}}}}\,.$$

Remarks:

- Note that a rocket in real cases is *not* launched from earth's surface without an own propulsion!
- The required kinetic energy to reach the 'escape velocity' is $T_0 = mgR$.

Problem 2.11 Which minimum initial velocity v_0 in position ① is necessary, such that the body with mass m reaches position ② if **P2.11**

a) it slides along a frictionless circular path (radius l),

b) it is fixed at a rigid massless rod (length l) ?

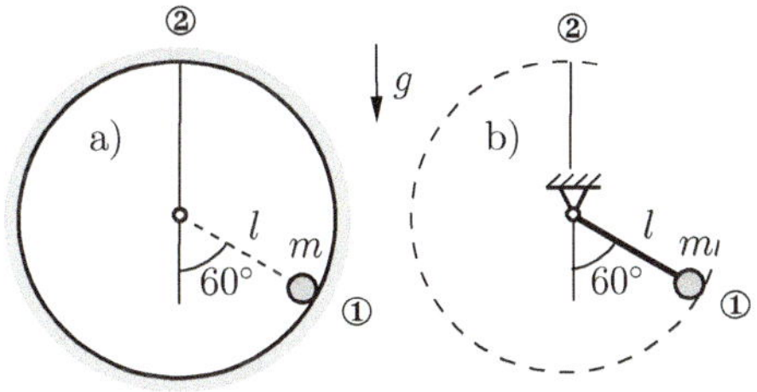

Solution In both cases the initial velocity v_0 is connected with v_2 at position ② by the energy-conservation law $T_2 + V_2 = T_1 + V_1$. Choosing zero potential energy at position ①, we obtain with $V_1 = 0$

$$\frac{m}{2}\,v_2^2 + mg(l + l\cos 60°) = \frac{m}{2}\,v_0^2 \quad \rightsquigarrow \quad v_0^2 = v_2^2 + 3gl\,.$$

a) The necessary velocity v_2 is obtained from the condition that the normal force N between the path and the body is non-negative (otherwise the body looses contact with the path). From the law of motion

$$\nearrow: \quad ma_n = N - mg\cos\varphi$$

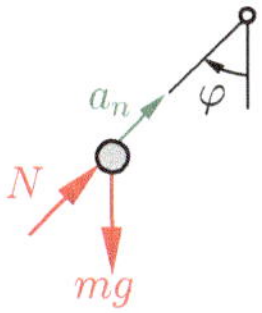

with $a_n = v^2/l$ we obtain at position ② ($\varphi = \pi$) for the limit case $N = 0$:

$$\frac{mv_2^2}{l} = mg \quad \rightsquigarrow \quad v_2^2 = gl\,.$$

Hence, it follows

$$v_0^2 = gl + 3gl \quad \rightsquigarrow \quad \underline{\underline{v_0 = 2\sqrt{gl}}}\,.$$

b) The initial velocity for the mass fixed at the rod will take a minimum if it comes to rest in position ②. For $v_2 = 0$, the energy-conservation law directly yields

$$\underline{\underline{v_0 = \sqrt{3}\,\sqrt{gl}}}\,.$$

Remark: In case b) the force S in the rod may get negative. For example, in ② ($v_2 = 0$) the force is $S = -mg$.

P2.12 **Problem 2.12** Determine the velocity $v(\varphi)$ of mass m of a simple pendulum in dependence on the maximum amplidude φ_0.

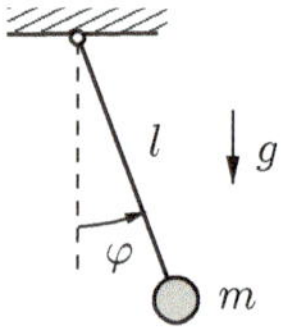

Discuss the result for characteristic angles φ.

Solution Since the velocity shall be determined in dependence on the position, the energy-conservation law is the first choice as solution method. As reference position for the potential energy, we choose the horizontal position $\varphi = \pi/2$ and find from

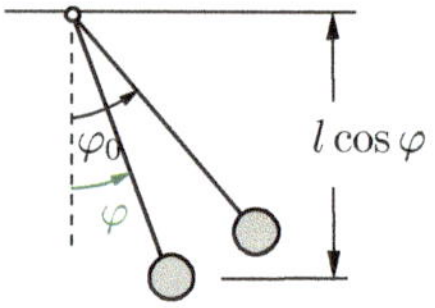

$$T(\varphi) + V(\varphi) = T_0 + V_0$$

with $v(\varphi_0) = 0$, i.e. $T_0 = 0$:

$$\frac{m}{2}\,v^2 - mgl\cos\varphi = 0 - mgl\cos\varphi_0 \quad \rightsquigarrow \quad \underline{\underline{v = \pm\sqrt{2gl(\cos\varphi - \cos\varphi_0)}}}\,.$$

The same result can be found by integrating the law of motion in tangential direction. From

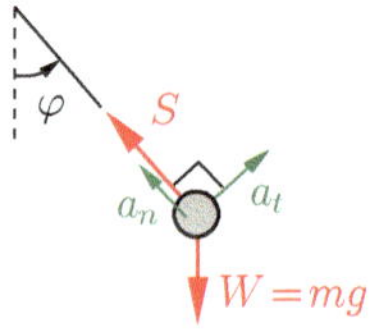

$$\nearrow: \quad ma_t = ml\ddot{\varphi} = -mg\sin\varphi$$

with $\ddot{\varphi}\mathrm{d}\varphi = \dot{\varphi}\mathrm{d}\dot{\varphi}$ and the initial condition $\dot{\varphi}(\varphi_0) = 0$, we obtain

$$\int_{\dot{\varphi}}^{0} \dot{\bar{\varphi}}\,\mathrm{d}\dot{\bar{\varphi}} = -\frac{g}{l}\int_{\varphi}^{\varphi_0} \sin\bar{\varphi}\,\mathrm{d}\bar{\varphi} \quad \rightsquigarrow \quad -\frac{\dot{\varphi}^2}{2} = \frac{g}{l}(\cos\varphi_0 - \cos\varphi)$$

or with $v = l\dot{\varphi}$ again

$$\underline{\underline{v^2 = 2gl(\cos\varphi - \cos\varphi_0)}}\,.$$

The maximum speed occurs for $\cos\varphi = 1$, i.e. at $\varphi = 0$:

$$v_{\max} = \sqrt{2gl(1-\cos\varphi_0)} = \sqrt{2gl2\sin^2\frac{\varphi_0}{2}} = 2\sqrt{gl}\sin\frac{\varphi_0}{2}\,.$$

For a *small* maximum amplitude φ_0 also the angle φ remains small, and we obtain by truncated series approximation

$$\cos\varphi \approx 1 - \varphi^2/2\,, \qquad \sin(\varphi_0/2) \approx \varphi_0/2$$
$$\rightsquigarrow \quad v^2 = gl(\varphi_0^2 - \varphi^2)\,, \qquad v_{\max} = \sqrt{gl}\,\varphi_0\,.$$

Problem 2.13 In a clamped *frictionless* pipe elbow (radius R) glides a sphere (weight $W = mg$) with zero initial velocity downwards from the top. **P2.13**

Determine the support reactions at the clamping in dependence on the position φ of the sphere.

At which φ the reactions take extreme values?

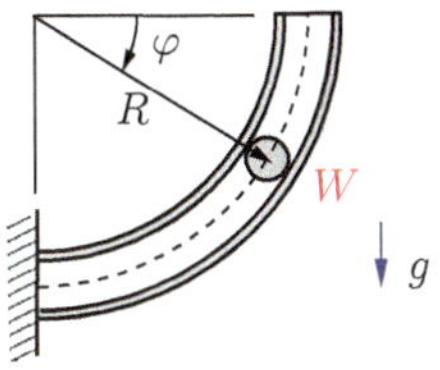

Solution NEWTON'S law yields in components:

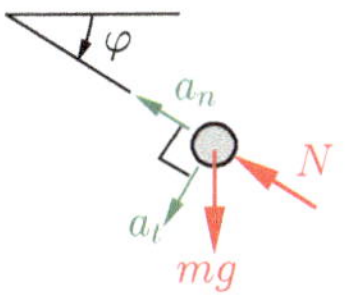

$$\swarrow:\quad ma_t = mg\cos\varphi\,,$$

$$\nwarrow:\quad ma_n = N - mg\sin\varphi\,.$$

With $a_t = R\ddot{\varphi}$, $a_n = R\dot{\varphi}^2$ and $\ddot{\varphi}\mathrm{d}\varphi = \dot{\varphi}\mathrm{d}\dot{\varphi}$, it follows from the 1st equation by integration

$$\int_0^{\dot{\varphi}} \dot{\bar{\varphi}}\,\mathrm{d}\dot{\bar{\varphi}} = \int_0^{\varphi} \frac{g}{R}\cos\bar{\varphi}\,\mathrm{d}\bar{\varphi} \quad\leadsto\quad \frac{\dot{\varphi}^2}{2} = \frac{g}{R}\sin\varphi\,.$$

Therewith, we obtain the normal force from the 2nd equation as

$$N(\varphi) = mg\sin\varphi + mR\dot{\varphi}^2 = 3W\sin\varphi\,.$$

The equilibrium conditions for the elbow lead to the support reactions:

$$\uparrow:\quad \underline{\underline{A_V}} = N\sin\varphi = \underline{\underline{3W\sin^2\varphi}}\,,$$

$$\leftarrow:\quad \underline{\underline{A_H}} = N\cos\varphi = 3W\sin\varphi\cos\varphi$$

$$= \underline{\underline{-\frac{3}{2}W\sin 2\varphi}}\,,$$

$$\curvearrowright A:\quad \underline{\underline{M_A}} = -NR\cos\varphi = \underline{\underline{-\frac{3}{2}WR\sin 2\varphi}}\,.$$

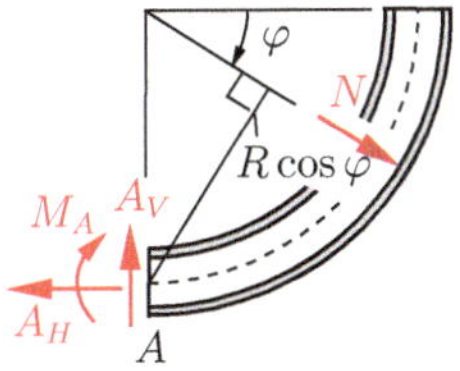

The clamping moment M_A and the horizontal force A_H take their maximum, when the sphere is at $\varphi = \pi/4$. The vertical force A_V is maximal for $\varphi = \pi/2$:

$$M_{A\,\max} = -\frac{3}{2}WR\,,\qquad A_{H\,\max} = \frac{3}{2}W\,,\qquad A_{V\,\max} = 3W\,.$$

P2.14 **Problem 2.14** A hockey puck (mass m) in an ideally smooth ice field (no friction) is shot with speed v_0 into the half-circular part of the boards and slides along the boards. At the end of the curved part, the speed is measured to be $0.6\,v_0$.

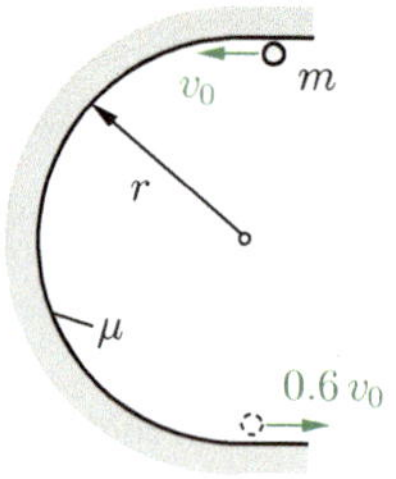

Determine the friction coefficient μ of the boards.

Solution With aid of the free body diagram, we obtain the equations of motion

$$\swarrow: \quad ma_t = -R\,, \qquad \searrow: \quad ma_n = N\,.$$

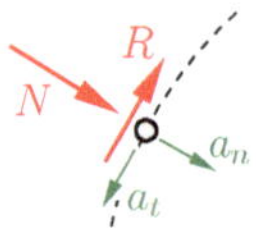

Introducing the kinematic relations $a_t = \dot{v}$, $a_n = v^2/r$ and the friction law $R = \mu N$ yields

$$m\dot{v} = -\mu m\,\frac{v^2}{r}\,.$$

Separation of variables and integration leads to the velocity $v(t)$:

$$\int_{v_0}^{v} \frac{d\bar{v}}{\bar{v}^2} = -\int_0^t \frac{\mu}{r}\,d\bar{t} \quad \leadsto \quad v(t) = \frac{v_0}{1 + \frac{\mu v_0}{r}\,t}\,.$$

From repeated integration, the path $s(t)$ is determined:

$$\int_0^s d\bar{s} = v_0 \int_0^t \frac{d\bar{t}}{1 + \mu\frac{v_0}{r}\,\bar{t}} \quad \leadsto \quad s(t) = \frac{r}{\mu}\ln\left(1 + \frac{\mu v_0}{r}\,t\right)\,.$$

The time t_1, until the speed is reduced to $0.6\,v_0$, is calculated as

$$0.6\,v_0 = \frac{v_0}{1 + \frac{\mu v_0}{r}\,t_1} \quad \leadsto \quad t_1 = \frac{2}{3}\,\frac{r}{\mu v_0}\,.$$

Thus, we obtain for the corresponding path length at the end of the curved boards

$$s_1 = s(t_1) = \frac{r}{\mu}\ln\left(1 + \frac{\mu v_0}{r}\,t_1\right) = \frac{r}{\mu}\,\ln\frac{5}{3}\,.$$

Equalizing it with the length of the half-circle yields

$$s_1 = r\pi \quad \leadsto \quad \underline{\underline{\mu = \frac{\ln(5/3)}{\pi} = 0.16}}\,.$$

Problem 2.15 A car (weight $W = mg$) passes a bend (static friction coefficient μ_0), whose curvature $1/\rho$ increases proportional to the covered distance s, i.e. $s = A^2/\rho$ (clothoid). At time $t_0 = 0$, the car is at $s_0 = 0$ and has an initial speed v_0. **P2.15**

At which speed v, where and when the car 'skids off the bend' if
a) it moves with constant speed,
b) it brakes with constant deceleration a_0?

Given.: $A = 35\,\text{m}$, $\mu_0 = 0.6$, $a_0 = g/4$, $v_0 = 72\,\text{km/h}$.

Solution **a)** For $a_t = 0$, Newton's law yields with $a_n = v_0^2/\rho$ the friction force

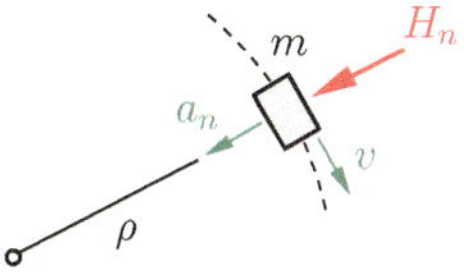

$$\swarrow:\quad H = H_n = m a_n = m\frac{v_0^2}{\rho}\,.$$

The car leaves its path when the limit friction force is attained:

$$H = \mu_0 mg \quad \leadsto \quad m\frac{v_0^2}{\rho_1} = \mu_0 mg \quad \leadsto \quad \rho_1 = \frac{v_0^2}{\mu_0 g}\,.$$

This leads with $s = A^2/\rho = v_0 t$ to $(v_1 = v_0)$

$$\underline{\underline{s_1 = \frac{\mu_0 g A^2}{\rho_1} = 18\,\text{m}}}\,, \qquad \underline{\underline{t_1 = \frac{s_1}{v_0} = 0.9\,\text{s}}}\,.$$

b) When the motion is decelerated, an additional force acts in tangential direction. With $a_t = -a_0$ follows

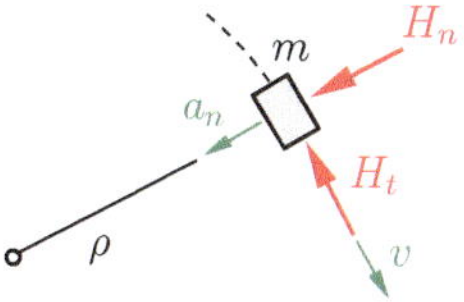

$$\searrow:\quad m a_t = -m a_0 = -H_t\,,$$

$$\swarrow:\quad m a_n = m\frac{v^2}{\rho} = H_n\,.$$

Static limit friction is attained for

$$H = \sqrt{H_n^2 + H_t^2} = \mu_0 mg \quad \leadsto \quad \sqrt{\left(\frac{mv^2}{\rho_2}\right)^2 + (m a_0)^2} = \mu_0 mg$$

$$\leadsto \quad \frac{v^2}{\rho_2} = \sqrt{\mu_0^2 g^2 - a_0^2}\,.$$

Thus, with $v = \sqrt{v_0^2 - 2a_0 s} = v_0 - a_0 t$ (constant deceleration) and $s = A^2/\rho$, we obtain

$$\underline{\underline{s_2}} = \frac{v_0^2}{4a_0} \overset{(+)}{-} \sqrt{\left(\frac{v_0^2}{4a_0}\right)^2 - \frac{A^2}{2a_0}\sqrt{\mu_0^2 g^2 - a_0^2}} = \underline{\underline{22.7\,\text{m}}}\,,$$

$$\underline{\underline{v_2}} = \sqrt{v_0^2 - 2a_0 s_2} = \underline{\underline{61.2\,\text{km/h}}}\,, \qquad \underline{\underline{t_2}} = \frac{v_0 - v_2}{a_0} = \underline{\underline{1.22\,\text{s}}}\,.$$

P2.16 **Problem 2.16** The potential of a free movable point mass m in a horizontal plane is given by $V(x, y) = k(x^2 + y^2)/2$.

a) Determine the acting forces and formulate the equations of motion.

b) Determine the path of the point mass in parameter and implicit representation for the initial conditions $x(0) = a$, $\dot{x}(0) = 0$, $y(0) = 0$, $\dot{y}(0) = v_0$.

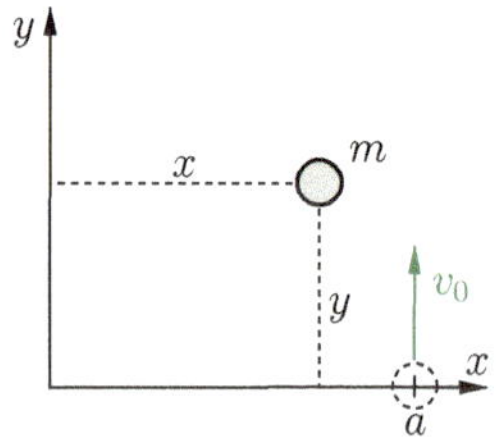

Solution **a)** The forces follow from the derivatives of the potential as

$$F_x = -\frac{\partial V}{\partial x} = -kx\,, \qquad F_y = -\frac{\partial V}{\partial y} = -ky\,,$$

which leads to the equations of motion

$$\rightarrow: \; m\ddot{x} = F_x = -kx\;, \qquad \uparrow: \; m\ddot{y} = F_y = -ky$$

or

$$\underline{\underline{\ddot{x} + \omega^2 x = 0}}\,, \qquad \underline{\underline{\ddot{y} + \omega^2 y = 0}}\,, \qquad \text{where} \quad \omega^2 = k/m\;.$$

b) Both differential equations describe free undamped vibrations, whose solutions are given by (cf. chapter 7)

$$x = A\cos\omega t + B\sin\omega t\;, \qquad y = C\cos\omega t + D\sin\omega t\;,$$

where A, B, C, D are constants. They are determined by using the initial conditions:

$$x(0) = a \quad \rightsquigarrow \quad A = a\;, \quad y(0) = 0 \quad \rightsquigarrow \; C = 0\;,$$

$$\dot{x}(0) = 0 \quad \rightsquigarrow \quad B = 0\;, \quad \dot{y}(0) = v_0 \quad \rightsquigarrow \; D = \frac{v_0}{\omega}\;.$$

Thus, in parameter representation, the path is described by

$$\underline{\underline{x(t) = a\cos\omega t}}\;, \qquad \underline{\underline{y(t) = \frac{v_0}{\omega}\sin\omega t}}\;.$$

The implicit representation is found by eliminating t through squaring and adding, resulting in

$$\underline{\underline{\left(\frac{x}{a}\right)^2 + \left(\frac{y}{v_0/\omega}\right)^2 = 1}}\;.$$

Accordingly, the point mass moves counterclockwise along an ellipse with half-axes a and v_0/ω.

Problem 2.17 A space vessel shall be lifted from a circular path with radius r_1 around the earth E (earth's radius R) to a more distant circular orbit of radius r_2. The transition is carried out by changing in A and in B suddenly the magnitude of the velocity of the vessel. P2.17

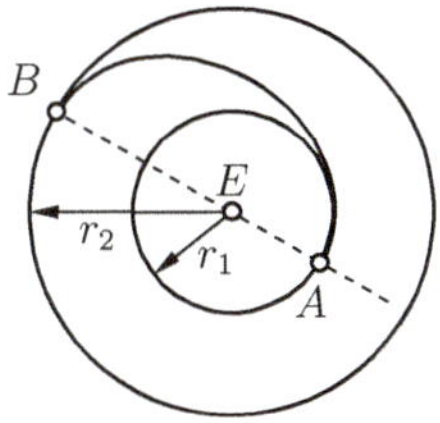

Determine the necessary velocity change in A.

Solution According to Kepler's 1st law, the vessel moves between A and B along an ellipse, whose one focus coincides with the earth. Since in A and B only the magnitude of velocity and not its direction shall be changed, the transition ellipse in A and in B must be tangential to the circles. From this condition the ellipse parameters follow as

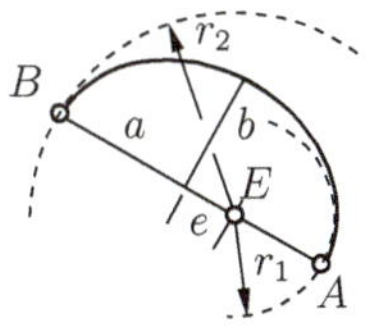

$$a = \frac{r_1 + r_2}{2}\,, \quad e = a - r_1 = \frac{r_2 - r_1}{2}\,, \quad b^2 = a^2 - e^2 = r_1 r_2\,,$$

and the curvature radius at A (vertex of the ellipse) yields

$$\rho = \frac{b^2}{a} = \frac{2 r_1 r_2}{r_1 + r_2}\,.$$

In A, the gravitational force has the magnitue $F = mg(R/r_1)^2$. At this location, the law of motion normal to the circular path (before velocity change) leads to

$$m\,\frac{v_1^2}{r_1} = mg\left(\frac{R}{r_1}\right)^2 \quad \leadsto \quad v_1 = R\sqrt{\frac{g}{r_1}}$$

and normal to the elliptic orbit (after velocity change) to

$$m\,\frac{v_A^2}{\rho} = mg\left(\frac{R}{r_1}\right)^2 \quad \leadsto \quad v_A = R\sqrt{\frac{g}{r_1}}\sqrt{\frac{2r_2}{r_1 + r_2}}\,.$$

Thus, the necessary velocity change is given by

$$\underline{\underline{\Delta v_A}} = v_A - v_1 = \underline{\underline{R\sqrt{\frac{g}{r_1}}\left\{\sqrt{\frac{2r_2}{r_1 + r_2}} - 1\right\}}}\,.$$

P2.18 **Problem 2.18** A barge K is towed in a channel by a haul engine L. In the towing rope acts a force $S = 9\,\text{kN}$, which is inclined by an angle $\alpha = 28°$ with respect to the rail track.

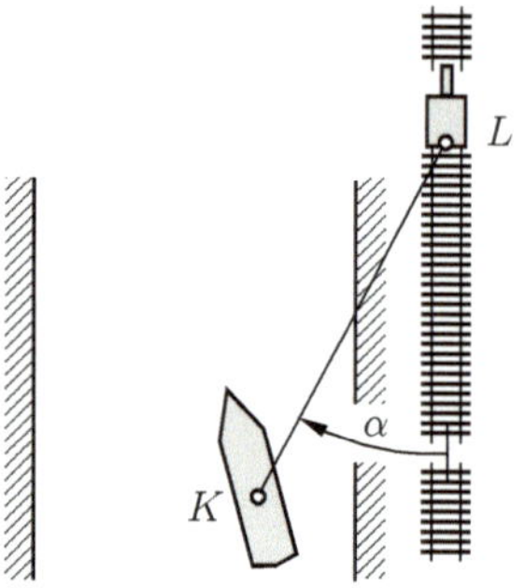

Determine

a) the work for a covered distance $s = 3\,\text{km}$,

b) the power for a towing speed $v = 9\,\text{km/h}$.

Solution **a)** For the work

$$U = \int \boldsymbol{F} \cdot \mathrm{d}\boldsymbol{r} = \int |\boldsymbol{F}| \cos\alpha \, |\mathrm{d}\boldsymbol{r}| \, ,$$

we obtain with $|\boldsymbol{F}| = \text{const} = S$, $\cos\alpha = \text{const} = \cos 28°$ and $|\mathrm{d}\boldsymbol{r}| = \mathrm{d}s$:

$$\underline{\underline{U}} = S \cos\alpha \; s = 9 \cdot 0.883 \cdot 3000 = 23800\,\text{kNm} = 23800\,\text{kJ} = \underline{\underline{23.8\,\text{MJ}}} \, .$$

b) The power is given by

$$\underline{\underline{P}} = \boldsymbol{F} \cdot \boldsymbol{v} = S \cos\alpha \, v = 9 \cdot 0.883 \cdot \frac{9}{3.6} = 19.9\,\text{kJ/s} = \underline{\underline{19.9\,\text{kW}}} \, .$$

P2.19 **Problem 2.19** Determine the necessary work for lifting a body of weight $W = 1\,\text{N}$ from erth's surface (earth's radius R) into the distance r_0 of the moon ($r_0 = 60\,R$).

Solution According to the gravitational law, the gravitational force varies inverse to the squared distance from the earth's surface. Thus, the 'weight' in distance r is

$$F = W \left(\frac{R}{r}\right)^2 \, .$$

Therewith, the work follows as

$$U = W \int_R^{60R} \left(\frac{R}{r}\right)^2 \mathrm{d}r = \frac{59}{60}\, W R \, .$$

With $R = 6370\,\text{km}$ and $W = 1\,\text{N}$, we obtain

$$\underline{\underline{U}} = \frac{59}{60} \cdot 6370\ \text{Nkm} = 6264\ \text{kJ} = \underline{\underline{6.3\ \text{MJ}}} \, .$$

Problem 2.20 A motor winch M tows a body of weight $W = mg$ with constant speed v_0 upwards a rough inclined plane (coefficient of kinetic friction μ). P2.20

Determine the necessary electric power P_A of the winch if its efficiency η is known.

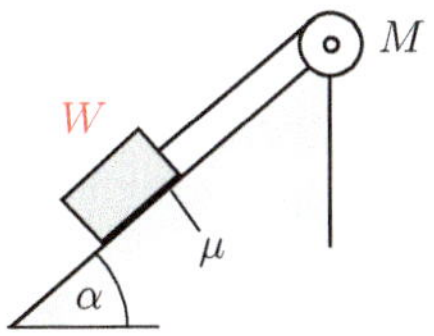

Solution For uniform motion ($\dot{v} = 0$), the force in the rope S follows from the equilibrium conditions

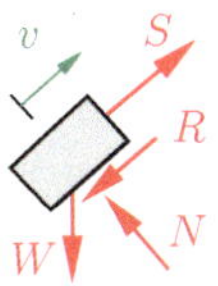

$$\nearrow: \quad S = W \sin\alpha + R\,, \qquad \nwarrow: \quad N = W\cos\alpha$$

and the friction law $R = \mu N$ as

$$S = W(\sin\alpha + \mu\cos\alpha)\ .$$

Thus, the power generated by the winch is

$$P = S\,v_0 = W(\sin\alpha + \mu\cos\alpha)v_0\ .$$

The power absorbed by the winch having an effienciency η is given by

$$\underline{\underline{P_A}} = \frac{P}{\eta} = \underline{\underline{\frac{W}{\eta}(\sin\alpha + \mu\cos\alpha)v_0}}\ .$$

Problem 2.21 A big container vessel with a drive power of 80000 kW covers in 7 days 4000 nautical miles. P2.21

Determine the average drag force F_d

Solution Using the conversion 1 nautical mile = 1.852 km and 1 kW = 1 kNm/s, we obtain from

$$P = F_d\,v \quad \text{with} \quad v = \frac{4000 \cdot 1852}{7 \cdot 24 \cdot 3600} = 12.25\ \frac{\text{m}}{\text{s}}$$

the drag force

$$\underline{\underline{F_d}} = \frac{P}{v} = \frac{80000\ \text{kNm/s}}{12.25\ \text{m/s}} = 6531\ \text{kN} = \underline{\underline{6.53\ \text{MN}}}\ .$$

P2.22 **Problem 2.22** In a centrifuge of radius r, rotating with constant angular velocity ω_0, a body (point mass m) is accelerated by dynamic friction (friction coefficient μ) from its initial angular velocity $\omega(0) = \omega_0/2$ to the final angular velocity ω_0.

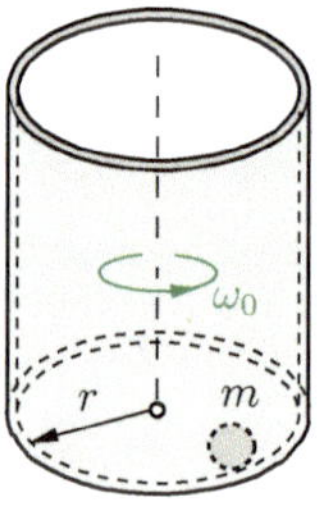

Determine the required acceleration time t_r, the drive torque $M(t)$, the power $P(t)$ and the work U done by the centrifuge.

Solution During acceleration, the point mass rotates with angular velocity $\omega(t)$. With the accelerations $a_t = r\dot{\omega}$, $a_n = r\omega^2$, the equations of motion are given by

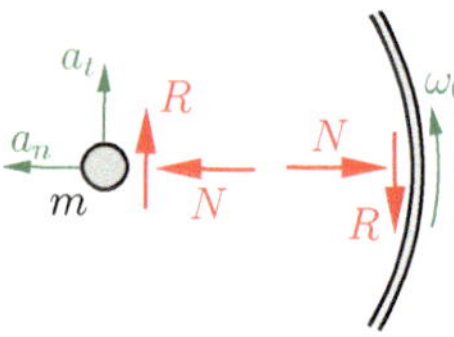

$$\uparrow: \quad mr\dot{\omega} = R\,, \qquad \leftarrow: \quad mr\omega^2 = N\,.$$

Introducing the friction law $R = \mu N$, eliminating N and using $\omega(0) = \omega_0/2$ leads to

$$\dot{\omega} = \mu\omega^2 \quad \rightsquigarrow \quad \int \frac{\mathrm{d}\omega}{\omega^2} = \mu \int \mathrm{d}t \quad \rightsquigarrow \quad \frac{2}{\omega_0} - \frac{1}{\omega} = \mu t\,.$$

The acceleration time t_r is obtained from the condition $\omega(t_r) = \omega_0$:

$$\frac{2}{\omega_0} - \frac{1}{\omega_0} = \mu t_r \quad \rightsquigarrow \quad \underline{\underline{t_r = \frac{1}{\mu\omega_0}}}\,.$$

Since the centrifuge is not accelerated, the driving torque is given by the moment of the friction force R:

$$\underline{\underline{M(t)}} = rR = mr^2\dot{\omega} = \mu m r^2\omega^2 = \underline{\underline{\frac{\mu m r^2\omega_0^2}{(2-\mu\omega_0 t)^2}}}\,.$$

Because M and ω_0 are coaxial, the required power is given by

$$\underline{\underline{P(t)}} = \boldsymbol{M}\cdot\boldsymbol{\omega}_0 = M\omega_0 = \underline{\underline{\frac{\mu m r^2\omega_0^3}{(2-\mu\omega_0 t)^2}}}\,.$$

The total work U done by the centrifuge (strictly speaking, the friction force) is calculated easiest from the difference of kinetic energies:

$$\underline{\underline{U}} = \frac{1}{2}\,mv^2(t_r) - \frac{1}{2}\,mv^2(0) = \frac{1}{2}\,m(r\omega_0)^2 - \frac{1}{2}\,m(r\omega_0/2)^2 = \underline{\underline{\frac{3}{8}\,mr^2\omega_0^2}}\,.$$

Problem 2.23 A soccer player kicks the ball from a distance e with a kick-off angle $\alpha = 45°$ against a vertical wall. The impact at the wall is assumed to be ideal-elastic. **P2.23**

What initial velocity v_0 of the ball is necessary if

a) it shall bounce back exactly to the foot of the player,

b) it shall bounce back to the head (height H) of the player?

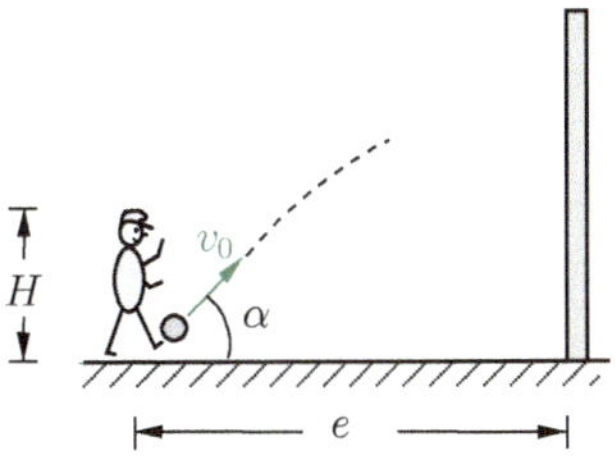

Solution Since no energy gets lost when the impact against the wall is ideal-elastic, the magnitudes of impact velocity v_1 and rebound velocity v_2 must be equal. Then, from the impulse law follows (reflection law)

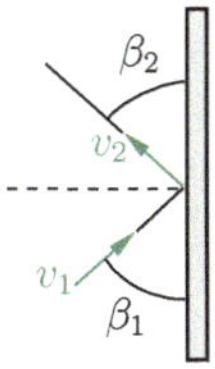

$$\uparrow: \quad mv_2 \cos\beta_2 - mv_1 \cos\beta_1 = 0 \quad \rightsquigarrow \quad \underline{\underline{\beta_1 = \beta_2}} \, .$$

Hence, we can replace the problem of reflection at the wall by a mirroring problem, where we imagine the trajectory being continued through the wall.

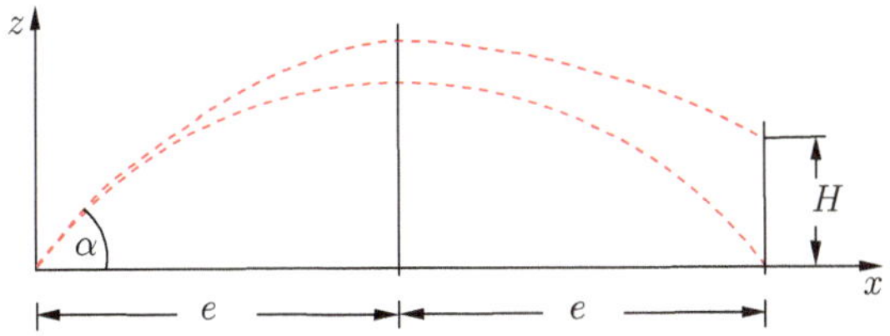

a) The 'flight distance' $d = 2e$ follows with $\alpha = 45°$ and $z_0 = 0$ as (see page 32)

$$d = 2e = v_0^2 \, \frac{\sin 2\alpha}{g} \quad \rightsquigarrow \quad \underline{\underline{v_0}} = \sqrt{\frac{2ge}{\sin 2\alpha}} = \underline{\underline{\sqrt{2ge}}} \, .$$

b) We introduce the coordinates of the kick-off point $x_0 = z_0 = 0$ and end point $x = 2e$, $z = H$ into the parabolic trajectory of motion (page 32) and obtain

$$H = -\frac{g}{2}\left(\frac{2e}{v_0 \cos 45°}\right)^2 + 2e \tan 45° \quad \rightsquigarrow \quad \underline{\underline{v_0 = 2e\sqrt{\frac{g}{2e - H}}}} \, .$$

P2.24 **Problem 2.24** A body (mass m) is driven along a rough horizontal path (kinetic friction coefficient μ) by a periodically acting force, such that $v(t_0) = v_0$ and $v_1 = v(t_1) = v_0/2$. During the driving phase Δt, the force profile $F(t)$ is triangular.

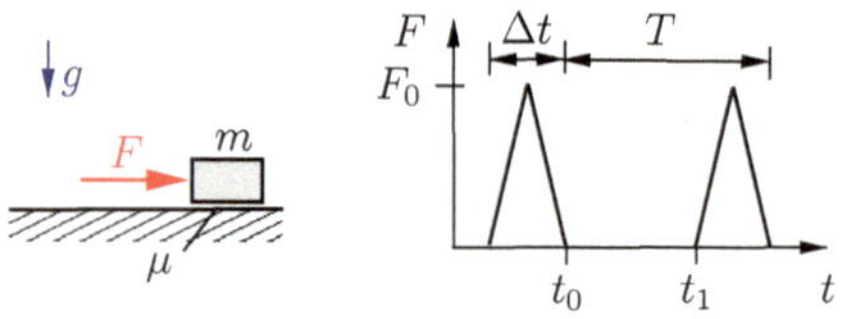

a) Determine the period T and the required peak force F_0 for a given Δt.

b) Calculate the work U done by F during a period T.

Solution a) We consider one period and start counting time at the end of the driving phase. First we apply the impulse law over the full period T. With $R = \mu N = \mu mg = \text{const}$ and the given $F(t)$ profile, the linear impulses of the friction force and the driving force are given by

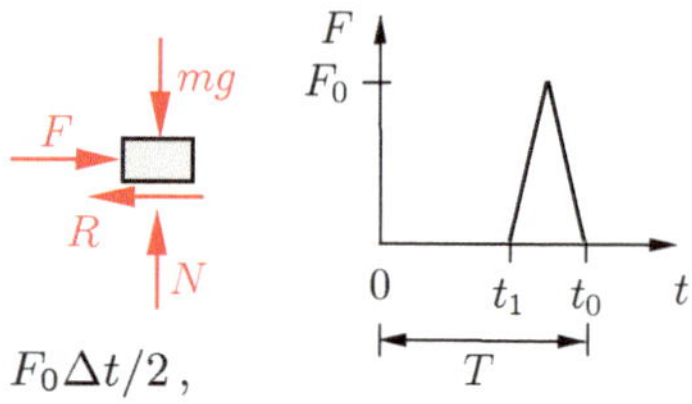

$$\widehat{R} = \mu mgT\,, \qquad \widehat{F} = \int_{\Delta t} F(t)\mathrm{d}t = F_0\Delta t/2\,,$$

Thus, with $v(0) = v(t_0) = v_0$, the impulse law leads to

$$\rightarrow: \quad m\,v_0 - m\,v_0 = 0 = -\mu mgT + \frac{1}{2}F_0\Delta t\,.$$

In the same way, we obtain with $v(t_0) = v_0$ and $v(t_1) = v_0/2$ from the impulse law applied over the driving phase Δt

$$\rightarrow: \quad m\,v_0 - m\,v_0/2 = -\mu mg\Delta t + \frac{1}{2}F_0\Delta t\,.$$

From these two equations for the two unknowns T and F_0, it follow

$$\underline{\underline{T = \Delta t + \frac{v_0}{2\mu g}}}\,, \qquad \underline{\underline{F_0 = 2\mu mg\left[1 + \frac{v_0}{2\mu mg\Delta t}\right]}}\,.$$

b) The work U done by $F(t)$ during time T is determined by using the work-energy theorem, i.e. from the difference of kinetic energies at t_0 and t_1:

$$\underline{\underline{U}} = T(t_0) - T(t_1) = \frac{1}{2}\,mv_0^2 - \frac{1}{2}\,m(v_0/2)^2 = \underline{\underline{\frac{3}{8}\,mv_0^2}}\,.$$

Remark: The work of F and of R during T are equal but have opposite signs. This easily allows calculating the covered distance: $l = 3v_0^2/(8\mu g)$.

Problem 2.25 A point mass, fixed at a massless thread, rotates along a horizontal circular path. At time $t = 0$, the radius is r_0 and the angular velocity is ω_0. P2.25

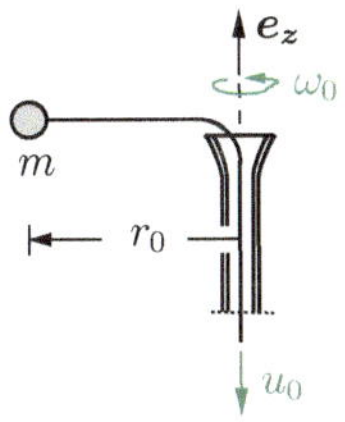

a) Determine $r(t)$ and $\omega(t)$, when the thread is pulled with constant speed u_0 downwards through the sketched vertical pipe.

b) At what time t_1 the angular velocity has doubled and how big is the associated radius r_1?

c) Determine for this case the change of kinetic energy ΔT of the point mass.

Solution a) Because there acts no external moment on the point mass with respect to the center of its path, the angular momentum remains conserved:

$$\boldsymbol{L} = \boldsymbol{r} \times m\,\boldsymbol{v} = \text{const} \ .$$

With $\boldsymbol{r} \times \boldsymbol{v} = r v_\varphi\, \boldsymbol{e}_z$ and $v_\varphi = r\omega$, it follows

$$L = m r^2 \omega = m r_0^2 \omega_0 \quad \rightsquigarrow \quad \omega = \omega_0 \, \frac{r_0^2}{r^2} \ .$$

The dependence of $r(t)$ on time is given by the constant thread speed $\dot{r} = -u_0$:

$$\underline{\underline{r(t) = r_0 - u_0 t}} \ .$$

Inserting into ω leads to

$$\underline{\underline{\omega(t) = \frac{\omega_0 r_0^2}{(r_0 - u_0 t)^2}}} \ .$$

b) From the condition $\omega(t_1) = 2\omega_0$, it follows

$$\underline{\underline{t_1 = \frac{r_0}{u_0}\left(1 - \frac{1}{2}\sqrt{2}\right)}} \quad \text{and} \quad \underline{\underline{r_1 = r_0 - u_0 t_1 = \frac{\sqrt{2}}{2}\, r_0}} \ .$$

c) The energy change is calculated as

$$\begin{aligned} \Delta T &= \frac{m}{2}(v_{\varphi_1}^2 + u_0^2) - \frac{m}{2}(v_{\varphi_0}^2 + u_0^2) \\ &= \frac{m}{2}\left(\frac{\sqrt{2}}{2}\, r_0 2\omega_0\right)^2 - \frac{m}{2}(r_0\omega_0)^2 = \underline{\underline{\frac{1}{2}\, m r_0^2 \omega_0^2}} \ . \end{aligned}$$

The kinetic energy has doubled.

P2.26 **Problem 2.26** In an upright standing frictionless hollow cylinder (radius R), a little sphere (point mass m) is inserted at point A with a horizontal initial speed v_0.

a) What angle α to the horizontal plane has the velocity v_B at point B lying in height distance h below A?

b) What speed v_0 is necessary such that the sphere impinges on ground at C with an angle of 45° and what magnitude has v_C in this case?

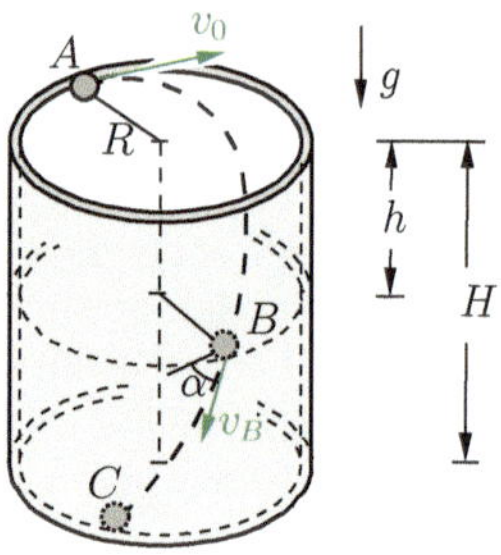

Solution a) The speed of the sphere in point B follows from the energy conservation law $T_A + V_A = T_B + V_B$:

$$\frac{1}{2}mv_0^2 + mgh = \frac{1}{2}mv_B^2 \quad \rightsquigarrow \quad v_B = \sqrt{v_0^2 + 2gh}\,.$$

Since there acts no external moment with respect to the cylinder axis, the angular momentum (moment of momentum) with respect to this axis is conserved:

$$L = \text{const} \quad \rightsquigarrow \quad L_A = L_B\,.$$

With $L_A = R(mv_0)$ and $L_B = R(mv_B\cos\alpha)$, the angle α follows as

$$\underline{\underline{\cos\alpha}} = \frac{v_0}{v_B} = \underline{\underline{\frac{v_0}{\sqrt{v_0^2 + 2gh}}}}\,.$$

b) With $\alpha = 45°$ and $h = H$, we obtain at C

$$\cos 45° = \frac{1}{2}\sqrt{2} = \frac{v_0}{\sqrt{v_0^2 + 2gH}}$$

or after squaring and solving for v_0

$$\underline{\underline{v_0 = \sqrt{2gH}}}\,.$$

Thus, after falling the height distance H, the velocity is

$$\underline{\underline{v_C}} = \sqrt{v_0^2 + 2gH} = \sqrt{2gH + 2gH} = \underline{\underline{2\sqrt{gH}}}\,.$$

Remark: Because on the sphere acts only the weight (vertical) and the normal force from the cylinder wall (normal to the wall) the horizontal velocity component remains unchanged v_0.

Problem 2.27 A cosine-shaped arch of a roller coaster is in A and in B pin-supported. The arch is passed without friction by a car (mass m) that has in point A the initial velocity $v_0 = \sqrt{ga/10}$. P2.27

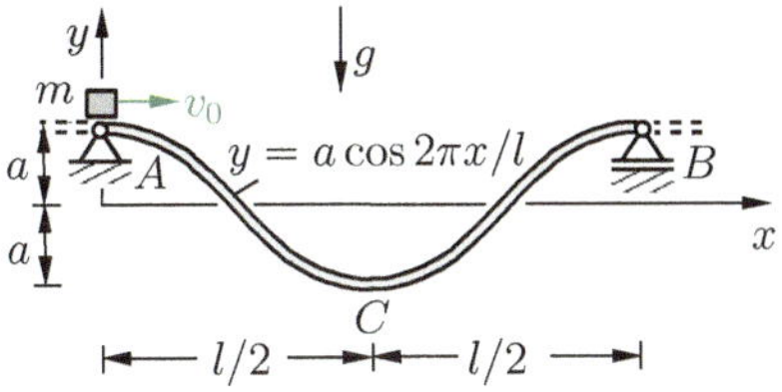

Determine the support reactions and the bending moment at C when the car just passes point C. The weight of the arch shall be disregarded.

Solution The velocity of the car at point C follows from the energy conservation law $T_A + V_A = T_C + V_C$:

$$\frac{1}{2}\,mv_0^2 + mga = \frac{1}{2}\,mv_C^2 - mga \quad \rightsquigarrow \quad v_C^2 = v_0^2 + 4ga = \frac{41}{10}\,ga\,.$$

The derivatives $y' = -(2\pi a/l)\sin 2\pi x/l$ and $y'' = -(4\pi^2 a/l^2)\cos 2\pi x/l$ yield $y'(l/2) = 0$ and $y''(l/2) = 4\pi^2 a/l^2$. Hence, the curvature radius ρ of the path at C is given by

$$\frac{1}{\rho_C} = \frac{y''(l/2)}{[1 + y'^2(l/2)]^{3/2}} = \frac{4\pi^2 a}{l^2}\,.$$

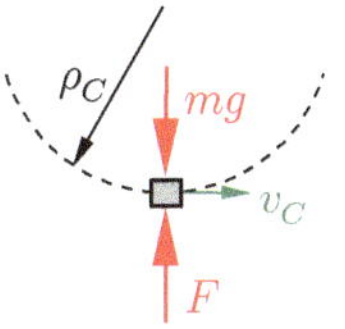

With the normal acceleration $a_n = v^2/\rho$, the law of motion allows to determine the force F at C, which acts from the arch onto the car:

$$\uparrow: \quad m\,\frac{v_C^2}{\rho_C} = F - mg \quad \rightsquigarrow \quad F = mg + m\,\frac{v_C^2}{\rho_C} = mg\Big(1 + \frac{164\,\pi^2 a^2}{10\,l^2}\Big).$$

Knowing F, the support reactions and the bending moment in C can be calculated:

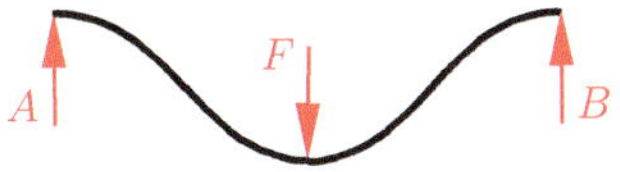

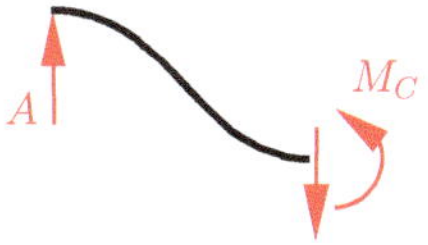

$$\underline{\underline{A = B = \frac{mg}{2}\Big(1 + \frac{164\,\pi^2 a^2}{10\,l^2}\Big)}}, \qquad \underline{\underline{M_C}} = A\,\frac{l}{2} = \underline{\underline{\frac{mgl}{4}\Big(1 + \frac{164\,\pi^2 a^2}{10\,l^2}\Big)}}.$$

Remark: When the results are evaluated for the data $m = 500\,\mathrm{kg}$, $2a = 10\,\mathrm{m}$, $l = 50\,\mathrm{m}$, we obtain $v_0 = 7{,}97\,\mathrm{km/h}$, $v_C = 51{,}05\,\mathrm{km/h}$, $\rho_C = 12{,}66\,\mathrm{m}$, $F = 12{,}84\,\mathrm{kN}$, $A = B = 6{,}42\,\mathrm{kN}$, $M_C = 160{,}55\,\mathrm{kNm}$.

Chapter 3

Dynamics of a System of Point Masses

3

System of Point Masses: The mass m_i is subject to the *external* force $\boldsymbol{F}_i$ and the *internal* forces $\boldsymbol{F}_{ij} = -\boldsymbol{F}_{ji}$. If the distances r_{ij} between the masses remain constant, the system of point masses is *rigid*.

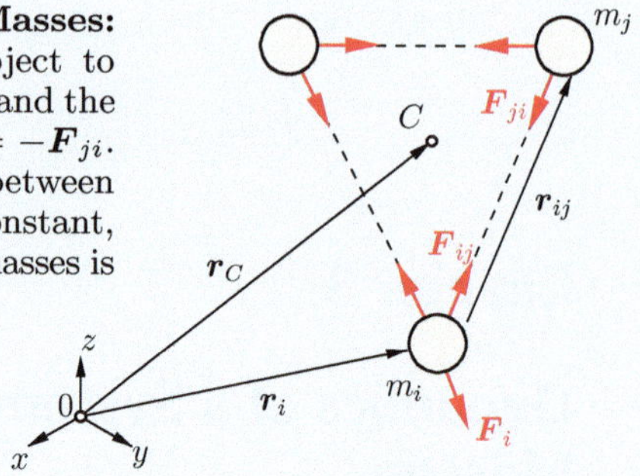

Law of Motion for the Center of Mass: The center of mass C moves such as a point mass with the same total mass subject to the resultant of all external forces acting on the system:

$$m\ddot{\boldsymbol{r}}_C = \dot{\boldsymbol{p}}_C = \sum \boldsymbol{F}_i$$

$$m = \sum m_i \quad \widehat{=} \quad \text{total mass,}$$

$$\boldsymbol{p}_C = m\dot{\boldsymbol{r}}_C = \sum m_i \dot{\boldsymbol{r}}_i = \sum \boldsymbol{p}_i \quad \widehat{=} \quad \text{total linear momentum.}$$

Impulse Law: Time integration of the law of motion yields

$$m\boldsymbol{v}_C - m\boldsymbol{v}_{C0} = \widehat{\boldsymbol{F}}\,,$$

with the linear impulse

$$\widehat{\boldsymbol{F}} = \int_{t_0}^{t} \sum \boldsymbol{F}_i \,\mathrm{d}\tau\,.$$

Conservation of Linear Momentum: In case that the resultant external force is zero, the linear momentum is conserved:

$$m\boldsymbol{v}_C = \sum m_i \boldsymbol{v}_i = \text{const}\,.$$

Angular Momentum Theorem: The time rate of change of the total angular momentum with respect to a fixed point 0 is equal to the resultant moment of all external forces about the same point:

$$\frac{\mathrm{d}\boldsymbol{L}^{(0)}}{\mathrm{d}t} = \boldsymbol{M}^{(0)}$$

$$\boldsymbol{L}^{(0)} = \sum \boldsymbol{r}_i \times m_i \boldsymbol{v}_i = \sum \boldsymbol{L}_i^{(0)} \quad \widehat{=} \quad \text{total angular momentum,}$$

$$\boldsymbol{M}^{(0)} = \sum \boldsymbol{r}_i \times \boldsymbol{F}_i \quad \widehat{=} \quad \text{total external moment.}$$

Special Case: For the rotation of a *rigid* system of point masses about a fixed axis a-a follows

$$\Theta_a \, \ddot{\varphi} = M_a \, ,$$

$$\Theta_a = \sum r_i^2 \, m_i \;\widehat{=}\; \text{mass moment of inertia relative to axis } a\text{-}a,$$

$$r_i \quad \widehat{=} \quad \text{orthogonal distance between mass } m_i \text{ and axis } a\text{-}a.$$

Work-Energy Theorem: The change of kinetic energy is equal to the sum of the work $U^{(e)}$ of all external and $U^{(i)}$ of all internal forces:

$$T - T_0 = U^{(e)} + U^{(i)}$$

$$T = \sum m_i v_i^2 / 2 \quad \widehat{=} \quad \text{kinetic energy,}$$

$$U^{(e)} = \sum \boldsymbol{F}_i \cdot \mathrm{d}\boldsymbol{r}_i \quad \widehat{=} \quad \text{work of external forces,}$$

$$U^{(i)} = \sum \boldsymbol{F}_{ij} \cdot \mathrm{d}\boldsymbol{r}_{ji} \quad \widehat{=} \quad \text{work of internal forces.}$$

For rigid constraints ($\mathrm{d}\boldsymbol{r}_{ji} = 0$) holds $W^{(i)} = 0$.

Conservation of Energy Law: If the external forces can be derived from a potential $V^{(e)}$ and the internal forces from a potential $V^{(i)}$, the work-energy theorem results in the conservation of energy law

$$T + V^{(i)} + V^{(e)} = T_0 + V_0^{(i)} + V_0^{(e)} = \text{const} \, .$$

Bodies with variable mass: The motion of a body with variable mass (e.g. a rocket) is described by

$$m(t) \, \boldsymbol{a} = \boldsymbol{F} + \boldsymbol{T}$$

where $\boldsymbol{F}$ = external force,

$m(t)$ = time dependent mass,

$\boldsymbol{T} = -\mu \, \boldsymbol{w}$ = thrust, where

$\mu = -\dot{m}$ = rate of mass change (mass flow),

$\boldsymbol{w}$ = mass flow velocity relative to the body.

P3.1 **Problem 3.1** On a frictionless horizontal plane, two wedges of masses m_1 and m_2 are placed on top of each other. The wedges can slide frictionless against each other.

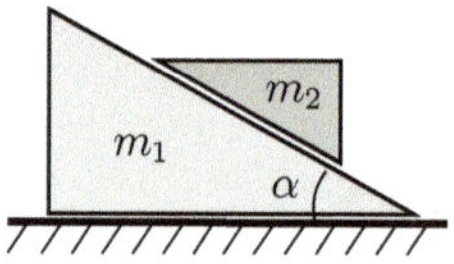

Determine the accelerations of both wedges. Check the result by considering the limit cases $m_1 \to \infty$ and $\alpha \to \pi/2$.

Solution We separate the two bodies and formulate the equations of motion in x- and in y-direction:

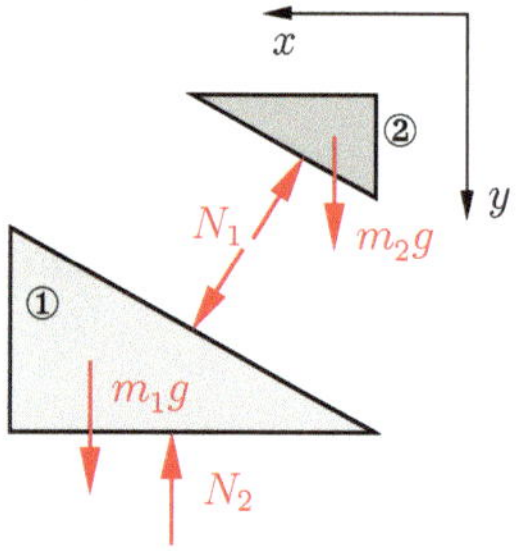

$$
\begin{aligned}
&① \leftarrow: && m_1\ddot{x}_1 = N_1 \sin\alpha\,,\\
&\quad\ \downarrow: && m_1\ddot{y}_1 = m_1 g - N_2 + N_1\cos\alpha\,,\\
&② \leftarrow: && m_2\ddot{x}_2 = -N_1 \sin\alpha\,,\\
&\quad\ \downarrow: && m_2\ddot{y}_2 = m_2 g - N_1\cos\alpha\,.
\end{aligned}
$$

Since wedge ① moves horizontally and at its top side wedge ② slides downwards, the kinematic relations read

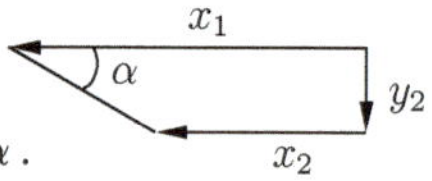

$$
\ddot{y}_1 = 0\,,
$$

$$
y_2 = (x_1 - x_2)\tan\alpha \quad \rightsquigarrow \quad \ddot{y}_2 = (\ddot{x}_1 - \ddot{x}_2)\tan\alpha\,.
$$

Thus, we have six equations for the six unknowns ($\ddot{x}_1$, $\ddot{y}_1$, $\ddot{x}_2$, $\ddot{y}_2$, N_1, N_2). By eliminating N_1 and N_2 it follows

$$
\underline{\underline{\ddot{x}_1 = \frac{\frac{m_2}{m_1}\, g\tan\alpha}{1 + \left(1 + \frac{m_2}{m_1}\right)\tan^2\alpha}}}\,, \qquad \underline{\underline{\ddot{y}_1 = 0}}\,,
$$

$$
\underline{\underline{\ddot{x}_2 = -\frac{g\tan\alpha}{1 + \left(1 + \frac{m_2}{m_1}\right)\tan^2\alpha}}}\,, \qquad \underline{\underline{\ddot{y}_2 = \frac{\left(1 + \frac{m_2}{m_1}\right) g\tan^2\alpha}{1 + \left(1 + \frac{m_2}{m_1}\right)\tan^2\alpha}}}\,.
$$

For the two limit cases we obtain:

$$
a)\ m_1 \to \infty: \quad \ddot{x}_1 \to 0 \quad \text{and} \quad |\ddot{y}_2/\ddot{x}_2| \to \tan\alpha \quad \text{(inclined plane)},
$$

$$
b)\ \alpha \to \frac{\pi}{2}: \quad \ddot{x}_1 = \ddot{x}_2 \to 0 \ \text{und}\ \ddot{y}_2 \to g \quad \text{(free fall)}.
$$

Remark: Adding the equations of motion in x-direction and time-integration confirms conservation of linear momentum: $m_1\ddot{x}_1 + m_2\ddot{x}_2 = 0$ $\rightsquigarrow$ $m_1\dot{x}_1 + m_2\dot{x}_2 = C$, i.e. the total linear momentum is constant!

Problem 3.2 The two massless pulleys are connected by a massless inextensible rope. The system is subject to the weights $W_1 = m_1 g$ and $W_2 = m_2 g$. **P3.2**

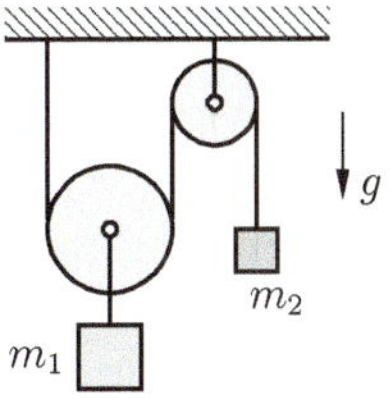

Determine the force in the rope, the acceleration of the mass m_1 and its velocity in dependence of the covered distance.

Solution We first separate the system, draw the free body diagrams including all acting forces and introduce the coordinates x_1 and x_2 (counted from an arbitrary initial position). From the equilibrium of moments about the centers of the pulleys follows

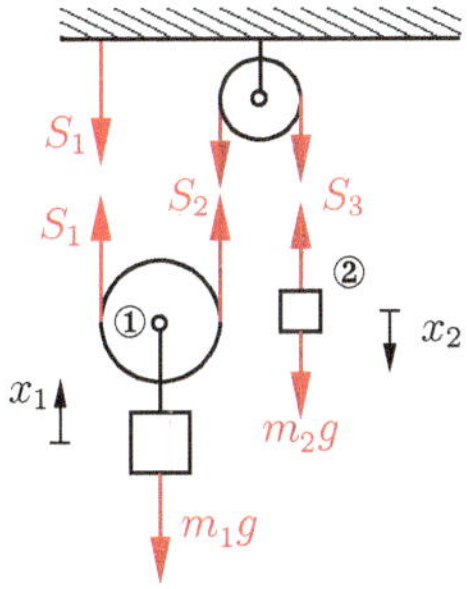

$$S_1 = S_2 = S_3 = S\ .$$

The equations of motion yield

$$① \uparrow: \quad m_1 \ddot{x}_1 = 2S - m_1 g\ , \qquad ② \downarrow: \quad m_2 \ddot{x}_2 = m_2 g - S\ .$$

When the body ② moves downwards by x_2 the body ① is lifted by $x_1 = x_2/2$. Therefore, with $\ddot{x}_1 = \ddot{x}_2/2$ we otain by solving for the unknows

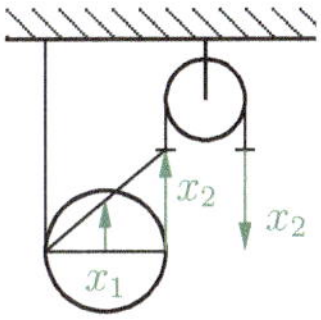

$$\underline{\underline{\ddot{x}_1 = g\,\frac{2m_2 - m_1}{m_1 + 4m_2}}}\ , \qquad \underline{\underline{S = \frac{3m_1 m_2 g}{m_1 + 4m_2}}}\ .$$

The relationship between velocity and covered distance is most easily determined from the energy conservation law $T + V = T_0 + V_0$. With $T_0 = 0$, $V_0 = 0$ follows

$$\frac{1}{2}\,m_1 \dot{x}_1^2 + \frac{1}{2}\,m_2 \dot{x}_2^2 + m_1 g x_1 - m_2 g x_2 = 0$$

or

$$\frac{1}{2}\,m_1 \dot{x}_1^2 + \frac{1}{2}\,m_2 (2\dot{x}_1)^2 = 2m_2 g x_1 - m_1 g x_1$$

$$\rightsquigarrow \qquad \underline{\underline{v_1(x_1) = \dot{x}_1 = \sqrt{2g x_1 \frac{2m_2 - m_1}{m_1 + 4m_2}}}}\ .$$

Remark: In the special case $2m_2 = m_1$, we obtain $\ddot{x}_1 \equiv 0$ and $S = W_2$. In this case, the system is in equilibrium!

P3.3 **Problem 3.3** A hinged pendulum is modeled as a massless rigid rod with two attached point masses of weights $W_1 = m_1 g$ and $W_2 = m_2 g$.

Formulate the equation of motion.

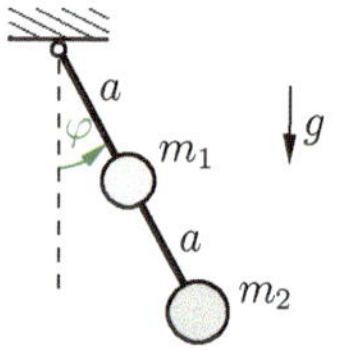

Solution We will solve the problem by two different approaches.

1st approach: Because the distance of the masses is constant, the pendulum represents a rigid system of point masses. Therefore, the angular momentum theorem with respect to the fixed hinge A

$$\Theta_A \ddot{\varphi} = M_A$$

can be applied. With

$$\Theta_A = a^2 m_1 + (2a)^2 m_2 = a^2(m_1 + 4m_2)$$

the equation of motion follows as (notice the positive sense of rotation!)

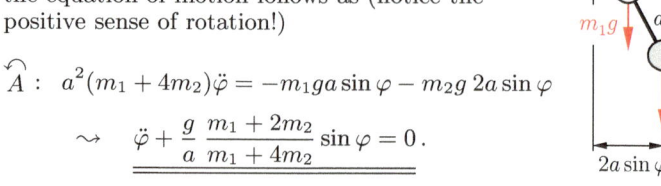

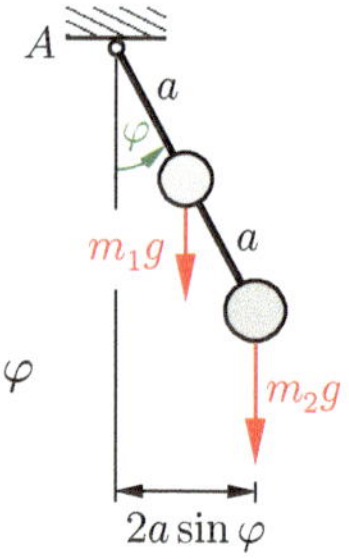

$$\overset{\curvearrowleft}{A}: \quad a^2(m_1 + 4m_2)\ddot{\varphi} = -m_1 g a \sin\varphi - m_2 g \, 2a \sin\varphi$$

$$\rightsquigarrow \quad \underline{\underline{\ddot{\varphi} + \frac{g}{a}\frac{m_1 + 2m_2}{m_1 + 4m_2}\sin\varphi = 0}}\,.$$

2nd approach: We start from the conservation of energy law $T + V =$ const, where the zero level for the potential energy is chosen at $\varphi = \pi/2$ (hinge A):

$$\frac{1}{2} m_1 v_1^2 + \frac{1}{2} m_2 v_2^2 - m_1 g a \cos\varphi - m_2 g \, 2a \cos\varphi = \text{const}\ .$$

With $v_1 = a\dot{\varphi}$ and $v_2 = 2a\dot{\varphi}$ follows

$$\frac{1}{2}(m_1 + 4m_2)a^2\dot{\varphi}^2 - ag\cos\varphi(m_1 + 2m_2) = \text{const}$$

and differentiation with respect to time leads to

$$(m_1 + 4m_2)a^2\ddot{\varphi}\dot{\varphi} + ag\sin\varphi(m_1 + 2m_2)\dot{\varphi} = 0\,.$$

Since $\dot{\varphi}$ is nonzero for all times t, it remains

$$\underline{\underline{\ddot{\varphi} + \frac{g}{a}\frac{m_1 + 2m_2}{m_1 + 4m_2}\sin\varphi = 0}}\,.$$

Problem 3.4 The constant acceleration a_0 of a truck (mass M) is such high that the box (weight $W = mg$) starts to slide on the loading area (coefficients of static and kinetic friction μ_0, μ). **P3.4**

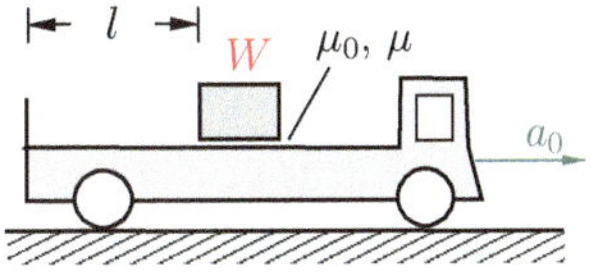

a) Determine the minimum acceleration a_0^* for the onset of sliding and the corresponding horizontal force F^* excerted by the truck to the pavement.

b) Calculate the time T, the box requires to bounce against the rear wall.

Solution a) As long as the box is not sliding, the equations of motion for truck and box are given by

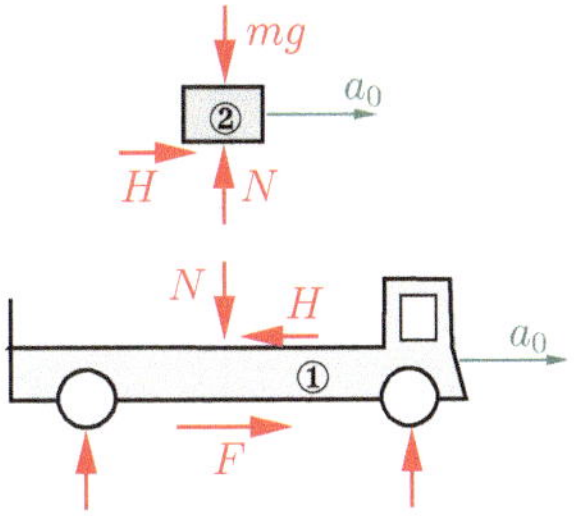

$$① \rightarrow: \quad Ma_0 = F - H\,,$$

$$② \rightarrow: \quad ma_0 = H\,,$$

where the static friction force H is limited by $H_0 = \mu_0 N = \mu_0 mg$. Introducing $H = H_0$ leads to the limit values of acceleration and force:

$$\underline{\underline{a_0^* = \mu_0 g}}\,, \qquad \underline{\underline{F^* = (M+m)\mu_0 g}}$$

b) During sliding, the friction force is $R = \mu N = \mu mg$, which for the box leads to the equation of motion

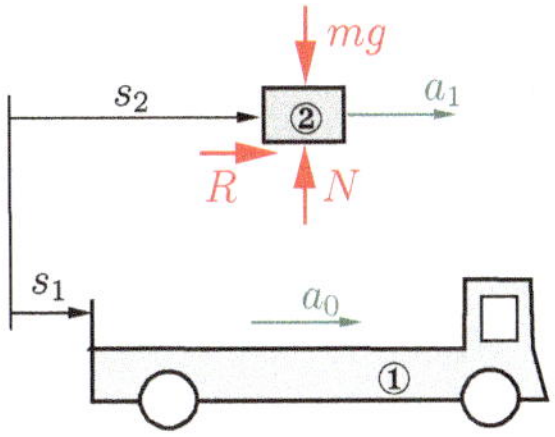

$$② \rightarrow: \quad ma_2 = R \quad \rightsquigarrow \quad a_2 = \mu g\,.$$

Integration of the two constant accelerations $a_1 = a_0$ and a_2 of truck and box, considering the initial conditions $s_1(0) = s_2(0) = s_0$, $v_1(0) = v_2(0) = v_0$, yields

$$v_1 = a_0 t + v_0\,, \qquad v_2 = \mu g t + v_0\,,$$

$$s_1 = a_0 \frac{t^2}{2} + v_0 t + s_0\,, \qquad s_2 = \mu g\,\frac{t^2}{2} + v_0 t + s_0\,.$$

The box bounces at time T against the rear wall, when

$$\Delta s = s_1 - s_2 = l \quad \rightsquigarrow \quad (a_0 - \mu g)T^2 = 2l \quad \rightsquigarrow \quad \underline{\underline{T = \sqrt{\frac{2l}{a_0 - \mu g}}}}\,.$$

P3.5 **Problem 3.5** A big block (mass M), initially resting on a frictionless ground, starts to move due to a little block of mass m, which is shot with speed v_0 to its rough upper surface. The kinetic friction coefficient between the two bodies is μ.

a) Determine the time T, that is required, until both bodies move with the common speed v^*.

b) Determine the covered distance of M, until the common speed is reached.

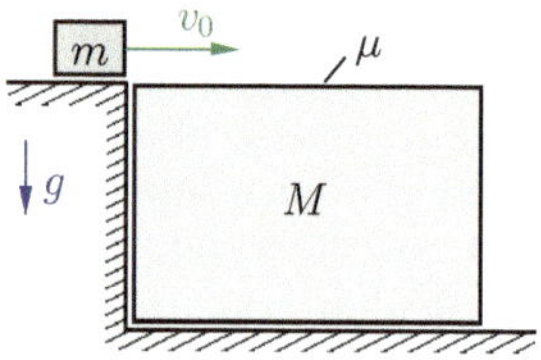

Solution We separate the two bodies, sketch the free body diagrams and introduce position coordinates. Then, the equations of motion read

$$
\begin{aligned}
&① \rightarrow: \quad M\ddot{x}_1 = R\,,\\
&② \rightarrow: \quad m\ddot{x}_2 = -R\,.
\end{aligned}
$$

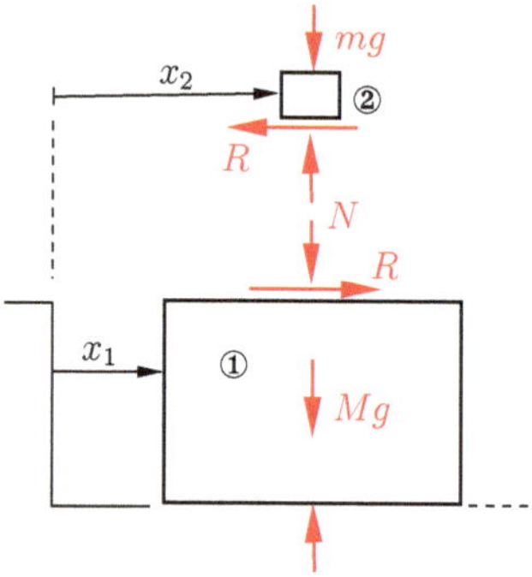

Introducing the friction law $R = \mu N = \mu m g$ leads to

$$\ddot{x}_1 = \frac{m}{M}\,\mu g\,, \qquad \ddot{x}_2 = -\mu g\,.$$

By integration, considering the initial conditions $\dot{x}_1(0) = 0$, $x_1(0) = 0$, $\dot{x}_2(0) = v_0$, $x_2(0) = 0$, we obtain

$$
\begin{aligned}
\dot{x}_1 &= \frac{m}{M}\,\mu g t\,, \qquad & x_1 &= \frac{1}{2}\frac{m}{M}\,\mu g t^2\,,\\
\dot{x}_2 &= v_0 - \mu g t\,, \qquad & x_2 &= v_0 t - \frac{1}{2}\,\mu g t^2\,.
\end{aligned}
$$

a) The condition $\dot{x}_1(T) = \dot{x}_2(T) = v^*$ leads to

$$\frac{m}{M}\,\mu g T = v_0 - \mu g T \quad \rightsquigarrow \quad \underline{\underline{T = \frac{v_0}{\mu g(1 + m/M)}}}$$

and

$$\underline{\underline{v^*}} = \dot{x}_1(T) = \frac{m}{M}\,\mu g T = \underline{\underline{v_0\frac{1}{1 + M/m}}}\,.$$

b) The covered distance of M is given by

$$\underline{\underline{x_1(T)}} = \frac{1}{2}\frac{m}{M}\,\mu g T^2 = \underline{\underline{\frac{1}{2}\,\frac{v_0^2\,m/M}{\mu g(1 + m/M)^2}}}\,.$$

P3.6

Problem 3.6 In an experiment, a particle of mass m is shot with speed v_0 against a second resting particle of mass $2m$ After impact, three particles are observed, where the sketched directions and the following masses and velocities are detected during measurements: $m_1 = m$, $v_1 = 2v_0$, $v_2 = v_0/2$.

Determine m_2, m_3 and v_3.

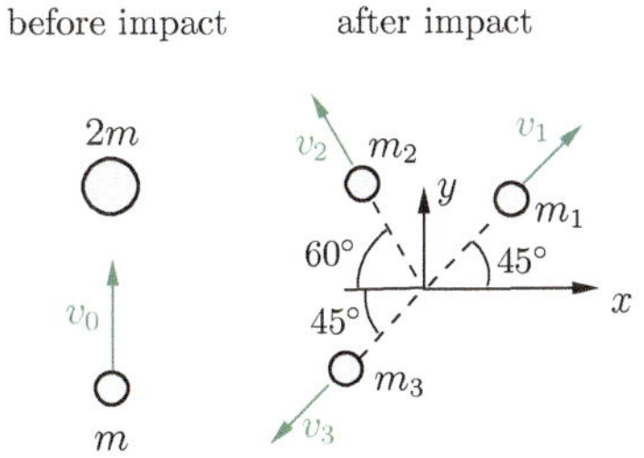

Solution The momentum of the particle system is unchanged after impact. Thus, with $\cos 45° = \sin 45° = \sqrt{2}/2$, $\cos 60° = 1/2$, $\sin 60° = \sqrt{3}/2$ we obtain in components

$$\uparrow: \quad mv_0 = m_1 v_1 \sqrt{2}/2 + m_2 v_2 \sqrt{3}/2 - m_3 v_3 \sqrt{2}/2 \,,$$
$$\rightarrow: \quad 0 = m_1 v_1 \sqrt{2}/2 - m_2 v_2 \, 1/2 - m_3 v_3 \sqrt{2}/2 \,.$$

In conjunction with

$$m_1 + m_2 + m_3 = 3m \,,$$

we now have three equations for the three unknows m_2, m_3, v_3. To solve them, it is advantageous to introduce the given quantities:

$$mv_0 = mv_0\sqrt{2} + m_2 v_0 \sqrt{3}/4 - m_3 v_3 \sqrt{2}/2 \,,$$
$$0 = mv_0\sqrt{2} - m_2 v_0 1/4 - m_3 v_3 \sqrt{2}/2 \,,$$
$$m_2 + m_3 = 2m \,.$$

Subtracting the 2nd from the 1st equation provides directly

$$mv_0 = m_2 v_0 \frac{1+\sqrt{3}}{4} \quad \rightsquigarrow \quad \underline{\underline{m_2 = m\,\frac{4}{1+\sqrt{3}} = 1.464\,m\,.}}$$

Thus, from the 3rd equation follows

$$\underline{\underline{m_3}} = 2m - m_2 = m\Big(2 - \frac{4}{1+\sqrt{3}}\Big) = \underline{\underline{\frac{2\sqrt{3}-2}{1+\sqrt{3}}\,m = 0.536\,m\,.}}$$

Finally, introducing m_2, m_3 into the 2nd equation yields

$$\underline{\underline{v_3 = v_0 \frac{m}{m_3}\Big(2 - \frac{1}{2\sqrt{2}}\,\frac{m_2}{m}\Big) = 2.765\,v_0\,.}}$$

Remark: Note, that the center of mass stays on the y-axis!

P3.7 **Problem 3.7** On a frictionless plane, a body with mass m_1 hits with velocity v a second body at rest (mass m_2) and connects to it. Subsequently, the composite body hits via a spring (spring constant k) a third body at rest (mass m_3).

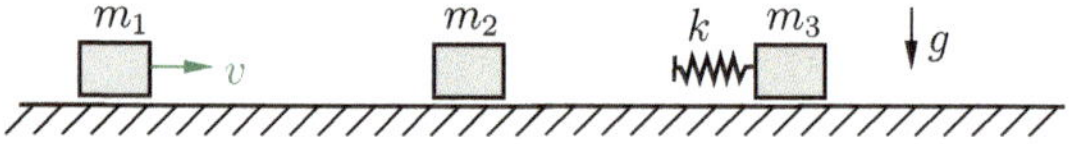

a) Determine the speed v of the mass m_1 such that m_3 remains at rest, if the plane solely at the location of m_3 is rough (coefficient of static friction μ_0).

b) Determine the speed of m_3 after collision if the plane at the location of m_3 is also frictionless.

Solution **a)** When m_1 and m_2 are connecting, the momentum remains conserved. Thus, the velocity $\bar{v}_{12}$ of the composite is given by

$$\rightarrow: \quad m_1 v = (m_1 + m_2)\bar{v}_{12} \quad \rightsquigarrow \quad \bar{v}_{12} = \frac{m_1}{m_1 + m_2}\, v\,.$$

If m_3 is at rest, the maximum compression x of the spring is reached, when during collision, the velocity of the composite body has come to zero. Then, energy conservation yields

$$\frac{1}{2}(m_1 + m_2)\bar{v}_{12}^2 = \frac{1}{2}k\,x^2 \quad \rightsquigarrow \quad x^2 = \frac{m_1 + m_2}{k}\,\bar{v}_{12}^2\,,$$

and the horizontal maximum force acting on m_3 follows as $F_k = k\,x$. In order that m_3 remains at rest, the equilibrium condition $H = F_c$ and the condition fo static friction $H < H_0 = \mu_0 m_3 g$ must be fulfilled. This leads to

$$k\,x < \mu_0 m_3 g \quad \rightsquigarrow \quad \underline{\underline{v < \frac{\mu_0 m_3 g}{m_1}\sqrt{\frac{m_1 + m_2}{k}}}}$$

b) The velocity $\bar{v}_3$ of m_3 after collision with the composite body can be determined from conservation of linear momentum and energy conservation:

$$\rightarrow: (m_1 + m_2)\bar{v}_{12} = (m_1 + m_2)\bar{\bar{v}}_{12} + m_3\bar{v}_3\,,$$
$$\frac{1}{2}(m_1 + m_2)\bar{v}_{12}^2 = \frac{1}{2}(m_1 + m_2)\bar{\bar{v}}_{12}^2 + \frac{1}{2}m_3\bar{v}_3^2\,.$$

Hence, introducing the already know velocity $\bar{v}_{12}$ of the composite body, we obtain

$$\underline{\underline{\bar{v}_3 = \frac{2m_1}{m_1 + m_2 + m_3}\, v\,.}}$$

Problem 3.8 Two swimmers (mass $m_1 = 75\,\text{kg}$, $m_2 = 60\,\text{kg}$) jump into the water from an initially resting rowboat (mass $m_B = 150\,\text{kg}$), which can ideally glide without any drag. The first swimmer jumps horizontally from the stern with speed $v_1 = 2\,\text{m/s}$ relative to the boat. Subsequently, the second swimmer jumps in the same direction with speed $v_2 = 3\,\text{m/s}$ relative to the boat. **P3.8**

a) Determine the velocity of the boat after the jumps.

b) What speed has the boat, after both swimmers jumped simultaneously with the speed $v_3 = 2.5\,\text{m/s}$?

Solution **a)** We introduce as positive direction the jump direction. Since the momentum of the system initially is zero, the sum of momenta after the first jump must stay zero. Thus, with the absolute velocities v_{B_1} and $v_{B_1} + v_1$ of the boat and the first jumper, we obtain

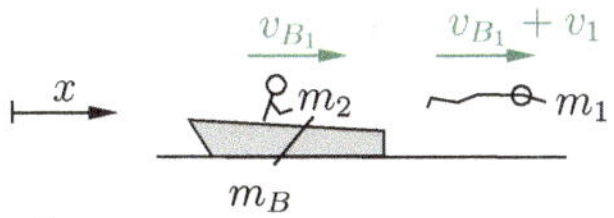

$$\rightarrow: \quad (m_B + m_2)v_{B_1} + m_1(v_{B_1} + v_1) = 0$$

$$\rightsquigarrow \quad \underline{\underline{v_{B_1} = -\frac{m_1}{m_B + m_1 + m_2}\, v_1 = -0.52\,\text{m/s}\,.}}$$

The sign indicates that boat and jumper move in opposite directions. Before the second jump, the boat including second jumper has the momentum $(m_B + m_2)v_{B_1}$, which is conserved after the jump:

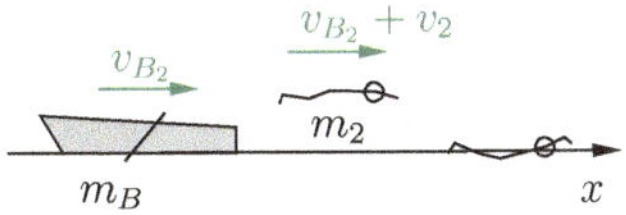

$$\rightarrow: \quad (m_B + m_2)v_{B_1} = m_B v_{B_2} + m_2(v_{B_2} + v_2)$$

$$\rightsquigarrow \quad \underline{\underline{v_{B_2} = \frac{(m_B + m_2)v_{B_1} - m_2 v_2}{m_B + m_2} = -1.38\,\text{m/s}\,.}}$$

b) If both swimmers jump simultaneously, the situation is analogeous to the first jump, but with changed masses and velocities:

$$\rightarrow: \quad m_B v_{B_3} + (m_1 + m_2)(v_{B_3} + v_3) = 0$$

$$\rightsquigarrow \quad \underline{\underline{v_{B_3} = -\frac{m_1 + m_2}{m_B + m_1 + m_2}\, v_3 = -1.18\,\text{m/s}\,.}}$$

P3.9 **Problem 3.9** A rope without any bending stiffness (mass m, length l), moves freely on a frictionless plane, such that its bend K changes its position continuously (similar as it does at a horsewhip). In the initial state, K is in the middle of the rope; the lower part ① has the velocity $v_{10} = v_0$ and the upper part ② the velocity $v_{20} = 0$.

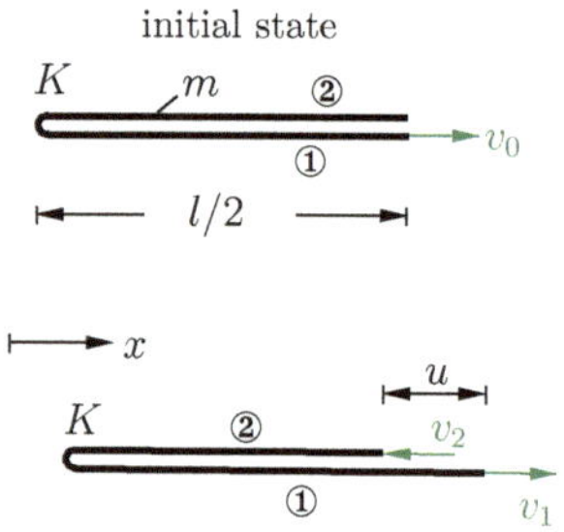

a) Determine the velocities v_2 and v_1 in dependence on the distance u of the ends of the rope.

b) What are the velocities of the parts ① and ② at the instant when the bend K passes the end of the rope? Sketch the functions $v_1(u)$ and $v_2(u)$.

a) Because no external forces are acting, the linear momentum and the total energy at initial state must be conserved. Thus, with the lengths $(l+u)/2$ and $(l-u)/2$ of the parts ① and ② of the rope, it follows

momentum conservation: $$\frac{m}{2}v_0 = \frac{m}{2}\frac{l+u}{l}v_1 + \frac{m}{2}\frac{l-u}{l}v_2\,,$$

energy conservation: $$\frac{1}{2}\frac{m}{2}v_0^2 = \frac{1}{2}\frac{m}{2}\frac{l+u}{l}v_1^2 + \frac{1}{2}\frac{m}{2}\frac{l-u}{l}v_2^2$$

respectively with $y = u/l$

$$v_0 = (1+y)v_1 + (1-y)v_2\,, \qquad (a)$$

$$v_0^2 = (1+y)v_1^2 + (1-y)v_2^2\,.$$

Solving for v_2 leads to the quadratic equation

$$2y(1-y)v_2^2 - 2(1-y)v_0\,v_2 - y\,v_0^2 = 0$$

with the solution

$$v_2 = \frac{v_0}{2y}\left[1 \pm \sqrt{1+\frac{2y^2}{1-y}}\right].$$

Introducing this result into (a), yields

$$v_1 = \frac{v_0}{2y(1+y)}\left[3y - 1 \mp (1-y)\sqrt{1+\frac{2y^2}{1-y}}\right].$$

To find the correct sign in front of the root, we consider the initial state $v_1 = v_0$, $v_2 = 0$. By determining the limit values for $y \to 0$ it can be seen that the 'lower' sign leads to the correct initial state. Thus, the solution reads

$$\underline{\underline{v_1 = \frac{v_0}{2y(1+y)}\left[3y - 1 + (1-y)\sqrt{1+\frac{2y^2}{1-y}}\right]}},$$

$$\underline{\underline{v_2 = \frac{v_0}{2y}\left[1 - \sqrt{1+\frac{2y^2}{1-y}}\right]}}.$$

b) The bend K reaches the end of the rope for $u \to l$ or for $y \to 1$, respectively. Determining the limit values leads to the velocities

$$\underline{\underline{\lim_{y\to 1} v_1(y) = \frac{v_0}{2}}}, \qquad \underline{\underline{\lim_{y\to 1} v_2(y) = -\infty}}.$$

c) In order to sketch $v_1(y)$ and $v_2(y)$, it is advisable first to approximate the functions near $y = 0$ and $y = 1$. By series expansion and neglecting higher order terms, we find

$$y \ll 1: \quad v_1(y) \approx v_0\,(1 - y/2), \qquad v_2(y) \approx -v_0\,y/2,$$

$$(1-y) \ll 1: \quad v_1(y) \approx \frac{v_0}{2}\left(1 + \frac{\sqrt{1-y}}{\sqrt{2}}\right), \quad v_2(y) \approx -\frac{v_0}{2}\,\frac{\sqrt{2}}{\sqrt{1-y}}.$$

In conjunction with the results from a), b), we obtain the displayed courses of the functions:

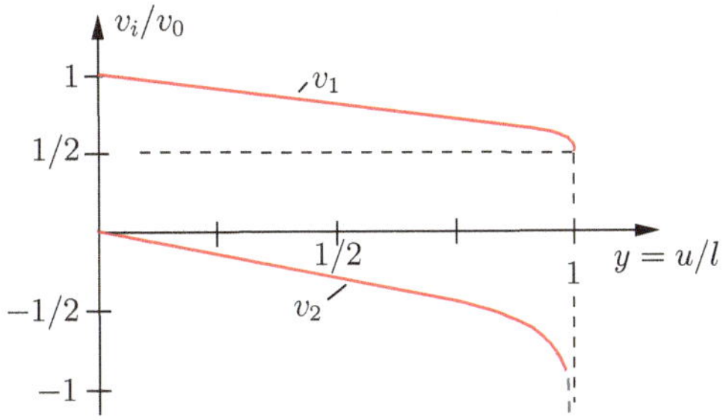

Remark: The infinite final velocity of the rope end ② (supersonic speed) may explain the so-called 'crack of the whipe'.

P3.10 **Problem 3.10** A rocket of initial mass m_0 (including propellant mass m_T) is vertically launched at time $t = 0$. Assume that the mass flow μ and ejection speed w are constant and that at thrust cut-off the propellant is fully consumed.

a) Determine the velocity $v(t)$ of the rocket, assuming that there is no air drag and the gravity g is *constant.*

b) What is the velocity at thrust cut-off for $m_T = 0.8\, m_0$, thrust duration $t_T = 2\,\text{min}$ and $w = 2000\,\text{m/s}$?

c) What are the accelerations at lift-off and just before thrust cut-off?

Solution **a)** The rocket motion is described by

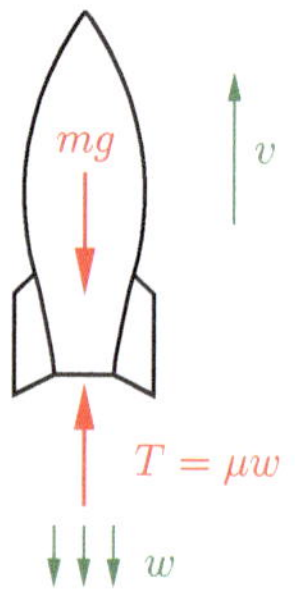

$$\uparrow: \quad m(t)\frac{\mathrm{d}v}{\mathrm{d}t} = -m(t)g + T\,.$$

With

$$T = \mu w\,, \qquad m(t) = m_0 - \mu t$$

follows the acceleration

$$a = \frac{\mathrm{d}v}{\mathrm{d}t} = \frac{\mu w}{m_0 - \mu t} - g\,.$$

Integration and considering the initial condition $v(0) = 0$ leads to the velocity

$$\underline{\underline{v(t) = w \ln \frac{1}{1 - \frac{\mu}{m_0}t} - g\,t}}\,.$$

b) The mass at thrust cut-off is $m(t_T) = m_0 - m_T$, resulting in

$$\mu = \frac{m_0 - m(t_T)}{t_T} = \frac{m_T}{t_T}\,,$$

Thus, the velocity at thrust cut-off is given by

$$\underline{\underline{v(t_T)}} = w \ln \frac{1}{1 - \frac{m_T}{m_0}} - g\,t_T = 2000 \ln \frac{1}{1 - 0.8} - 9.81 \cdot 120 = 2042 \frac{\text{m}}{\text{s}} = \underline{\underline{7350\,\frac{\text{km}}{\text{h}}}}.$$

c) The accelerations at lift-off ($t = 0$) and just before thrust-off ($t = t_T$) follow as

$$\underline{\underline{a(0)}} = \frac{\mu w}{m_0} - g = \frac{m_T\, w}{m_0\, t_T} - g = \frac{0,8 \cdot 2000}{120} - 9.81 = \underline{\underline{3.52\,\frac{\text{m}}{\text{s}^2}}}$$

$$\underline{\underline{a(t_T)}} = \frac{\mu w}{m_0 - m_T} - g = \frac{m_T\, w}{0.2\, m_0\, t_T} - g = \underline{\underline{56.84\,\frac{\text{m}}{\text{s}^2}}}\,.$$

Problem 3.11 At point A of a drum, a rope (length $l_0 > h$, total mass m_0) is attached, the lower part of which initially rests on ground C. At time $t = 0$ the drum starts to rotate such that the rope is lifted with constant acceleration a_0 and reeled. For $t = 0$ point A is at $s = 0$.

P3.11

Determine the force H in the rope at the drum inlet B. Assume that the rope hangs vertical at all times.

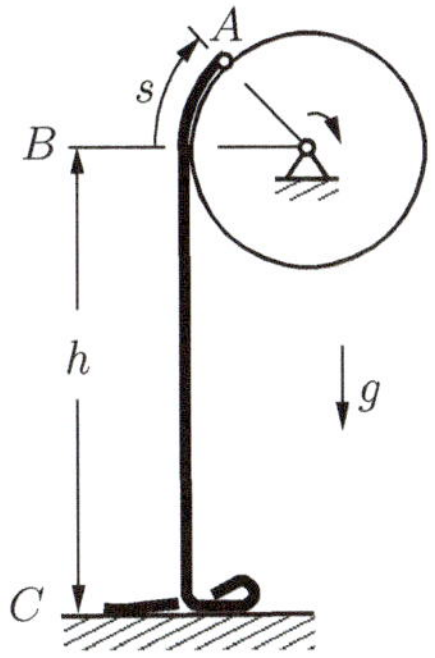

Solution We first determine the reeled rope length s from the known acceleration and the initial conditions $s(0) = 0$, $\dot{s}(0) = 0$:

$$\ddot{s} = a_0\,, \qquad \dot{s} = v = a_0 t\,, \qquad s = \frac{1}{2} a_0 t^2\,.$$

In what follows, two cases must be distinguished.

Case a: One part of the rope still rests at the ground ($s + h \leq l_0$). In this case, the hanging part of the rope is regarded as a body with variable mass, in which at the ground C mass flows in and at B mass flows out. Then, the mass m of the hanging part is

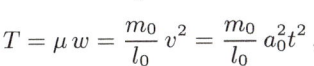

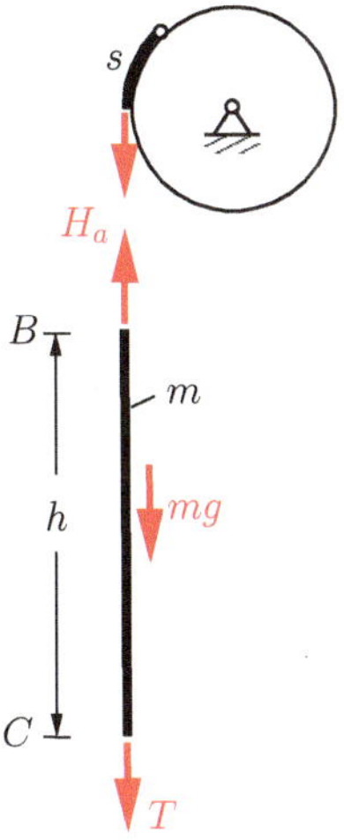

$$\frac{m_0}{l_0} = \frac{m}{h} \quad \rightsquigarrow \quad m = m_0\,\frac{h}{l_0}$$

and the mass flow follows as

$$\mu = \frac{\mathrm{d}m}{\mathrm{d}t} = \frac{(\mathrm{d}s/l_0)m_0}{\mathrm{d}t} = \frac{m_0}{l_0}\,v\,.$$

At the lower end C of the hanging part, the 'ejection velocity' w (there is no mass outflow but a mass inflow!) has the magnitude $w = v$, at the upper end at B it is zero (velocity of the outflowing mass = velocity of the hanging part). Accordingly, there exists only at C a thrust of magnitude

$$T = \mu\, w = \frac{m_0}{l_0}\,v^2 = \frac{m_0}{l_0}\,a_0^2 t^2\,,$$

which is downwards directed. Hence, the equation of motion reads

$$\uparrow: \quad m\, a_0 = H_a - m\, g - T\,.$$

It leads, after inserting T and with $v^2 = 2a_0 s$, to the force

$$\underline{\underline{H_a = m_0 \frac{h}{l_0}\left(a_0 + g + 2a_0 \frac{s}{h}\right)}}\,.$$

From this result, for $s = 0$ and for the limit case $s = l_0 - h$ (just before the end of the rope lifts-off from ground), we obtain

$$H_a(0) = m_0 \frac{h}{l_0}\,(a_0 + g), \qquad H_a(l_0 - h) = m_0 \frac{h}{l_0}\left(-a_0 + g + 2a_0 \frac{l_0}{h}\right).$$

Case b: The rope no longer touches the ground ($s + h \geq l_0$).
In this case, the mass of the hanging part is

$$\frac{m_0}{l_0} = \frac{m}{l_0 - s} \quad \rightsquigarrow \quad m = m_0\,\frac{l_0 - s}{l_0}\,.$$

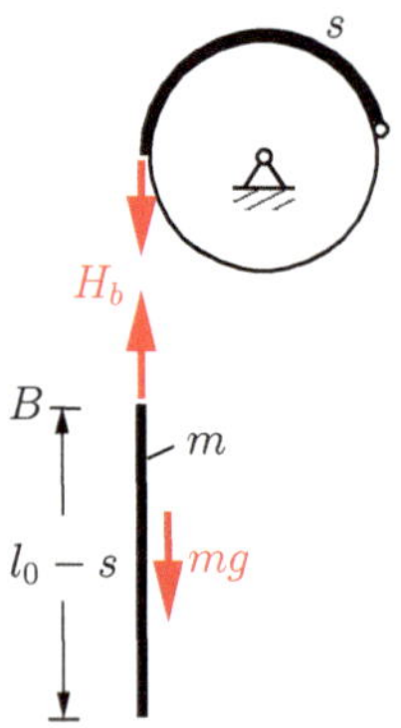

Since there is no mass flow at the lower end and the 'ejection velocity' at B is zero, there is no thrust anywhere. The equation of motion now reads

$$\uparrow: \quad m\,a_0 = H_b - m\,g\,.$$

Thus, the force in the rope follows as

$$\underline{\underline{H_b = m_0 \frac{l_0 - s}{l_0}\,(a_0 + g)}}\,.$$

For the limit case $s = l_0 - h$ (just after lifting the rope from ground) we obtain

$$H_b(l_0 - h) = m_0 \frac{h}{l_0}\,(a_0 + g)\,.$$

Remark: When comparing the limit cases of **a** and **b** for $s = l_0 - h$, it can be recognized that the force in the rope at lift-off experiences a jump of magnitude

$$\Delta H = H_a(l_0 - h) - H_b(l_0 - h) = 2\,\frac{l_0 - h}{l_0}\,m_0 a_0\,.$$

Problem 3.12 A whipcord (mass m, length l) is pulled at one end with the constant speed v_0. Given are the initial conditions $x_1(0) = l/2$, $x_2(0) = l/2$ and $\dot{x}_1(0) = v_0$, $\dot{x}_2(0) = 0$. P3.12

Determine the required force F in dependence on the distance u of the cord ends.

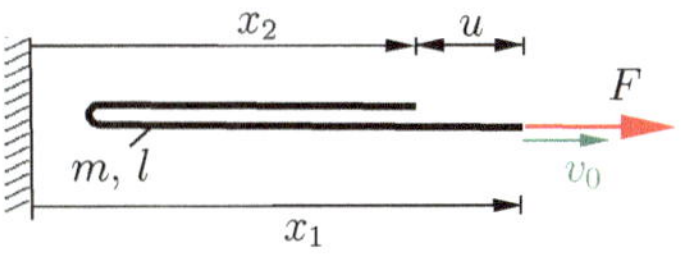

Solution Considering $\dot{x}_1 = v_0 = \text{const}$ and $\ddot{x}_1 = 0$, the kinematic relationships between u and x_1, x_2 are given by

$$u = x_1 - x_2\,, \qquad \dot{u} = \dot{x}_1 - \dot{x}_2 = v_0 - \dot{x}_2\,, \qquad \ddot{u} = -\ddot{x}_2$$

and the initial conditions are $u(0) = 0$ and $\dot{u}(0) = v_0$.

In what follows we consider part ① and part ② of the whipcord as bodies with variable mass. Hence, the equations of motion read

$$\text{①}: \qquad 0 = F - T\,, \quad (a)$$

$$\text{②}: \quad m\frac{l-u}{2l}\,\ddot{x}_2 = T\,. \quad (b)$$

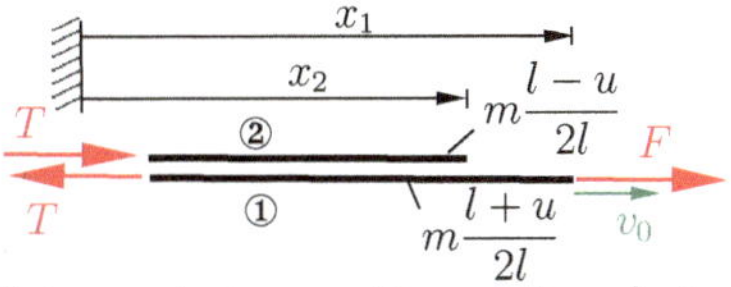

During the time increment dt the distance between the cord ends increases by du where part ② shortens by d$u/2$ and part ① elongates by d$u/2$. Accordingly, for part ② the mass flow and the relative ejection velocity are $\mu = \dot{u}m/2l$ and $w = \dot{x}_1 - \dot{x}_2 = \dot{u}$, respectively. Thus, the 'thrust' is given by

$$T = -\mu\, w = -\frac{m}{2l}\,\dot{u}^2 \qquad (c)\,.$$

Introducing T into (b) and considering $\ddot{x}_2 = -\ddot{u}$ yields

$$(l-u)\ddot{u} = \dot{u}^2$$

and with $\ddot{u} = \dot{u}\mathrm{d}\dot{u}/\mathrm{d}u$, separation of variables and integration

$$\frac{\mathrm{d}\dot{u}}{\dot{u}} = \frac{\mathrm{d}u}{l-u} \quad \rightsquigarrow \quad \ln\frac{\dot{u}}{\dot{u}(0)} = \ln\frac{l-u(0)}{l-u} \quad \rightsquigarrow \quad \dot{u} = v_0\,\frac{l}{l-u}\,.$$

This leads with (c) and (a) to

$$\underline{\underline{F = T = \frac{m\,v_0^2}{2l}\left(\frac{l}{l-u}\right)^2}}\,.$$

Chapter 4

Kinematics of Rigid Bodies

4

The **Motion of a Rigid Body** can be composed of a *Translation* and a *Rotation.*

Spatial Motion

The relations between the positions, the velocities and the accelerations of points A and P of a rigid body are given by

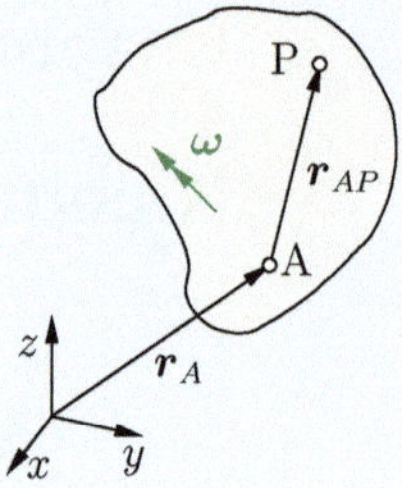

$$r_P = r_A + r_{AP} \ ,$$

$$v_P = v_A + \omega \times r_{AP} \ ,$$

$$a_P = a_A + \dot{\omega} \times r_{AP} + \omega \times (\omega \times r_{AP})$$

where $\omega \mathrel{\widehat{=}}$ angular velocity vector.

Remarks:

- ω has the direction of the current rotation axis.
- From the equations above follow v_P and a_P for any arbitrary point P of the body if v_A, a_A, ω and $\dot{\omega}$ are known.

Planar Motion

With $\omega = \omega e_z = \dot{\varphi} e_z$, $r_{AP} = r e_r$ follows

$$v_P = v_A + v_{AP} \ ,$$

$$a_P = a_A + a^t_{AP} + a^n_{AP} \ ,$$

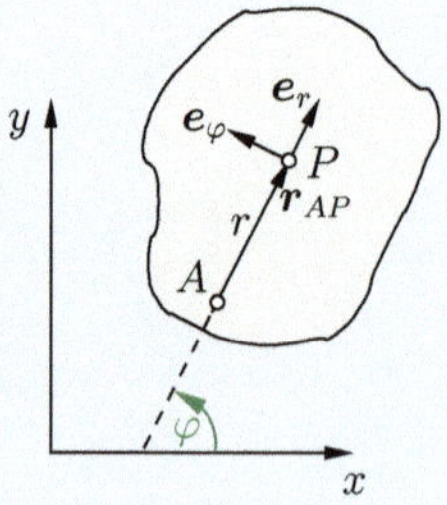

where

$$v_{AP} = \omega \times r_{AP} = \omega r e_\varphi \qquad (\perp \text{ to } r_{AP})$$

$$a^t_{AP} = \dot{\omega} \times r_{AP} = \dot{\omega} r e_\varphi \qquad (\perp \text{ to } r_{AP})$$

$$a^n_{AP} = \omega \times (\omega \times r_{AP}) = -\omega^2 r e_r \quad (\parallel \text{ to } r_{AP})$$

'The velocity (acceleration) of point P is equal to the velocity (acceleration) of point A plus the velocity (acceleration) caused by the rotation of P about A'.

Velocity and Acceleration Diagram

The graphical representation of the velocities and accelerations of plane kinematic problems and their graphical solution is done by using a velocity and a acceleration diagram. The directions of the velocity and acceleration components are taken from the *layout diagram.*:

layout diagram velocity diagram acceleration diagram

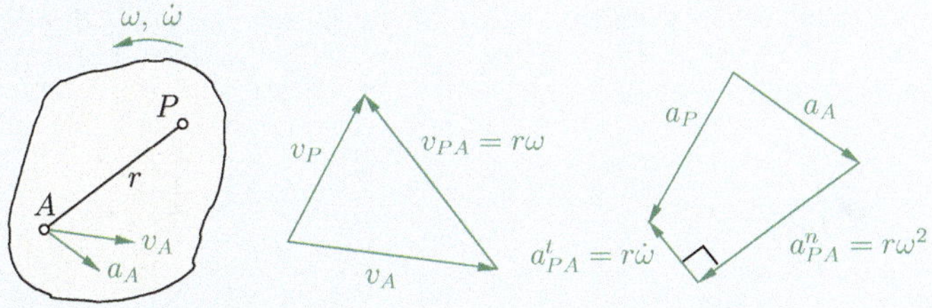

Instantaneous Center of Rotation

The plane motion of a rigid body having an instantaneous angular velocity ω can be considered at each instant as a pure rotation about the instantaneous center Π (instantaneous center of rotation, instantaneous center of zero velocity):

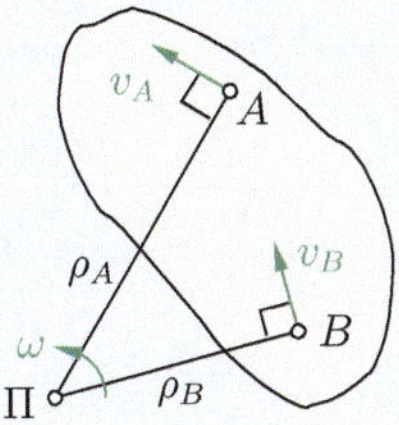

$$v_A = \rho_A \omega \ ,$$

$$v_B = \rho_B \omega \ .$$

The trajectory passed by the instantaneous center of rotation is called *centrode.*

Rolling Wheel

The locus of the center of rotation is given by:

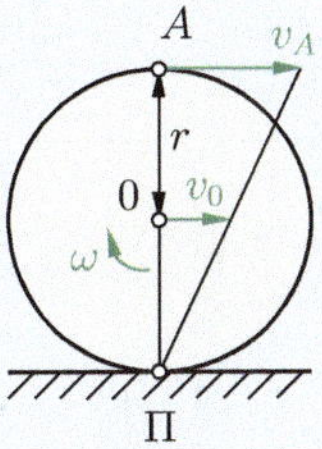

$$v_0 = r\omega \ ,$$

$$v_A = 2r\omega \ .$$

P4.1 **Problem 4.1** The end point A of a rigid bar moves in a horizontal channel with speed v_A and acceleration a_A.

Determine the velocity and acceleration of B as well as the angular velocity $\omega = \dot{\varphi}$ and the angular acceleration $\dot{\omega}$ of the bar.

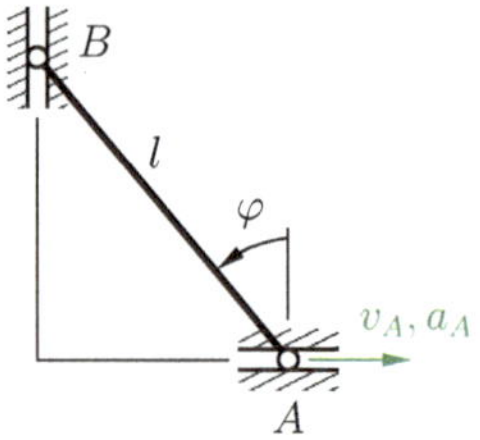

Solution The velocity and acceleration of B are vertically directed. With that information, the *velocity diagram* can be plotted and from the graph we find by inspection:

$$l\omega = \frac{v_A}{\cos\varphi}$$

$$\rightsquigarrow \quad \underline{\underline{\omega}} = \dot{\varphi} = \underline{\underline{\frac{v_A}{l\cos\varphi}}}\,,$$

$$\underline{\underline{v_B}} = \underline{\underline{v_A \tan\varphi}}\,.$$

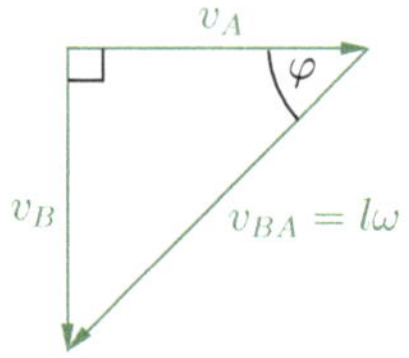

From the *acceleration diagram* follow in the same way

$$l\dot{\omega} = \frac{a_A}{\cos\varphi} + \omega^2 l \tan\varphi$$

$$\rightsquigarrow \quad \underline{\underline{\dot{\omega}}} = \underline{\underline{\frac{a_A}{l\cos\varphi} + \frac{v_A^2 \sin\varphi}{l^2\cos^3\varphi}}}\,,$$

$$\underline{\underline{a_B}} = a_A \tan\varphi + \frac{l\omega^2}{\cos\varphi}$$

$$= \underline{\underline{a_A \tan\varphi + \frac{v_A^2}{l\cos^3\varphi}}}\,.$$

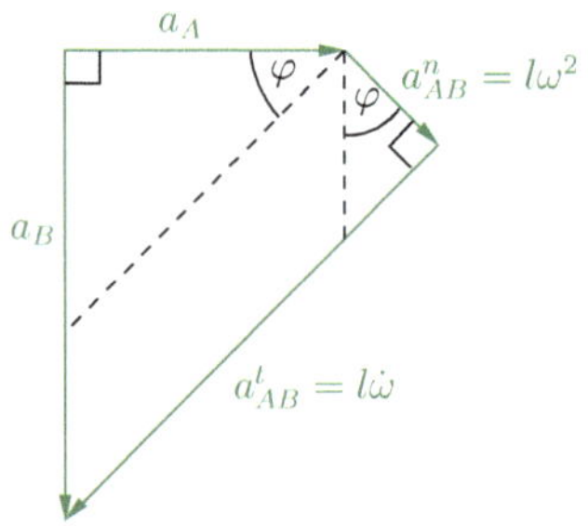

The problem can also be solved purely analytically. We introduce an appropriate coordinate system and from the coordinates

$$x_A = l\sin\varphi\,,$$

$$y_B = l\cos\varphi$$

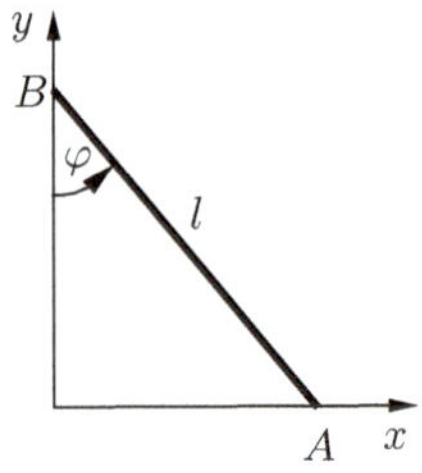

one obtains with $\dot{x}_A = v_A$ and $\ddot{x}_A = \dot{v}_A = a_A$

$$\dot{x}_A = l\dot{\varphi}\cos\varphi \quad \leadsto \quad \underline{\underline{\omega = \dot{\varphi} = \frac{v_A}{l\cos\varphi}}}\,,$$

$$\underline{\underline{\dot{\omega}}} = \frac{\dot{v}_A l\cos\varphi + v_A l\dot{\varphi}\sin\varphi}{l^2\cos^2\varphi} = \underline{\underline{\frac{a_A}{l\cos\varphi} + \frac{v_A^2\sin\varphi}{l^2\cos^3\varphi}}}\,,$$

$$\underline{\underline{\dot{y}_B}} = v_B = -l\dot{\varphi}\sin\varphi = \underline{\underline{-v_A\tan\varphi}}\,,$$

$$\underline{\underline{\ddot{y}_B}} = a_B = -\dot{v}_A\tan\varphi - v_A\frac{\dot{\varphi}}{\cos^2\varphi} = \underline{\underline{-a_A\tan\varphi - \frac{v_A^2}{l\cos^3\varphi}}}\,.$$

Note that since y is positive upwards directed, the quantities $\dot{y}_B$ and $\ddot{y}_B$ have a negative sign.

The velocity v_B and the angular velocity ω can also be determined by using the center of instantaneous rotation Π. This point Π is given by the intersection of the perpendiculars to v_A and v_B, the directions of which are known. In this way we obtain

$$v_A = \omega l\cos\varphi$$

$$\leadsto \quad \underline{\underline{\omega}} = \underline{\underline{\frac{v_A}{l\cos\varphi}}}\,,$$

$$\underline{\underline{v_B}} = \omega l\sin\varphi = \underline{\underline{v_A\tan\varphi}}\,.$$

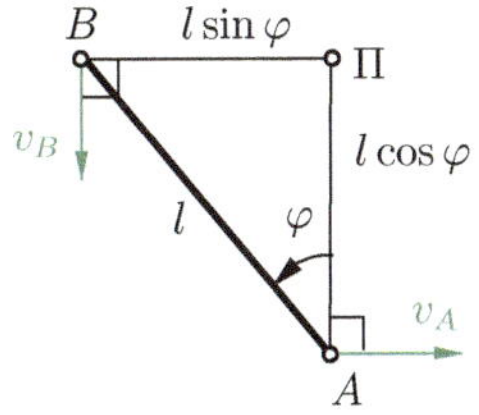

Remark: From the derived relations it can be seen that $\dot{\omega}$ and a_B are nonzero even when point A moves at *constant speed* ($a_A = 0$).

P4.2 **Problem 4.2** At a crank mechanism, the wheel rotates with constant angular velocity ω and point P moves along a vertical straight guide rail.

Determine analytically the velocity and acceleration of P.

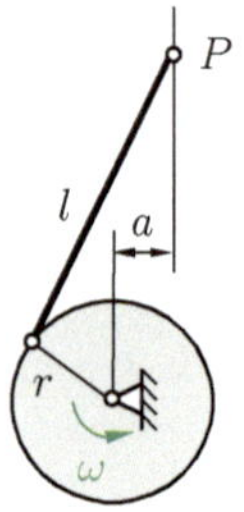

Solution We introduce a coordinate sytem and the two auxiliary angles φ and ψ, where $\varphi = \omega t$ and $\dot{\varphi} = \omega = \text{const}$. Then the position, velocity and acceleration of P are given by

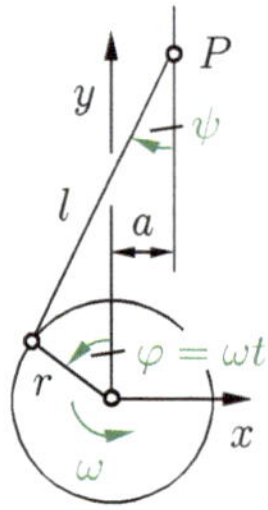

$$y_P = r\cos\varphi + l\cos\psi\,,$$

$$\dot{y}_P = -r\omega\sin\varphi - l\dot{\psi}\sin\psi\,,$$

$$\ddot{y}_P = -r\omega^2\cos\varphi - l\ddot{\psi}\sin\psi - l\dot{\psi}^2\cos\psi\,.$$

The still unknown quantities $\sin\psi$, $\cos\psi$, $\dot{\psi}$ and $\ddot{\psi}$ follow from the condition that P moves along a vertical line:

$$x_P = a = -r\sin\varphi + l\sin\psi\,,$$

$$\dot{x}_P = 0 = -r\omega\cos\varphi + l\dot{\psi}\cos\psi\,,$$

$$\ddot{x}_P = 0 = r\omega^2\sin\varphi + l\ddot{\psi}\cos\psi - l\dot{\psi}^2\sin\psi\,.$$

Solving for the unknowns yields

$$\sin\psi = \frac{a}{l} + \frac{r}{l}\sin\varphi\,, \qquad \cos\psi = \sqrt{1 - \left(\frac{a}{l} + \frac{r}{l}\sin\varphi\right)^2}\,,$$

$$\dot{\psi} = \omega\,\frac{r}{l}\,\frac{\cos\varphi}{\cos\psi}\,, \qquad \ddot{\psi} = -\omega^2\,\frac{r}{l}\,\frac{\sin\varphi}{\cos\psi} + \dot{\psi}^2\,\frac{\sin\psi}{\cos\psi}\,.$$

Introduction into $\dot{y}_P$ and $\ddot{y}_P$ finally leads to

$$\underline{\underline{\dot{y}_P = -r\omega\left\{\sin\varphi + \cos\varphi\,\frac{\sin\psi}{\cos\psi}\right\}}}\,,$$

$$\underline{\underline{\ddot{y}_P = -r\omega^2\left\{\cos\varphi - \sin\varphi\,\frac{\sin\psi}{\cos\psi} + \frac{r}{l}\,\frac{\cos^2\varphi}{\cos^3\psi}\right\}}}\,.$$

Problem 4.3 The boom with a pulley and a carrying rope is reeled in with speed v by a second horizontal cable. There is no sliding between pulley and carrying rope. **P4.3**

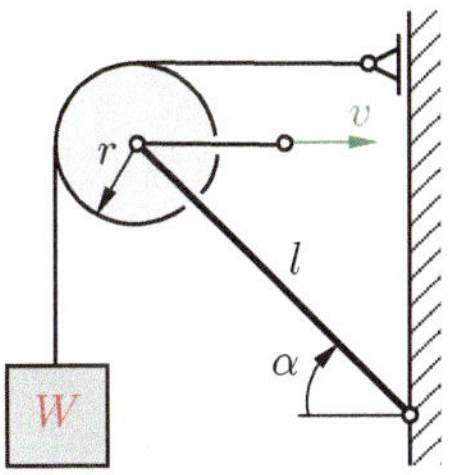

Determine the angular velocities of the boom and the pulley for $\alpha = 45°$?

Solution Because the motion of the boom is a pure rotation about A, the velocity v_P is perpendicular to l, and thus we have

$$l\dot{\alpha} = v_P$$

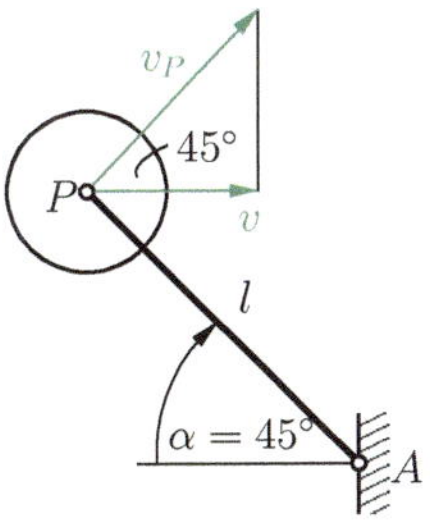

and

$$v_P = \sqrt{2}\,v\,.$$

Hence, the angular velocity of the boom is

$$\underline{\underline{\dot{\alpha} = \sqrt{2}\,\frac{v}{l}}}\,.$$

To determine the angular velocity ω of the pulley we use its instantaneous center of rotation Π. Its location is given by the intersection of the perpendiculars to v_P and v_B.

Notice: Due to the constraint by the rope, point B can only move in vertical direction.

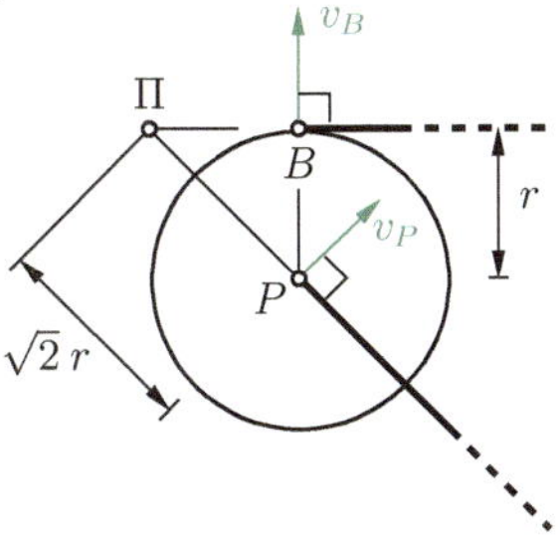

From

$$\sqrt{2}\,r\omega = v_P$$

we obtain

$$\underline{\underline{\omega}} = \frac{v_P}{\sqrt{2}\,r} = \underline{\underline{\frac{v}{r}}}\,.$$

P4.4 **Problem 4.4** A bar of length l moves in plane. In the displayed position, the velocity $\boldsymbol{v}_A$ of point A and the angular velocity ω are known.

a) Determine the magnitude and direction of the velocity of point B.

b) Where on the bar is point C located, whose instantaneous velocity has the x-direction?

c) Identify the location of the instantaneous center of rotation Π.

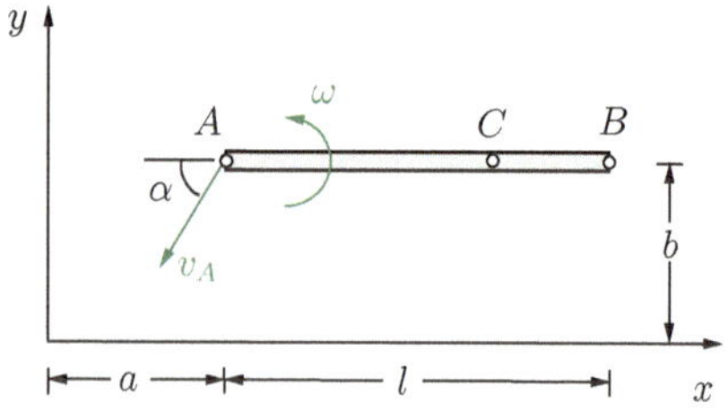

Given: $v_A = 3\,\text{m/s}$, $\alpha = 60°$, $\omega = 6/\text{s}$, $l = 1/2\text{m}$, $a = b = 1/4\,\text{m}$.

Solution **a)** From the relation $\boldsymbol{v}_B = \boldsymbol{v}_A + \boldsymbol{\omega} \times \boldsymbol{r}_{AB}$ and

$$\boldsymbol{v}_A = -v_A \cos\alpha\, \boldsymbol{e}_x - v_A \sin\alpha\, \boldsymbol{e}_y\,, \quad \boldsymbol{\omega} = \omega\, \boldsymbol{e}_z\,, \quad \boldsymbol{r}_{AB} = l\; \boldsymbol{e}_x\,,$$

$$\boldsymbol{v}_B = -v_A \cos\alpha\, \boldsymbol{e}_x + (\omega\, l - v_A \sin\alpha)\, \boldsymbol{e}_y \quad \rightsquigarrow \quad \begin{cases} v_{Bx} = -1.5\,\text{m/s}\,, \\ v_{By} = +0.402\,\text{m/s} \end{cases}$$

we obtain

$$\underline{\underline{v_B}} = \sqrt{v_{Bx}^2 + v_{Bx}^2} = \underline{\underline{1.55\,\text{m/s}}}\,, \quad \cos\beta = \frac{v_{Bx}}{v_B} \quad \rightsquigarrow \quad \underline{\underline{\beta = 165.4°}}\,.$$

b) The condition $v_{Cy} = 0$ leads with

$$\begin{aligned} \boldsymbol{v}_C &= \boldsymbol{v}_A + \boldsymbol{\omega} \times \boldsymbol{r}_{AC} \\ &= -v_A \cos\alpha\, \boldsymbol{e}_x \\ &\quad + (\omega\, c - v_A \sin\alpha)\, \boldsymbol{e}_y \end{aligned}$$

to

$$\underline{\underline{c = \frac{v_A \sin\alpha}{\omega} = 0.433\,\text{m}}}\,.$$

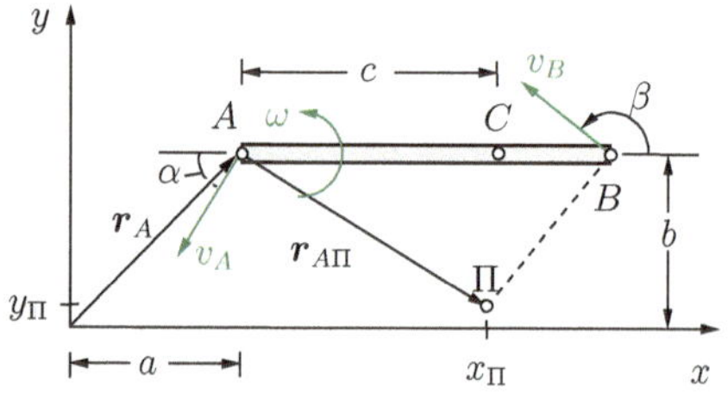

c) From the condition for the instantaneous center of rotation Π

$$\boldsymbol{v}_\Pi = \boldsymbol{v}_A + \boldsymbol{\omega} \times \boldsymbol{r}_{A\Pi} = 0 \quad \rightsquigarrow \quad \boldsymbol{\omega} \times \boldsymbol{v}_A = -\boldsymbol{\omega} \times (\boldsymbol{\omega} \times \boldsymbol{r}_{A\Pi}) = \omega^2 \boldsymbol{r}_{A\Pi}$$

follows with $\boldsymbol{r}_\Pi = \boldsymbol{r}_A + \boldsymbol{r}_{A\Pi}$ the location of Π:

$$\boldsymbol{r}_\Pi = \boldsymbol{r}_A + \frac{1}{\omega^2}\, \boldsymbol{\omega} \times \boldsymbol{v}_A = (a + \frac{v_A}{\omega} \sin\alpha)\, \boldsymbol{e}_x + (b - \frac{v_A}{\omega} \cos\alpha)\, \boldsymbol{e}_y\,,$$

$$\rightsquigarrow \quad \underline{\underline{x_\Pi = a + \frac{v_A}{\omega} \sin\alpha = 0.683\,\text{m}}}\,, \qquad \underline{\underline{y_\Pi = b - \frac{v_A}{\omega} \cos\alpha = 0}}\,.$$

Problem 4.5 In a gear, the wheel ① rolls along the fixed circular bearing ②. It is driven by a constant angular velocity Ω. P4.5

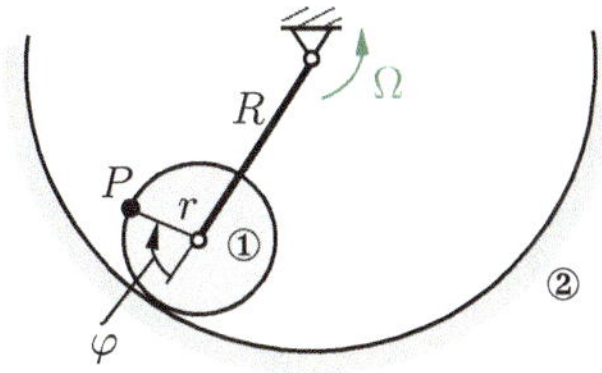

Determine analytically the magnitudes of velocity and acceleration for a point P of the wheel.

Solution We introduce a coordinate system and consider an initial and a displaced position of the wheel. With the angles α and β, where $\alpha + \beta = \varphi$, it follows for the covered arc length of P

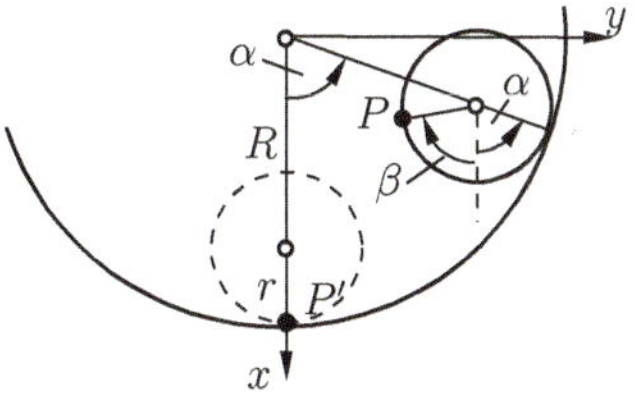

$$(R + r)\alpha = r(\beta + \alpha)$$
$$\rightsquigarrow \quad R\alpha = r\beta$$

Differentiation leads with $\dot{\alpha} = \Omega$ and $\dot{\beta} = \omega$ to

$$R\,\Omega = r\,\omega \quad \rightsquigarrow \quad \omega = \Omega\,R/r\,.$$

The velocity and acceleration components are derived from the position of P:

$$x_P = R\cos\alpha + r\cos\beta\,, \qquad y_P = R\sin\alpha - r\sin\beta\,,$$

$$\dot{x}_P = -R\Omega\sin\alpha - r\omega\sin\beta\,, \qquad \dot{y}_P = R\Omega\cos\alpha - r\omega\cos\beta\,,$$

$$\ddot{x}_P = -R\Omega^2\cos\alpha - r\omega^2\cos\beta\,, \quad \ddot{y}_P = -R\Omega^2\sin\alpha + r\omega^2\sin\beta\,.$$

Thus, with $R\,\Omega = r\,\omega$ and $\alpha + \beta = \varphi$ we obtain

$$\begin{aligned} v_P^2 &= \dot{x}_P^2 + \dot{y}_P^2 = R^2\Omega^2 + r^2\omega^2 + 2R\Omega r\omega(\sin\alpha\sin\beta - \cos\alpha\cos\beta) \\ &= 2R^2\Omega^2 - 2R^2\Omega^2\cos(\alpha+\beta) = 2R^2\Omega^2(1 - \cos\varphi) = 4R^2\Omega^2\sin^2\frac{\varphi}{2} \\ &\rightsquigarrow \quad \underline{\underline{v_P = 2R\Omega\sin\frac{\varphi}{2}}}\,, \end{aligned}$$

$$\begin{aligned} a_P^2 &= \ddot{x}_P^2 + \ddot{y}_P^2 = R^2\Omega^4 + r^2\omega^4 + 2R\Omega^2 r\omega^2(\cos\alpha\cos\beta - \sin\alpha\sin\beta) \\ &= (R\Omega^2)^2 + \left(\frac{R^2\Omega^2}{r}\right)^2 - 2R\Omega^2\,\frac{R^2\Omega^2}{r}\cos(\alpha+\beta) \\ &\rightsquigarrow \quad \underline{\underline{a_P = R\Omega^2\sqrt{1 + (R/r)^2 + 2(R/r)\cos\varphi}}}\,. \end{aligned}$$

P4.6

Problem 4.6 An idealized fair ride consists of a base plate and attached gondola crosses, both rotating with constant angular velocities ω_1 and $\omega_2 = 2\omega_1$, respectively.

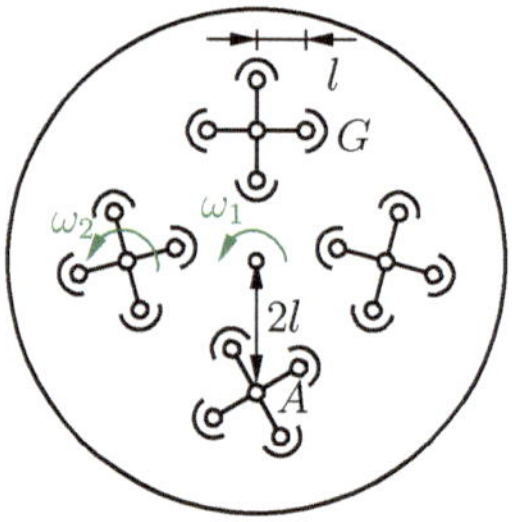

Determine the magnitudes of velocity and acceleration of a gondola G.

Solution We first solve the problem by sketching the velocity and acceleration diagrams. This can be done with the aid of the displayed layout diagram. Using the cosine rule, it can be seen

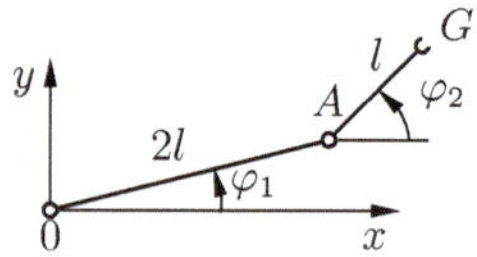

$$v_G^2 = (2l\omega_1)^2 + (l\omega_2)^2 - 2\cdot 2l\omega_1 l\omega_2 \cos(\pi - \varphi_2 + \varphi_1)$$

$$= 8l^2\omega_1^2[1 + \cos(\varphi_2 - \varphi_1)]$$

$$= 16l^2\omega_1^2 \cos^2\frac{\varphi_2 - \varphi_1}{2},$$

$$\rightsquigarrow \quad \underline{\underline{v_G = 4l\omega_1 \cos\frac{\varphi_2 - \varphi_1}{2}}},$$

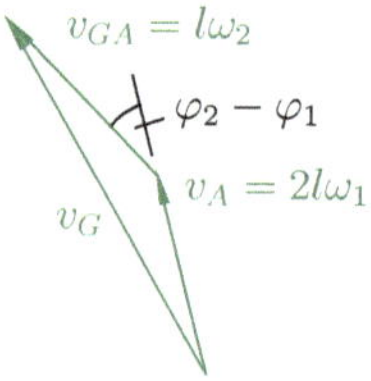

$$a_G^2 = (2l\omega_1^2)^2 + (l\omega_2^2)^2 - 2\cdot 2l\omega_1^2 l\omega_2^2 \cos(\pi - \varphi_2 - \varphi_1)$$

$$= 4l^2\omega_1^4[5 + 4\cos(\varphi_2 - \varphi_1)],$$

$$\rightsquigarrow \quad \underline{\underline{a_G = 2l\omega_1^2\sqrt{5 + 4\cos(\varphi_2 - \varphi_1)}}}.$$

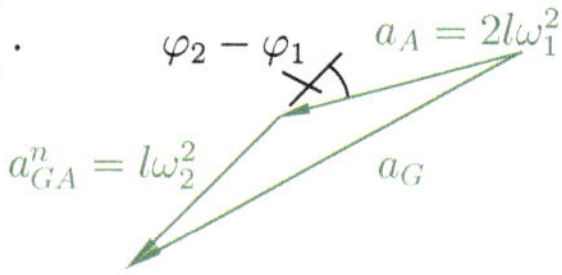

Remark: Because the cosine varies between +1 and −1, the maximum and minimum acceleration are given by $a_{G\max} = 6l\,\omega_1^2$ and $a_{G\min} = 2l\,\omega_1^2$, respectively.

The problem can also be solved analytically by considering the components of the position vector and by subsequent differentiation with respect to time. With $\dot\varphi_1 = \omega_1$ and $\dot\varphi_2 = \omega_2 = 2\omega_1$ we obtain

$$x_G = 2l\cos\varphi_1 + l\cos\varphi_2\ ,$$

$$\dot x_G = -2l\,\dot\varphi_1\sin\varphi_1 - l\,\dot\varphi_2\sin\varphi_2 = -2l\,\omega_1\sin\varphi_1 - l\,2\,\omega_1\sin\varphi_2\ ,$$

$$\ddot x_G = -2l\,\omega_1^2\cos\varphi_1 - l\,4\,\omega_1^2\cos\varphi_2\ ,$$

$$y_G = 2l\sin\varphi_1 + l\sin\varphi_2\ ,$$

$$\dot y_G = 2l\,\omega_1\cos\varphi_1 + l\,2\,\omega_1\cos\varphi_2\ ,$$

$$\ddot x_G = -2l\omega_1^2\sin\varphi_1 - l\,4\,\omega_1^2\sin\varphi_2\ .$$

Using the addition theorem it follows

$$\begin{aligned} v_G^2 = \dot x_G^2 + \dot y_G^2 &= (2l\,\omega_1)^2 + (l\,2\,\omega_1)^2 + \\ &\quad +8l^2\omega_1^2(\sin\varphi_1\sin\varphi_2 + \cos\varphi_1\cos\varphi_2) \\ &= 8l^2\omega_1^2[1+\cos(\varphi_2-\varphi_1)] \end{aligned}$$

$$\leadsto \quad \underline{\underline{v_G = 4l\omega_1\cos\frac{\varphi_2-\varphi_1}{2}}}\ ,$$

$$\begin{aligned} a_G^2 = \ddot x_G^2 + \ddot y_G^2 &= (2l\omega_1^2)^2 + (4l\omega_1^2) + \\ &\quad +16l^2\omega_1^4(\cos\varphi_1\cos\varphi_2 + \sin\varphi_1\sin\varphi_2) \\ &= 4l^2\omega_1^2[5+4\cos(\varphi_2-\varphi_1)] \end{aligned}$$

$$\leadsto \quad \underline{\underline{a_G = 2l\omega_1^2\sqrt{5+4\cos(\varphi_2-\varphi_1)}}}\ .$$

Remark: The instantaneous center of rotation Π of the gondola cross is located on the perpendicular to v_A. Because A rotates about 0 its velocity is $v_A = 2l\omega_1$. Due to the rotation of the gondola cross $\overline{AG}$ with the angular velocity $\omega_2 = 2\omega_1$ about Π at the same time the relation $v_A = \rho_A\omega_2$ is valid. Thus, it follows $\rho_A = l$ and therefore

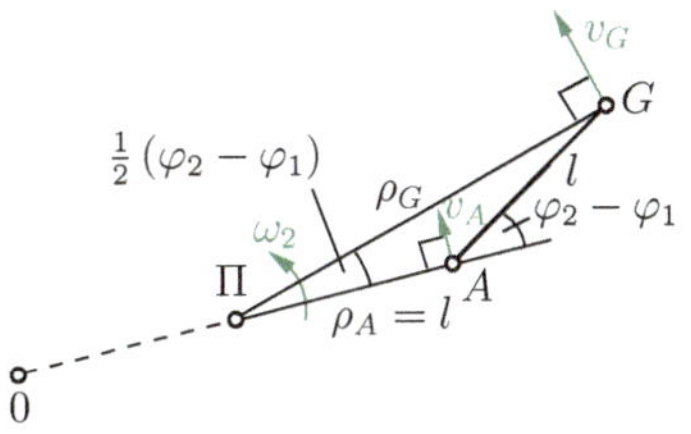

$$\underline{\underline{v_G}} = \omega_2\rho_G = \underline{\underline{\omega_2 2l\cos\frac{\varphi_2-\varphi_1}{2}}}\ .$$

P4.7 **Problem 4.7** The crank DC of a crank-rocker mechanism rotates with constant angular velocity ω.

Determine the vertical velocity of point B and the angular velocity $\dot{\psi}$ of the bar AB.

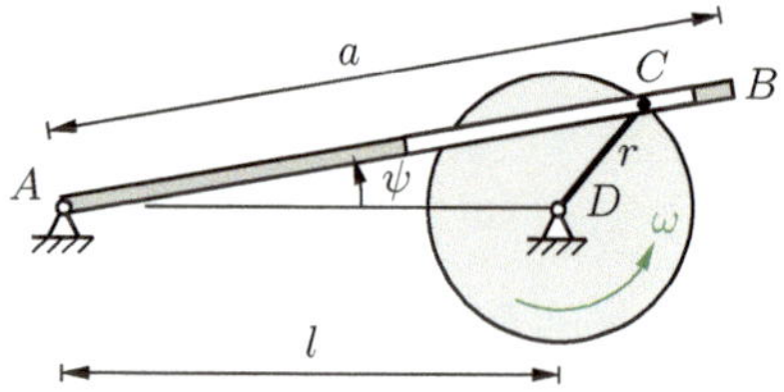

Solution From the sketch we obtain the geometric relation

$$r\sin\varphi = (l + r\cos\varphi)\tan\psi$$

$$\rightsquigarrow \quad \tan\psi = \frac{r\sin\varphi}{l + r\cos\varphi}\,.$$

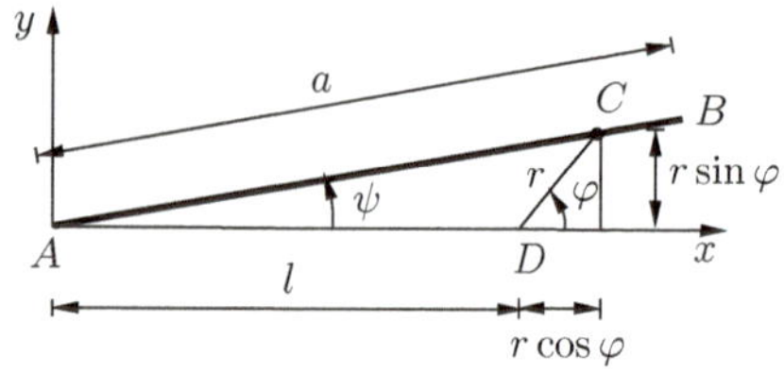

This leads to

$$y_B = a\sin\psi = a\,\frac{\tan\psi}{\sqrt{1+\tan^2\psi}} = ar\,\frac{\sin\varphi}{\sqrt{l^2+r^2+2rl\cos\varphi}}$$

and for the vertical velocity, using $\dot{\varphi} = \omega$, to

$$\underline{\underline{\dot{y}_B}} = ar\,\frac{\omega\cos\varphi\sqrt{l^2+r^2+2rl\cos\varphi} - \dfrac{\sin\varphi\,(-2rl\omega\sin\varphi)}{2\sqrt{l^2+r^2+2rl\cos\varphi}}}{(l^2+r^2+2rl\cos\varphi)}$$

$$= \underline{\underline{ar\omega\,\frac{(r+l\cos\varphi)(l+r\cos\varphi)}{(l^2+r^2+2rl\cos\varphi)^{3/2}}}}\,.$$

The angular velocity $\dot{\psi}$ is obtained by differentiating $\tan\psi$ with respect to time:

$$\frac{1}{\cos^2\psi}\,\dot{\psi} = \frac{r\cos\varphi\,(l+r\cos\varphi) - r\sin\varphi\,(-r\sin\varphi)}{(l+r\cos\varphi)^2}\,\omega$$

$$\rightsquigarrow \quad \underline{\underline{\dot{\psi}}} = \frac{r(r+l\cos\varphi)}{(l+r\cos\varphi)^2}\,\frac{1}{1+\tan^2\psi}\,\omega = \underline{\underline{\frac{r(r+l\cos\varphi)}{l^2+r^2+2rl\cos\varphi}\,\omega}}\,.$$

P4.8

Problem 4.8 In a part of a freight lift, three wheels are connected by unrolling vertical cables.

Determine the velocities and angular velocities of the wheels ② and ③ when wheel ① rotates with a given angular velocity ω_1.

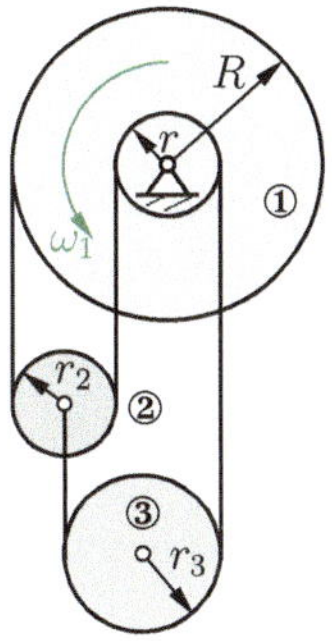

Solution Wheel ① rotates about the fixed point Π_1. Therefore, the velocities of points A and B are given by

$$v_A = R\omega_1 \,, \qquad v_B = r\omega_1 \,.$$

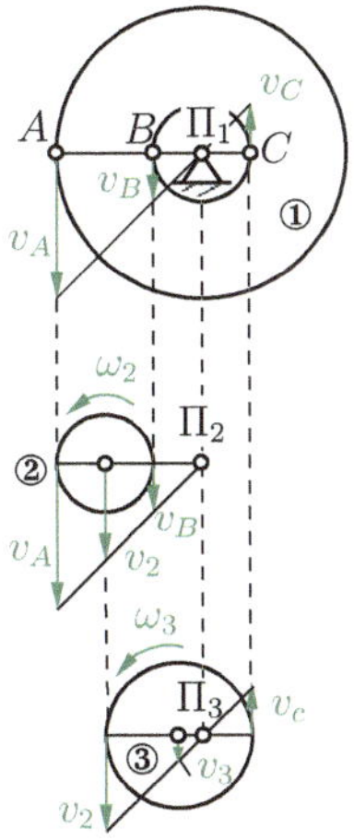

Since the cables unroll, v_A and v_B will be transferred unchanged to the wheel ②. Thus, from

$$v_A = v_2 + r_2\omega_2 \quad \text{and} \quad v_B = v_2 - r_2\omega_2$$

follows with $r_2 = (R - r)/2$

$$\underline{\underline{v_2}} = \frac{1}{2}(v_A + v_B) = \underline{\underline{\frac{1}{2}(R + r)\omega_1}} \,,$$

$$\underline{\underline{\omega_2}} = \frac{v_A - v_2}{r_2} = \underline{\underline{\omega_1}} \,.$$

In the same way we obtain for wheel ③ from

$$v_2 = v_3 + r_3\omega_3 \,, \quad v_C = v_3 - r_3\omega_3$$

with $r_3 = 2r + r_2 = (R + 3r)/4$ and $v_C = -r\omega_1$ the results

$$\underline{\underline{v_3}} = \frac{1}{2}(v_2 + v_C) = \underline{\underline{\frac{1}{4}(R - r)\omega_1}} \,,$$

$$\underline{\underline{\omega_3}} = \frac{v_2 - v_3}{r_3} = \underline{\underline{\omega_1}} \,.$$

Remark: The instantaneous centers of rotation Π_1, Π_2 and Π_3 are located on one and the same straight line.

P4.9

Problem 4.9 Point A of a bar of length l is guided horizonally with the constant velocity v_0 and slides at B across a post.

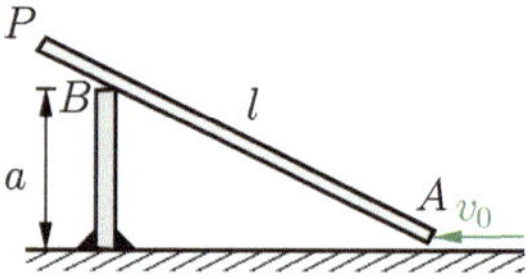

Determine the magnitudes of velocity and acceleration of point P of the bar.

Solution The angular velocity $\dot\varphi$ of the bar is determined from

$$\cot\varphi = \frac{x}{a}$$

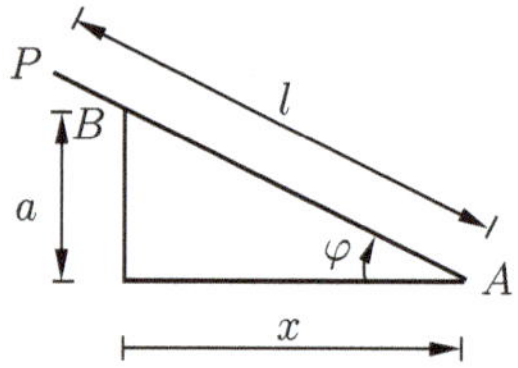

by time differentiation with $\dot x = -v_0$ as

$$-\frac{\dot\varphi}{\sin^2\varphi} = \frac{\dot x}{a} \quad\leadsto\quad \dot\varphi = \frac{v_0}{a}\sin^2\varphi\ .$$

Second differentiation leads to the angular acceleration

$$\ddot\varphi = \frac{v_0}{a}\,2\sin\varphi\,\cos\varphi\,\dot\varphi = 2\left(\frac{v_0}{a}\right)^2\sin^3\varphi\,\cos\varphi\ .$$

Then, by using the cosine rule, we obtain from the velocity diagram

$$v_P^2 = v_0^2 + (l\dot\varphi)^2 - 2v_0 l\dot\varphi\cos\left(\frac{\pi}{2}-\varphi\right)$$

$$\leadsto\quad \underline{\underline{v_P = v_0\sqrt{1+\left(\frac{l}{a}\right)^2\sin^4\varphi - 2\left(\frac{l}{a}\right)\sin^3\varphi}}}\ .$$

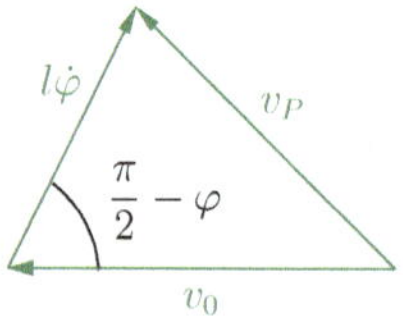

The acceleration diagram yields

$$\underline{\underline{a_P}} = \sqrt{l^2\dot\varphi^4 + l^2\ddot\varphi^2}$$

$$= l\left(\frac{v_0}{a}\right)^2\sqrt{\sin^8\varphi + 4\sin^6\varphi\cos^2\varphi}$$

$$= \underline{\underline{= l\left(\frac{v_0}{a}\right)^2\sin^3\varphi\sqrt{\sin^2\varphi + 4\cos^2\varphi}}}\ .$$

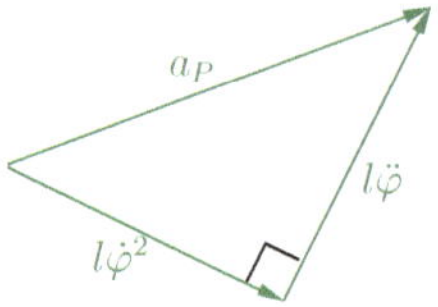

Remark: For $\varphi = \pi/2$ (point A arrives at the post) follows

$$v_P = v_0(l/a - 1) = (l-a)\dot\varphi|_{\pi/2}\ .$$

Problem 4.10 The wheel of a crank drive rotates with constant angular velocity ω.

Determine graphically the velocity and acceleration of the horizontally moving point B for the angle $\alpha = 30°$.

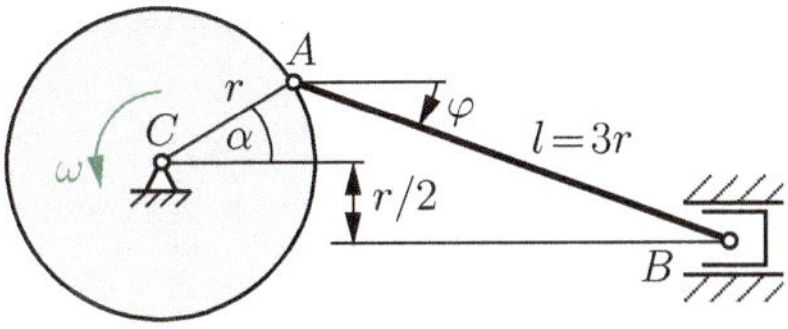

Solution From the *layout diagram* the directions of the different velocity and acceleration terms are taken: $v_A \perp$ to r, $v_{BA} \perp$ to l, v_B horizontal, a_A in direction of r, a^n_{BA} in direction of l, $a^t_{BA} \perp$ to l, a_B horizontal. With this knowledge, the velocity and acceleration diagrams can be constructed.

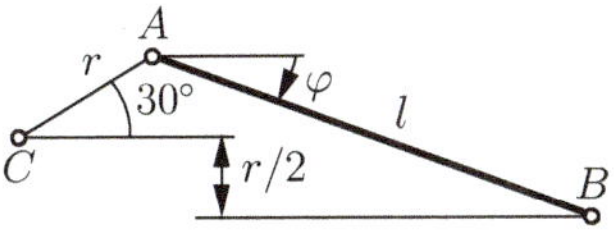

velocity diagram

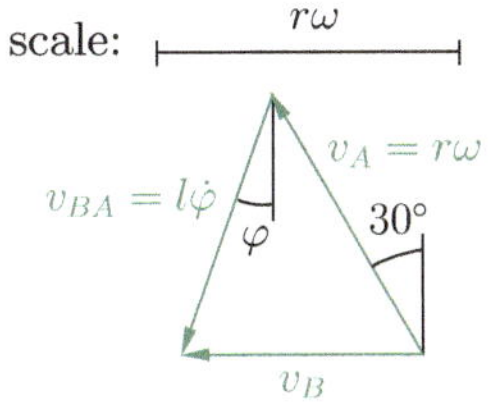

$$v_{BA} = 0.92\, r\omega = l\dot{\varphi}$$

$$\rightsquigarrow \quad \dot{\varphi} = \frac{0.92\, r}{l}\,\omega = 0.31\,\omega\ ,$$

$$\underline{\underline{v_B = 0.81\, r\omega}}\ .$$

acceleration diagram

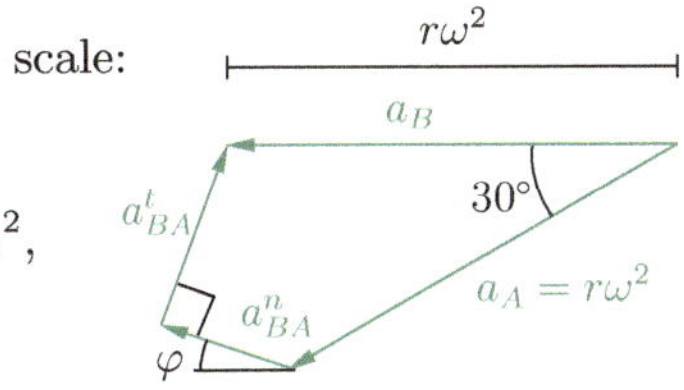

$$a^n_{BA} = l\dot{\varphi}^2 = 3r(0.31\,\omega)^2 = 0.29\, r\omega^2,$$

$$\underline{\underline{a_B = 0.99\, r\omega^2}}\ .$$

P4.11 **Problem 4.11** Point A of the mechanism has in the displayed position the velocity v_A.

Determine
a) for the 4 bars the instantaneous centers of rotation,

b) the velocities of points B and C as well as the angular velocities of all 4 bars.

Given: $a = 30$ cm, $b = 40$ cm, $v_A = 1$ m/s.

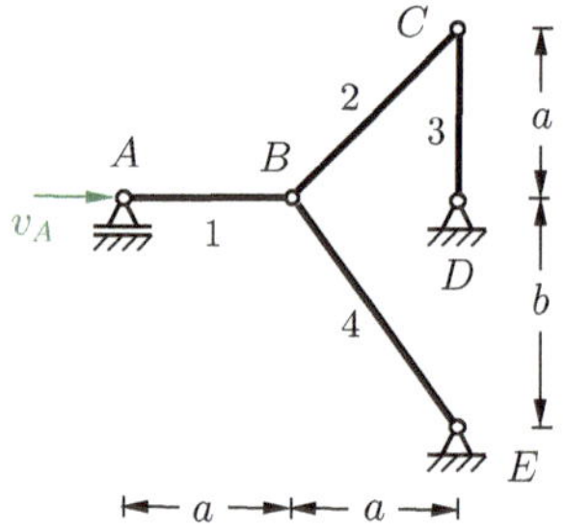

Solution **a)** The centers Π_3 (bar 3) and Π_4 (bar 4) are the hinged supports D and E, respectively. The centers Π_1 (bar 1) and Π_2 (bar 2) are given by the intersections of the perpendiculars to v_A and v_B ($v_B \perp EB$) and to v_B and v_C ($v_C \perp DC$), respectively (see figure).

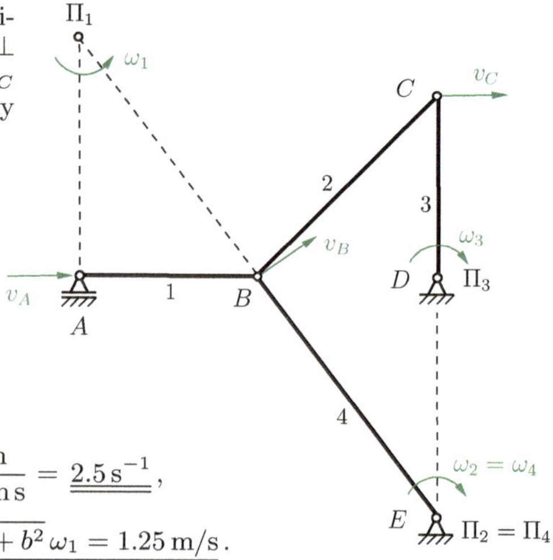

b) For bar 1 follows:

$$v_A = \overline{\Pi_1 A}\,\omega_1 = b\,\omega_1$$

$$\rightsquigarrow \quad \underline{\underline{\omega_1 = \frac{v_A}{b}}} = \frac{1\,\text{m}}{0.4\,\text{m}\,\text{s}} = \underline{\underline{2.5\,\text{s}^{-1}}},$$

$$\underline{\underline{v_B}} = \overline{\Pi_1 B}\,\omega_1 = \underline{\underline{\sqrt{a^2+b^2}\,\omega_1 = 1.25\,\text{m/s}}}.$$

For bar 4 and for bar 2 we obtain

$$v_B = \overline{\Pi_2 B}\,\omega_4 = \sqrt{a^2+b^2}\,\omega_4\,, \qquad \omega_4 = \omega_2$$

$$\rightsquigarrow \quad \underline{\underline{\omega_2 = \omega_4 = \omega_1 = 2.5\,\text{s}^{-1}}},$$

$$\underline{\underline{v_C}} = \overline{\Pi_2 C}\,\omega_2 = \underline{\underline{(a+b)\omega_2 = 1.75\,\text{m/s}}}.$$

Finally, the rotation of bar 3 about Π_3 leads to

$$v_C = \overline{\Pi_3 C}\,\omega_3 = a\,\omega_3 \quad \rightsquigarrow \quad \underline{\underline{\omega_3 = \frac{v_C}{a} = 5.83\,\text{s}^{-1}}},$$

Problem 4.12 A toggle lever is driven with the angular velocity ω. **P4.12**

Determine graphically the velocities of the points B, C and D for a certain angle β by using the centers of instaneous rotation.

Given.: $l = 30\,\text{cm}$, $a = 40\,\text{cm}$, $r = 15\,\text{cm}$,

$b = 50\,\text{cm}$, $\beta = 70°$, $\omega = 4\,\text{s}^{-1}$.

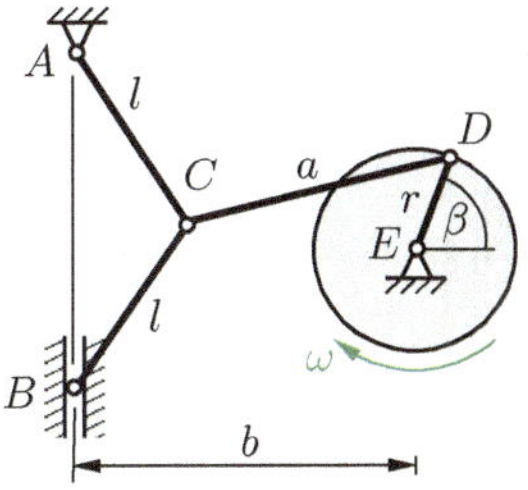

Solution The centers of rotation Π_1 (bar CD) and Π_2 (bar BC) are given by the intersections of the perpendiculars to $v_D = r\omega$ ($v_D \perp ED$) and v_C ($v_C \perp AC$) and to v_C and v_B, respectively. This leads in the layout diagram to the depicted representation.

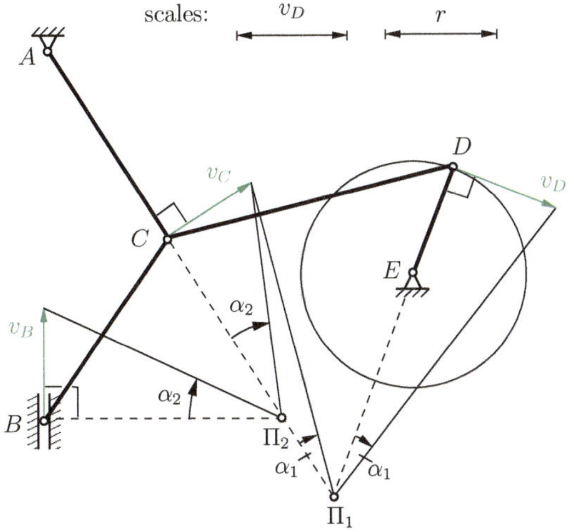

With the chosen scale, we read from the figure

$$v_C \cong 0.9\, v_D\,, \qquad v_B \cong 1.0\, v_D$$

and obtain with

$$\underline{\underline{v_D}} = r\omega = \underline{\underline{0.6\,\text{m/s}}}$$

the velocities

$$\underline{\underline{v_C \cong 0.5\,\text{m/s}}}\,, \qquad \underline{\underline{v_B \cong 0.6\,\text{m/s}.}}$$

P4.13 **Problem 4.13** A mechanism consists of three pin connected bars which are moved by a rope that is hauled in with constant speed v_S.

Determine graphically the velocities of the points B and C for $\varphi = 60°$.

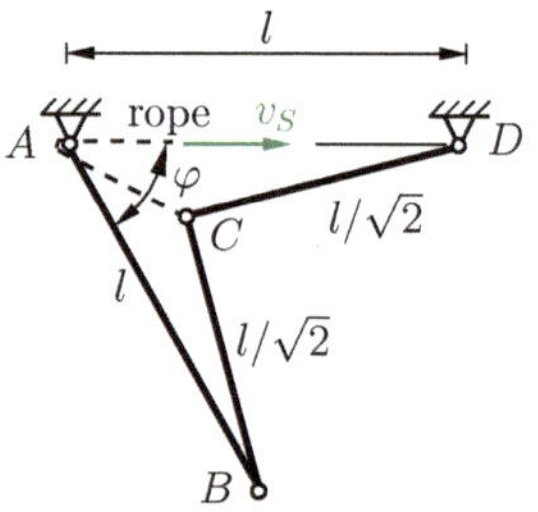

Solution We first solve the problem in the layout diagram by using the instantaneous center of rotation Π of bar CB. Π is given by the intersection of the perpendiculars to v_B $(v_B \perp AB)$ and v_C $(v_C \perp CD$, magnitude of v_C determinable from $v_S)$.

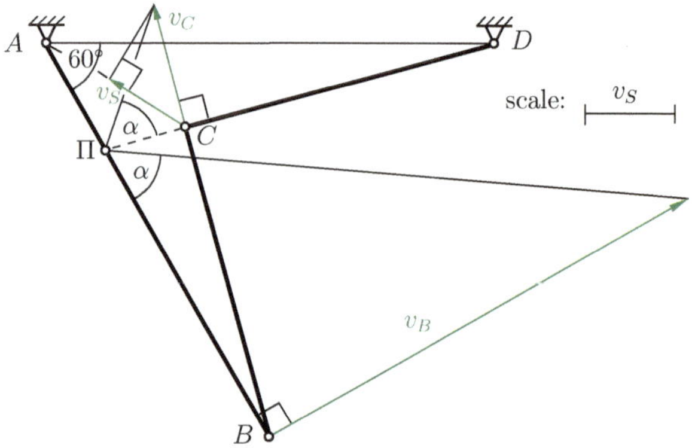

From the figure we obtain by inspection:

$$\underline{\underline{v_C \cong 1.4\, v_S}}\,, \qquad \underline{\underline{v_B \cong 5.4\, v_S}}\,.$$

The same result can be obtained by using the velocity diagram. Here, the directions of the velocities (e.g. $v_{BC} \perp CB$) are taken from the layout diagram:

$$\underline{\underline{v_C \cong 1.4\, v_S}}\,,$$

$$\underline{\underline{v_B \cong 5.4\, v_S}}\,.$$

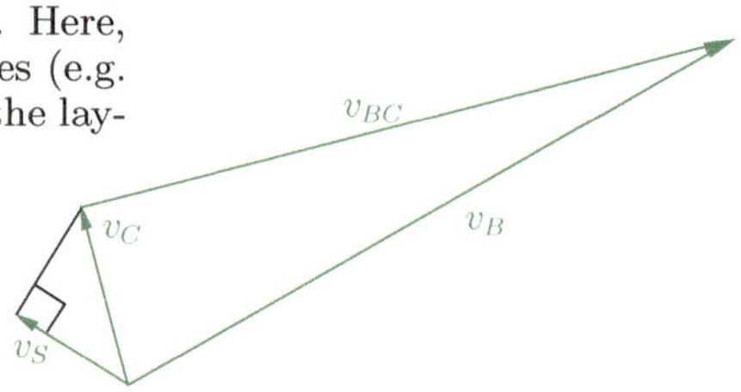

Problem 4.14 At the rolling wheel, a bar is connected by a hinge. The bar slides at B along an inclined plane. **P4.14**

Determine graphically the velocities and accelerations of B and C for the displayed position and the given constant speed v_0 of the wheel.

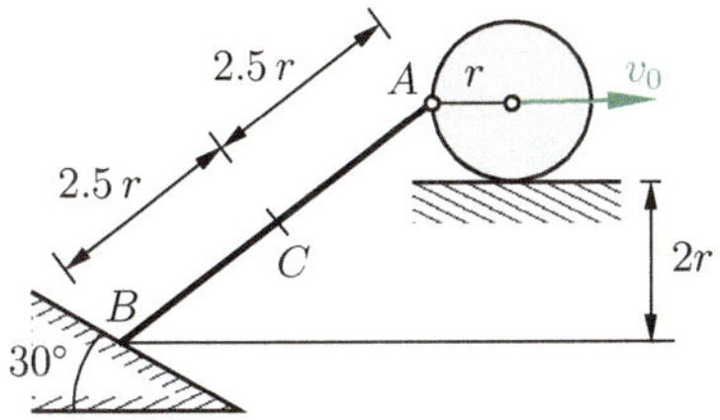

Solution From the *layout diagram*, using the centers of instantaneous rotation Π_1 (wheel) and Π_2 (bar), the directions of the velocities and accelerations can be obtained. With that information the **velocity diagram** is drawn ($v_A \perp \overline{\Pi_1 A}$, $v_{BA} \perp \overline{BA}$, $v_B \perp \overline{\Pi_2 B}$) and it follows by inspection

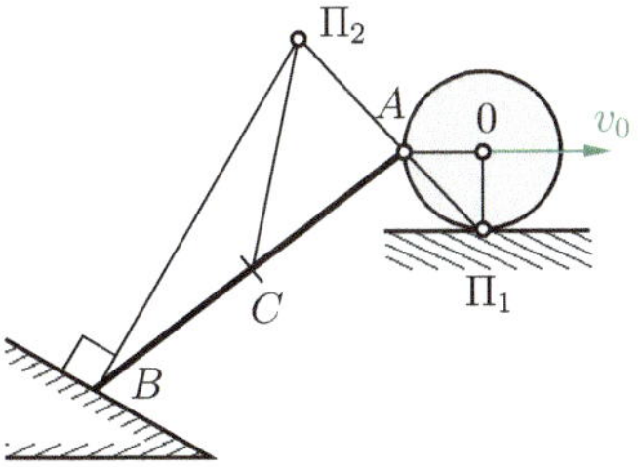

$$\underline{\underline{v_B \cong 3.6\, v_0}}\,, \qquad \underline{\underline{v_C \cong 2.1\, v_0}}\,,$$

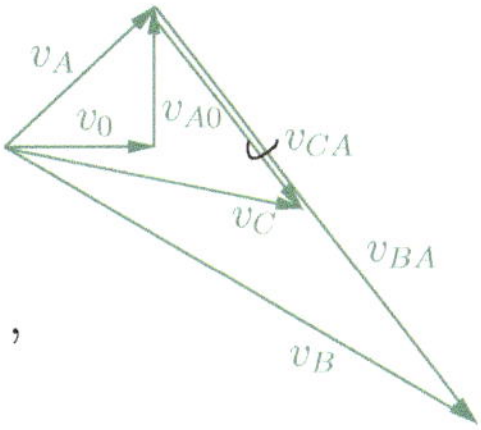

$$v_{BA} = 3.5\, v_0 \quad \rightsquigarrow \quad a^n_{BA} = \frac{v^2_{BA}}{5r} = 2.45\, \frac{v_0^2}{r}\,,$$

$$v_{CA} = 1.7\, v_0 \quad \rightsquigarrow \quad a^n_{CA} = \frac{v^2_{CA}}{2.5r} = 1.2\, \frac{v_0^2}{r}\,.$$

The **acceleration diagram** leads with $a_A = v_0^2/r$ and $a^t_{CA} = a^t_{BA}/2$ to

$$\underline{\underline{a_B = 8.4\, \frac{v_0^2}{r}}}\,,$$

$$\underline{\underline{a_C = 4.7\, \frac{v_0^2}{r}}}\,.$$

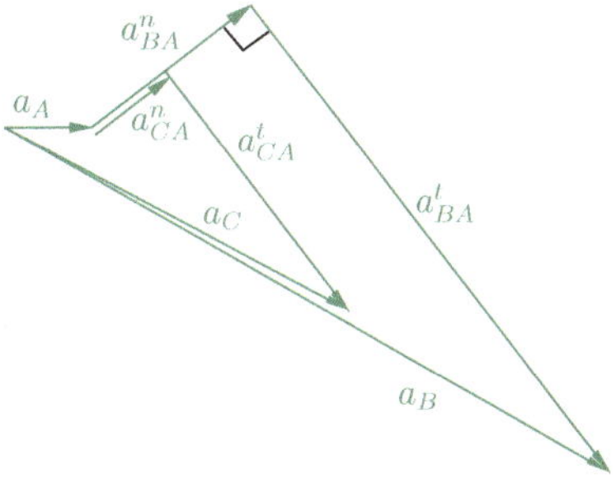

P4.15 **Problem 4.15** From the tire of a car (radius $r = 30\,\mathrm{cm}$, speed $v_0 = 108\,\mathrm{km/h}$) separates a stone at point P ($\alpha = 30°$). Determine

a) the velocity components of the stone at the instant of separation,
b) the maximum flight height and the flying distance of the stone,
c) the minimum distance of a following car with same speed so that it will not be hit by the stone.

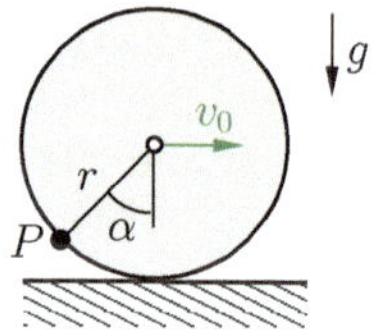

The height of the separation point above street level, the air drag and the length of the following car can be disregarded.

Solution **a)** The velocity components are determined by using the center of instantaneous rotation Π. We obtain with $\omega = v_0/r$, $a = 2r\sin\frac{\alpha}{2}$, $\beta = \pi/2 - \alpha/2 = 75°$ and the conversion $1\,\mathrm{m/s} = 3.6\,\mathrm{km/h}$

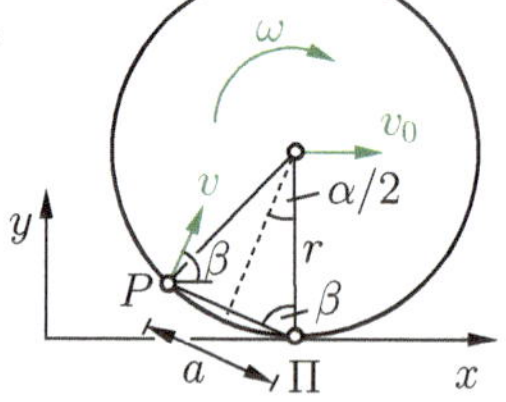

$$v = \omega a = 2v_0 \sin\tfrac{\alpha}{2} = 15.53\,\mathrm{m/s}\,,$$

$$\underline{\underline{v_x = v\cos\beta = 4.02\,\mathrm{m/s}}}\,,$$

$$\underline{\underline{v_y = v\sin\beta = v_0\sin\alpha = 15\,\mathrm{m/s}}}$$

b) The flight height h, flight distance d and flight time t_d follow from the equations for projectile motion (see page 32)

$$\underline{\underline{h}} = v^2\sin^2\beta/(2g) = \underline{\underline{v_0^2\sin^2\alpha/(2g) = 11.47\,\mathrm{m}}}\,,$$

$$\underline{\underline{d}} = v^2\sin 2\beta/g = \underline{\underline{4v_0^2\sin^2\frac{\alpha}{2}\sin\alpha/g = 12.29\,\mathrm{m}}}\,,$$

$$t_d = 2v\sin\beta/g = 2v_0\sin\alpha/g = 3.06\,\mathrm{s}.$$

c) The minimum distance c follows from the distances covered during the flight time t_d. The following car and the stone arrive at the same time at the same position if:

$$v_0 t_w = c + w\,, \quad \leadsto \quad \underline{\underline{c}} = v_0 t_w - w = \underline{\underline{79.45\,\mathrm{m}}}\,.$$

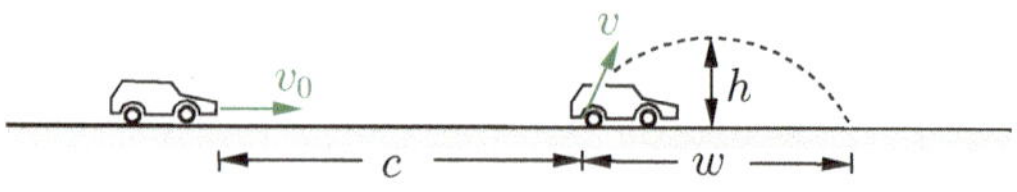

Problem 4.16 In a gear, the shaft ① (radius $3a$) and the ring ② (radius $6a$) rotate with constant angular velocities ω_1 and ω_2 about the point 0. They drive at contact points B and C without slip a stepped shaft ③ (radii a and $2a$). **P4.16**

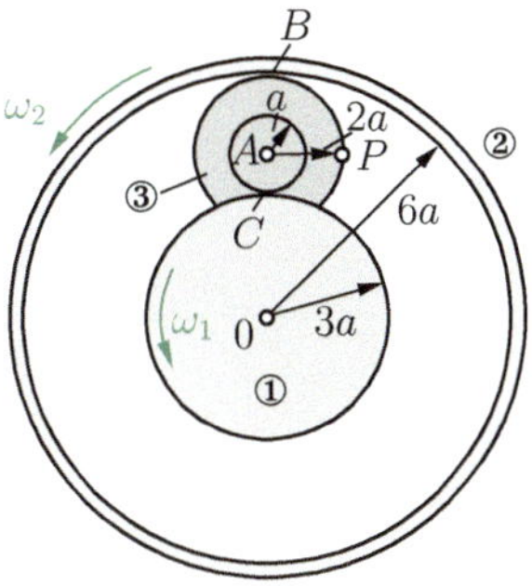

a) Determine the magnitude of velocity v_A and the acceleration ω_3 of the shaft ③.
b) Determine the velocity $\boldsymbol{v}_P(t)$ and the acceleration $\boldsymbol{a}_P(t)$ of point P.

Solution **a)** The velocities of shaft ① and ring ② at the contact points follow from the angular velocities:

$$v_C = 3a\,\omega_1\,, \quad v_B = 6a\,\omega_2\,.$$

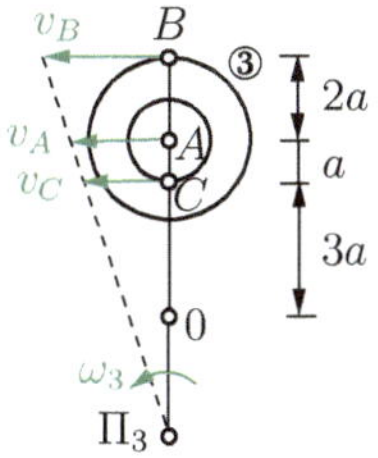

The shaft ③ carries out a plane motion. Accordingly, its velocities at the contact points can be described by

$$v_C = v_A - a\,\omega_3\,, \quad v_B = v_A + 2a\,\omega_3\,.$$

Equating the particular velocities leads to

$$\underline{\underline{v_A = 2a(\omega_1 + \omega_2)}}\,, \qquad \underline{\underline{\omega_3 = 2\omega_2 - \omega_1}}\,.$$

b) Point A undergoes a circular motion with the angular velocity

$$\omega_A = v_A/4a = (\omega_1 + \omega_2)/2\,.$$

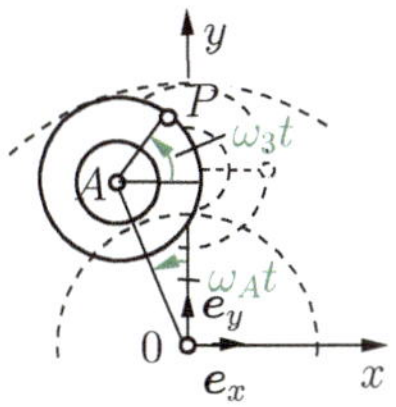

If we choose a fixed coordinate system with the origin 0, the position vector of P is given by

$$\boldsymbol{r}_P = [-4a\sin\omega_A t + 2a\cos\omega_3 t]\boldsymbol{e}_x + [4a\cos\omega_A t + 2a\sin\omega_3 t]\boldsymbol{e}_y\,.$$

From that, the velocity and acceleration are determined by differentiation:

$$\underline{\underline{\boldsymbol{v}_P = [-4a\omega_A\cos\omega_A t - 2a\omega_3\sin\omega_3 t]\boldsymbol{e}_x + [-4a\omega_A\sin\omega_A t + 2a\omega_3\cos\omega_3 t]\boldsymbol{e}_y}}\,,$$

$$\underline{\underline{\boldsymbol{a}_P = [4a\omega_A^2\sin\omega_A t - 2a\omega_3^2\cos\omega_3 t]\boldsymbol{e}_x + [-4a\omega_A^2\cos\omega_A t - 2a\omega_3^2\sin\omega_3 t]\boldsymbol{e}_y}}\,.$$

Chapter 5

Kinetics of a Rigid Body

5

Spatial Motion: The motion of a rigid body is described by the *Principle of Linear Momentum* and the *Principle of Angular Momentum.*

Principle of Linear Momentum:

$$\frac{\mathrm{d}\boldsymbol{p}}{\mathrm{d}t} = \boldsymbol{F} \qquad \text{or} \qquad m\boldsymbol{a}_c = \boldsymbol{F}$$

where $\boldsymbol{F}$ = sum of external forces,
m = total mass of the body,
$\boldsymbol{a}_c$ = acceleration of the center of mass,
$\boldsymbol{p} = m\,\boldsymbol{v}_c$ = linear momentum.

Principle of Angular Momentum:

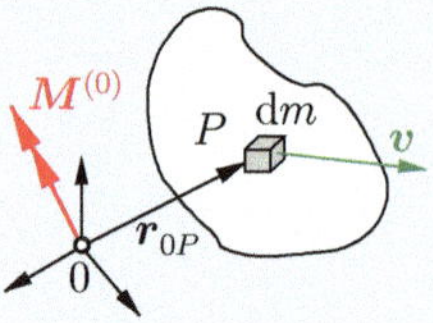

$$\frac{\mathrm{d}\boldsymbol{L}^{(0)}}{\mathrm{d}t} = \boldsymbol{M}^{(0)}$$

where $\boldsymbol{M}^{(0)}$ = sum of external moments with respect to 0,
$\boldsymbol{L}^{(0)} = \int \boldsymbol{r}_{0P} \times \boldsymbol{v}\,\mathrm{d}m$ = moment of momentum with respect to 0.

If the reference point 0 is space fixed or the center of mass, the angular momentum can be expressed by

$$\boldsymbol{L}^{(0)} = \boldsymbol{\Theta}^{(0)} \cdot \boldsymbol{\omega}$$

where $\boldsymbol{\omega}$ = angular velocity,

$$\boldsymbol{\Theta}^{0} = \begin{pmatrix} \Theta_x & \Theta_{xy} & \Theta_{xz} \\ \Theta_{yx} & \Theta_y & \Theta_{yz} \\ \Theta_{zx} & \Theta_{zy} & \Theta_z \end{pmatrix} = \text{inertia tensor}$$

Euler's Equations (Principle of angular momentum with respect to a body-fixed principal-axes system):

$$\Theta_1\,\dot{\omega}_1 - (\Theta_2 - \Theta_3)\,\omega_2\,\omega_3 = M_1\ ,$$
$$\Theta_2\,\dot{\omega}_2 - (\Theta_3 - \Theta_1)\,\omega_3\,\omega_1 = M_2\ ,$$
$$\Theta_3\,\dot{\omega}_3 - (\Theta_1 - \Theta_2)\,\omega_1\,\omega_2 = M_3$$

where $1, 2, 3$ = principal axes,
Θ_i = principal moments of inertia,
ω_i = components of angular velocity.

Moments of Inertia (see also volume 1, chapter 9)

Axial Moments of Inertia:

$$\Theta_x = \int (y^2 + z^2)\mathrm{d}m\ ,$$
$$\Theta_y = \int (z^2 + x^2)\mathrm{d}m\ ,$$
$$\Theta_z = \int (x^2 + y^2)\mathrm{d}m\ ,$$
$$\Theta_a = \int r^2\mathrm{d}m = m\,r_g^2\ ,$$

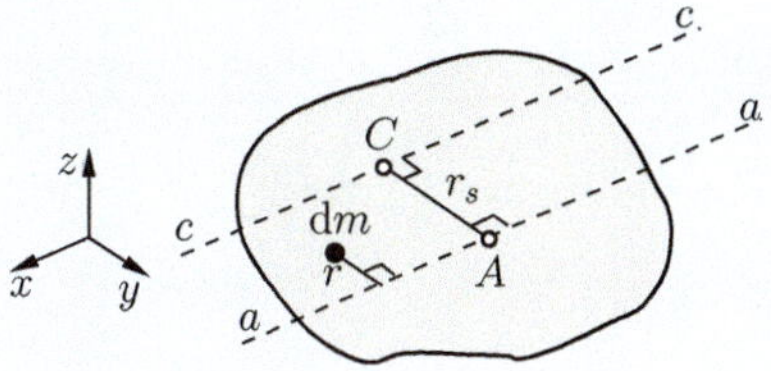

where r_g = radius of gyration.

Parallel-Axis Theorem: $$\Theta_a = \Theta_c + mr_c^2\ .$$

Products of Inertia:
$$\Theta_{xy} = \Theta_{yx} = -\int xy\mathrm{d}m\ ,$$
$$\Theta_{yz} = \Theta_{zy} = -\int yz\mathrm{d}m\ ,$$
$$\Theta_{zx} = \Theta_{xz} = -\int zx\mathrm{d}m\ .$$

Table of some moments of inertia:

slender rod	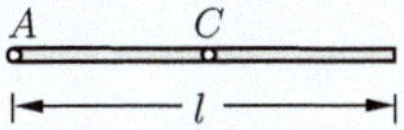	$\Theta_c = \frac{ml^2}{12},\quad \Theta_A = \frac{ml^2}{3}$
cylinder	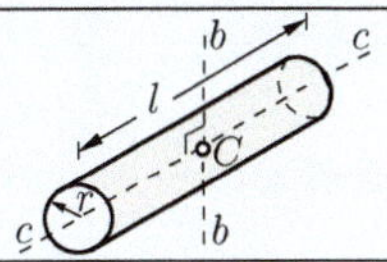	$\Theta_c = \frac{mr^2}{2},\quad \Theta_b = \frac{m}{12}(3r^2 + l^2)$
thin disc	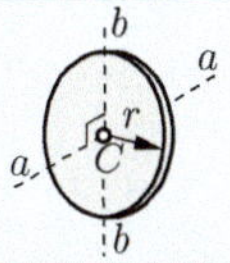	$\Theta_a = \frac{mr^2}{2},\quad \Theta_b = \frac{mr^2}{4}$
sphere	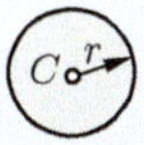	$\Theta_c = \frac{2}{5}\,mr^2$
cuboid	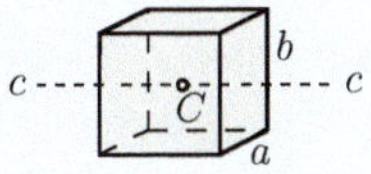	$\Theta_c = \frac{m}{12}(a^2 + b^2)$

Rotation about a Fixed Axis $a-a$:

Angular momentum $L_a = \Theta_a \omega \,,$

Principle of Angular Momentum $\Theta_a \,\dot{\omega} = \sum M_a \,.$

Time integration yields

$$\Theta_a \omega - \Theta_a \omega_0 = \int\limits_{t_0}^{t} \sum M_a \mathrm{d}\tau = \widehat{M}_a \,.$$

Plane Motion: Principles of linear and angular momentum (equations of motion)

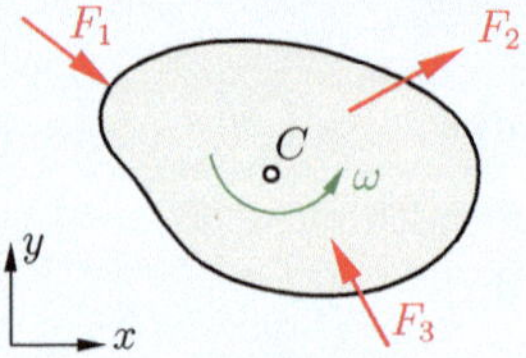

$$m\ddot{x}_c = \sum F_x \,,$$

$$m\ddot{y}_c = \sum F_y \,,$$

$$\Theta_A \dot{\omega} = \sum M_A$$

where A = fixed point or center of mass C. Time integration yields (*principles of linear and angular impulse and momentum*)

$$m\dot{x}_c - m\dot{x}_{c0} = \widehat{F}_x \,, \quad m\dot{y}_c - m\dot{y}_{c0} = \widehat{F}_y \,, \quad \Theta_A \omega - \Theta_A \omega_0 = \widehat{M}_A \,,$$

$$\text{where} \quad \widehat{F}_x = \int\limits_{t_0}^{t} \sum F_x \mathrm{d}\tau, \quad \widehat{F}_y = \int\limits_{t_0}^{t} \sum F_y \mathrm{d}\tau, \quad \widehat{M}_A = \int\limits_{t_0}^{t} \sum M_A \mathrm{d}\tau.$$

Work-Energy Theorem (see also chapter 2):

$$T - T_0 = U$$

Conservation of Energy Law (valid for conservative forces):

$$T + V = T_0 + V_0 = \text{const}$$

Kinetic Energy (plane motion):

$$T = \frac{1}{2} m v_c^2 + \frac{1}{2} \Theta_c \omega^2 \,.$$

Special case pure Translation: $T = \frac{1}{2}\, m v_C^2 \,,$

Special case Rotation about a fixed axis $a-a$: $T = \frac{1}{2}\, \Theta_a \omega^2 \,.$

Problem 5.1 A block (mass m) slides downwards an inclined rough plane. P5.1

Determine the acceleration. Under what circumstances tilt over is excluded?

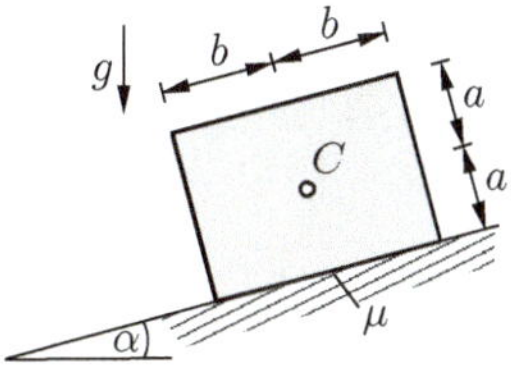

Solution As long as the block does not tilt, its motion is pure translation. With $\dot\omega = 0$, $\ddot y_C = 0$, the equations of motion are given by

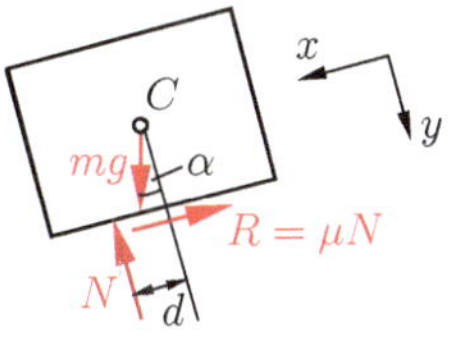

$$\swarrow: \quad m\ddot x_C = mg\sin\alpha - \mu N\,,$$
$$\searrow: \quad 0 = N - mg\cos\alpha\,,$$
$$\overset{\curvearrowright}{C}: \quad 0 = d\,N - a\mu N\,.$$

Thus, it follows

$$\underline{\underline{\ddot x_C = g\,(\sin\alpha - \mu\cos\alpha)}}\,, \qquad d = a\,\mu\,.$$

Tilt over is excluded for

$$d \le b \quad \leadsto \quad \underline{\underline{\mu \le b/a}}\,.$$

Problem 5.2 A homogeneous bar is hinged supported at point A of a vehicle and loosely rests at B. P5.2

Determine the reaction forces at A and B when the vehicle moves with an acceleration a.

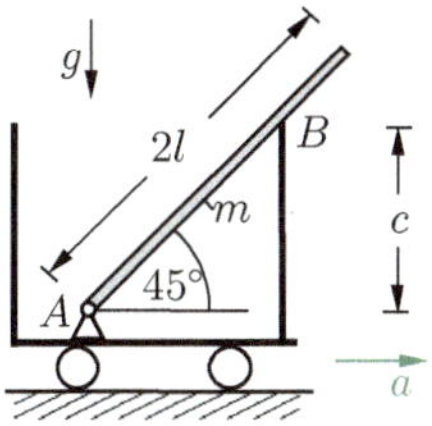

Solution For pure translation with $\ddot x_C = a$, $\ddot y_C = 0$ and $\dot\omega = 0$, the equations of motion read

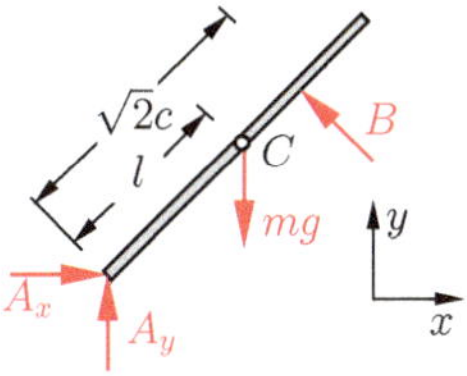

$$\rightarrow: \quad ma = A_x - \tfrac{1}{2}\sqrt{2}\,B\,,$$
$$\uparrow: \quad 0 = A_y + \tfrac{1}{2}\sqrt{2}\,B - mg\,,$$
$$\overset{\curvearrowright}{C}: \quad 0 = \tfrac{1}{2}\sqrt{2}\,l A_y - \tfrac{1}{2}\sqrt{2}\,l A_x - (\sqrt{2}\,c - l)B.$$

This leads to
$$\underline{\underline{B = mg\Big(1 - \frac{a}{g}\Big)\frac{l}{2c}}}\,,$$

$$\underline{\underline{A_x = mg\Big[\frac{a}{g} + \sqrt{2}\,\Big(1 - \frac{a}{g}\Big)\frac{l}{4c}\Big]}}, \quad \underline{\underline{A_y = mg\Big[1 - \sqrt{2}\,\Big(1 - \frac{a}{g}\Big)\frac{l}{4c}\Big]}}.$$

Remarks:
- For $a = g$ follows $B = 0$, $A_x = A_y = mg$.
- For $a > g$ the bar lifts off at B.

P5.3 **Problem 5.3** A homogeneous disk (weight $W = mg$) of quadratic parabola shape is fixed in the position α by three pin-supported bars.

Determine the acceleration of the disk and the forces in the bars immediately after the mounting by bar 3 is released.

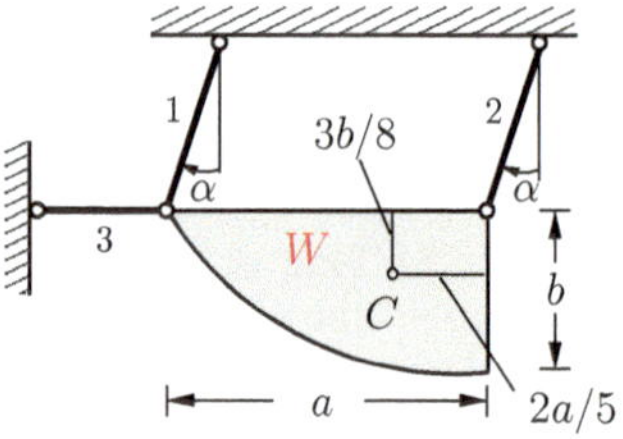

Solution When the mounting by bar 3 is released, the disk starts to move purely translationally. Therefore, the acceleration is the same for all points of the disk and it can be described by the acceleration of point B which rotates about the support A:

$$\boldsymbol{a} = \boldsymbol{a}_B = \dot{v}\,\boldsymbol{e}_t + \frac{v^2}{\rho_A}\boldsymbol{e}_n\ .$$

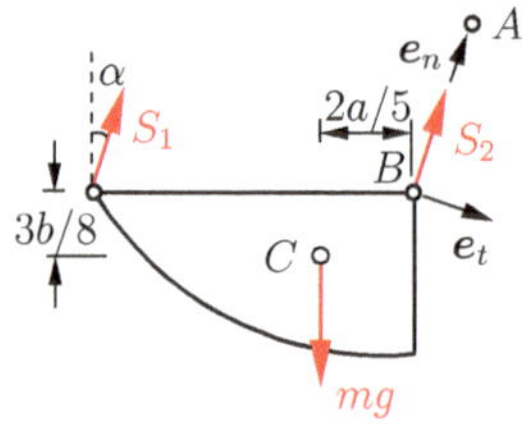

Immediately after the constraint is released, the velocity is still zero, i.e.,

$$v = 0 \quad \rightsquigarrow \quad \boldsymbol{a} = \dot{v}\,\boldsymbol{e}_t\ ,$$

and the equations of motion read

$$\searrow: \quad m\dot{v} = mg\sin\alpha\ ,$$

$$\nearrow: \quad 0 = S_1 + S_2 - mg\cos\alpha\ ,$$

$$\overset{\curvearrowright}{C}: \quad 0 = \frac{3}{5}aS_1\cos\alpha + \frac{3}{8}bS_1\sin\alpha - \frac{2}{5}aS_2\cos\alpha + \frac{3}{8}bS_2\sin\alpha\,.$$

From these three equations, we obtain

$$\underline{\underline{\dot{v} = g\sin\alpha}}\,,$$

$$\underline{\underline{S_1 = \frac{mg}{16}\Big(8\cos\alpha - 5\,\frac{b}{a}\,\sin\alpha\Big)}}\,, \qquad \underline{\underline{S_2 = \frac{mg}{16}\Big(8\cos\alpha + 5\,\frac{b}{a}\,\sin\alpha\Big)}}.$$

Remark: For $\tan\alpha = 8a/5b$ the force in bar 1 is $S_1 = 0$. In this case, the action line of S_2 passes through C.

Problem 5.4 A car with mass m brakes on a rough road without sliding. P5.4

Determine the possible braking decelerations a_b when the brakes are effective only to the front wheels or only to the rear wheels. The mass of the wheels shall be disregarded.

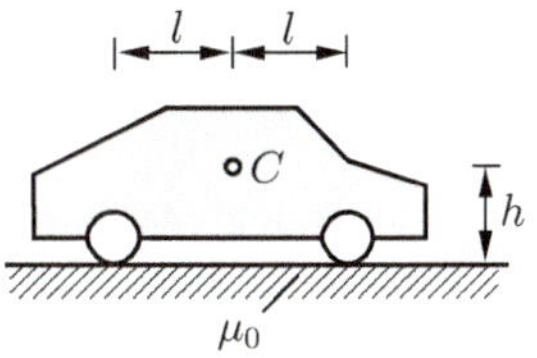

Solution When the brakes affect only the front wheels, we obtain with $\ddot{x}_C = -a_b$, $\ddot{y}_C = 0$, $\dot{\omega} = 0$ from

$$\rightarrow: \quad -ma_b = -H_1 \,,$$
$$\uparrow: \quad 0 = N_1 + N_2 - mg \,,$$
$$\overset{\curvearrowleft}{C}: \quad 0 = lN_1 - lN_2 - hH_1$$

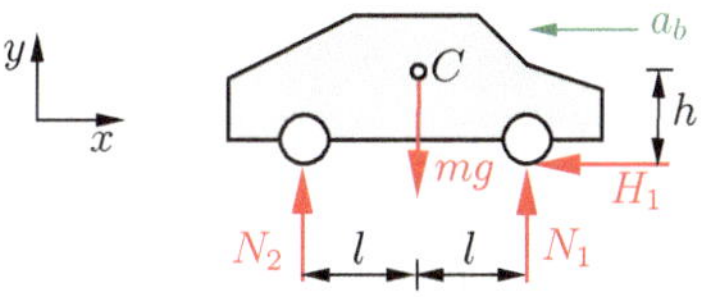

the forces

$$H_1 = ma_b \,, \quad N_1 = \frac{1}{2}m\,(g + \frac{h}{l}a_b)\,, \quad N_2 = \frac{1}{2}m\,(g - \frac{h}{l}a_b)\,.$$

The conditions that the front wheels do not slide and the rear wheels do not lift off lead to

$$H_1 \le \mu_0 N_1 \quad \rightsquigarrow \quad \underline{\underline{a_b \le g\,\frac{l}{h}\,\frac{1}{[2l/(\mu_0 h) - 1]}}} = \kappa_1 \,,$$

$$N_2 \ge 0 \quad \rightsquigarrow \quad \underline{\underline{a_b \le g\,\frac{l}{h}}} = \kappa_2 \,.$$

For $\mu_0 h/l \le 1$, we have $\kappa_1 \le \kappa_2$, i.e. it results $a_b \le \kappa_1$, while for $\mu_0 h/l \ge 1$ follows $a_b \le \kappa_2$.

When the brake is effective to the rear wheels, the force H_1 in the equations of motion must be replaced by H_2. The resulting forces N_1 and N_2 then remain unchanged and we obtain $H_2 = ma_b$. Thus, from $H_2 \le \mu_0 N_2$ and $N_2 \ge 0$ follows

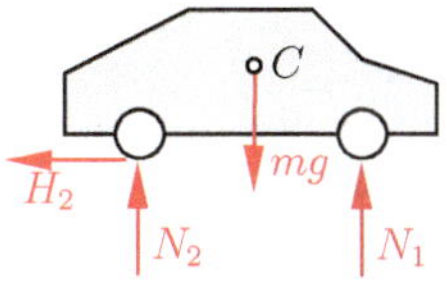

$$\underline{\underline{a_b \le g\,\frac{l}{h}\,\frac{1}{[2l/(\mu_0 h) + 1]}}} = \kappa_3 \,, \qquad a_b \le g\,\frac{l}{h} = \kappa_2 \,.$$

Because of $\kappa_3 < \kappa_2$ in this case always $a_b \le \kappa_3$ holds.

Remarks:

- The maximum deceleration is for front brakes always greater (κ_1, $\kappa_2 > \kappa_3$).
- The case $\mu_0 h/l \ge 1$ does not occur for an ordinary car under real conditions. Therefore, κ_2 is non-relevant.

P5.5 **Problem 5.5** The mass density of a graded shaft is given by $\rho = \rho_0(1 + \kappa r^2)$.

Determine the moments of inertia Θ_x and Θ_y.

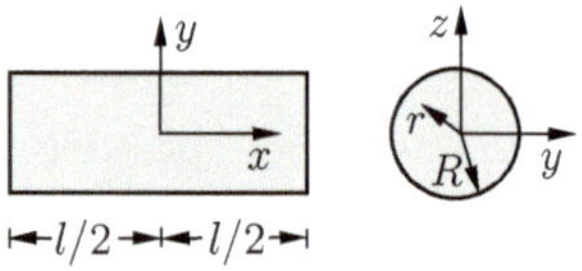

Solution We first consider a thin circular disk element of thickness dx. With the mass

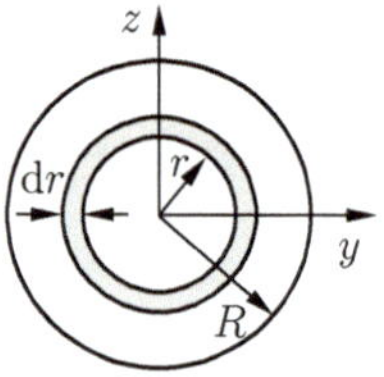

$$\mathrm{d}m = \rho\, 2\pi r\, \mathrm{d}r\, \mathrm{d}x = 2\pi\rho_0(r + \kappa r^3)\mathrm{d}r\, \mathrm{d}x$$

of a chosen ring, we obtain for its mass and its moment of inertia

$$\begin{aligned}
\mathrm{d}m^* &= \int_0^R \mathrm{d}m = \frac{\pi}{2}\rho_0 R^2 \mathrm{d}x(2 + \kappa R^2)\,,\\
\mathrm{d}\Theta_x^* &= \int_0^R r^2 \mathrm{d}m = 2\pi\rho_0 \mathrm{d}x \int_0^R (r^3 + \kappa r^5)\mathrm{d}r\\
&= 2\pi\rho_0 \mathrm{d}x\left(\frac{R^4}{4} + \kappa\,\frac{R^6}{6}\right) = \frac{\pi}{6}\rho_0 \mathrm{d}x\, R^4(3 + 2\,\kappa R^2)\,.
\end{aligned}$$

Because of the radial symmetry of the disk, its axial moments of inertia with respect to x, y and z (axes through the center of mass of the disk) are simply related by $\mathrm{d}\Theta_x^* = \mathrm{d}\Theta_y^* + \mathrm{d}\Theta_z^* = 2\mathrm{d}\Theta_y^*$, i.e.,

$$\mathrm{d}\Theta_y^* = \frac{\pi}{12}\rho_0 \mathrm{d}x\, R^4(3 + 2\,\kappa R^2)\,.$$

To calculate Θ_x and Θ_y, we now integrate over the length l, where we use the parallel-axis theorem to obtain Θ_y :

$$\begin{aligned}
\underline{\underline{\Theta_x}} &= \int_{-l/2}^{l/2} \mathrm{d}\Theta_x^* = \frac{\pi}{6}\rho_0 R^4(3 + 2\,\kappa R^2)\int_{-l/2}^{l/2} \mathrm{d}x = \underline{\underline{\frac{\pi}{6}\rho_0 l R^4(3 + 2\,\kappa R^2)}}\,,\\
\underline{\underline{\Theta_y}} &= \int_{-l/2}^{l/2} (\mathrm{d}\Theta_y^* + x^2 \mathrm{d}m^*)\\
&= \frac{\pi}{12}\rho_0 R^4(3 + 2\,\kappa R^2)\int_{-l/2}^{l/2} \mathrm{d}x + \frac{\pi}{2}\rho_0 R^2(2 + \kappa R^2)\int_{-l/2}^{l/2} x^2 \mathrm{d}x\\
&= \underline{\underline{\frac{\pi}{24}\,\rho_0 l R^2\Big[2R^2(3 + 2\kappa R^2) + l^2(2 + \kappa R^2)\Big]}}\,.
\end{aligned}$$

Remark: For $\kappa = 0$ follow with $m = \rho_0\pi R^2 l$ the results of the homogeneous shaft: $\Theta_x = mR^2/2\,,\quad \Theta_y = m(3R^2 + l^2)/12$.

Problem 5.6 Determine the moments of inertia Θ_a of rings (density ρ) with different circular and half-circular cross sections of radius $c = R/2$. P5.6

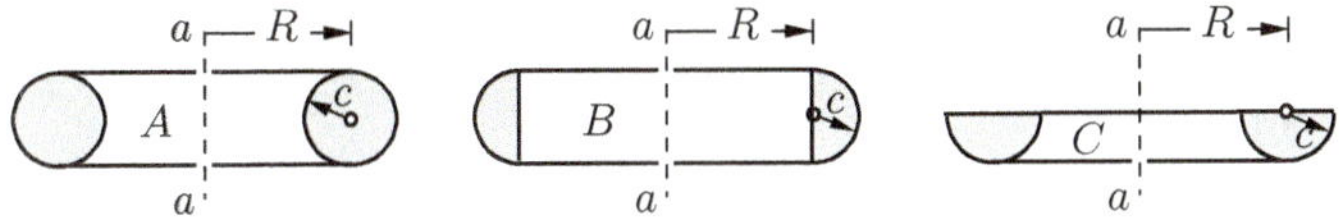

Solution The moment of inertia is defined as $\int r^2 \mathrm{d}m$. We first choose an appropriate ring-shaped mass element of radius r and thickness $\mathrm{d}r$. From the sketch follows

$$\mathrm{d}m = \rho\, 2\pi r 2c\, \cos\varphi\, \mathrm{d}r\ ,$$

$$r = R + c\,\sin\varphi\,,$$

$$\mathrm{d}r = c\,\cos\varphi\,\mathrm{d}\varphi\,,$$

$$\leadsto \quad \mathrm{d}m = 4\pi\rho c^2 (R + c\sin\varphi)\cos^2\varphi\,\mathrm{d}\varphi\ .$$

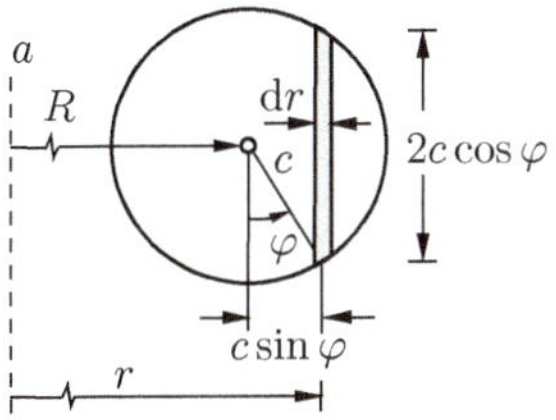

Depending on the cross section, we must integrate over different domains:

$$\begin{aligned}\Theta_a &= 4\pi\rho c^2 \int_{\varphi_1}^{\varphi_2} (R + c\sin\varphi)^3 \cos^2\varphi\mathrm{d}\varphi\\ &= 4\pi\rho c^2 \int_{\varphi_1}^{\varphi_2} [R^3\cos^2\varphi + 3R^2 c\sin\varphi\cos^2\varphi\\ &\qquad\qquad +3Rc^2\sin^2\varphi\cos^2\varphi + c^3\sin^3\varphi\cos^2\varphi]\mathrm{d}\varphi\\ &= 4\pi\rho c^2 \Big[R^3\tfrac{1}{4}(2\varphi + \sin 2\varphi) - R^2 c\,\cos^3\varphi\\ &\qquad +3Rc^2(\tfrac{1}{8}\varphi - \tfrac{1}{32}\sin 4\varphi) + c^3(-\tfrac{1}{3}\cos^3\varphi + \tfrac{1}{5}\cos^5\varphi)\Big]_{\varphi_1}^{\varphi_2}.\end{aligned}$$

In case A, we obtain with $\varphi_1 = -\pi/2$, $\varphi_2 = +\pi/2$, $m_A = 2\pi^2\rho c^2 R$

$$\underline{\underline{\Theta_a^A}} = 4\pi\rho c^2\Big(\frac{\pi}{2}\,R^3 + \frac{3\pi}{8}\,Rc^2\Big) = \underline{\underline{m_A\Big(R^2 + \frac{3}{4}c^2\Big) = 1.19\,m_A R^2}}\ .$$

Case B with $\varphi_1 = 0$, $\varphi_2 = +\pi/2$, $m_B = \pi^2\rho c^2(R + 4c/(3\pi))$ (note: $m_B \neq m_A/2\,$!) leads to

$$\underline{\underline{\Theta_a^B}} = 4\pi\rho c^2\Big(\frac{\pi}{4}\,R^3 + R^2 c + \frac{3\pi}{16}\,Rc^2 + \frac{2}{15}\,c^3\Big) = \underline{\underline{1.52\,m_B R^2}}\ .$$

Finally, in case C follows for symmetry reasons ($m_C = m_A/2$))

$$\underline{\underline{\Theta_a^C = \Theta^A/2 = m_C\Big(R^2 + \frac{3}{4}c^2\Big) = 1.19\,m_C R^2}}\ .$$

P5.7 **Problem 5.7** Determine the moments of inertia Θ_x and Θ_y for a homogeneous cone of mass m.

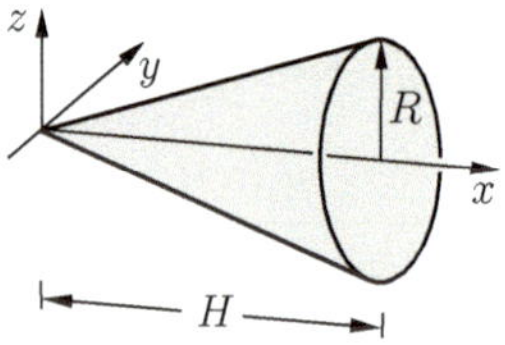

Solution It is practical to use as mass element the sketched circular disk. With its radius

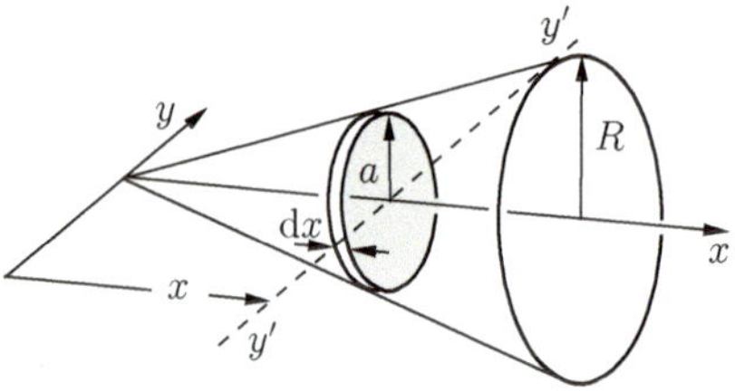

$$a = x\,\frac{R}{H}$$

and mass

$$dm = \rho\pi a^2 dx = \rho\pi\,\frac{R^2}{H^2}\,x^2 dx$$

its moment of inertia $d\Theta_x$ is given by

$$d\Theta_x = \frac{1}{2}\,a^2 dm = \rho\,\frac{\pi}{2}\left(\frac{R}{H}\right)^4 x^4 dx\,.$$

Thus, for the cone follows by integration

$$\Theta_x = \int d\Theta_x = \rho\frac{\pi}{2}\left(\frac{R}{H}\right)^4 \int_0^H x^4 dx = \rho\frac{\pi}{10}R^4 H\,.$$

The moment of inertia of the circular disk element with respect to the y'-axis is given by

$$d\Theta_{y'} = \frac{1}{4}a^2 dm = \rho\frac{\pi}{4}\left(\frac{R}{H}\right)^4 x^4 dx\,.$$

Therefore, we obtain for the cone by using the parallel-axis theorem

$$\begin{aligned}\Theta_y = \int [d\Theta_{y'} + x^2 dm] &= \rho\pi\left(\frac{R}{H}\right)^2 \int_0^H \left[\frac{1}{4}\left(\frac{R}{H}\right)^2 x^4 + x^4\right] dx\\ &= \rho\frac{\pi}{5}R^2 H\left(\frac{1}{4}R^2 + H^2\right)\,.\end{aligned}$$

Finally, introducing the mass $m = \rho\pi R^2 H/3$, the moments of inertia can be written as

$$\underline{\underline{\Theta_x = \frac{3}{10}\,mR^2}}\,, \qquad \underline{\underline{\Theta_y = \frac{3}{5}\,m\left(\frac{1}{4}R^2 + H^2\right)}}\,.$$

Problem 5.8 Determine the moments of inertia Θ_x and Θ_{xy} for a homogeneous rectangular thin plate (mass m, thickness $t \ll a, b$). **P5.8**

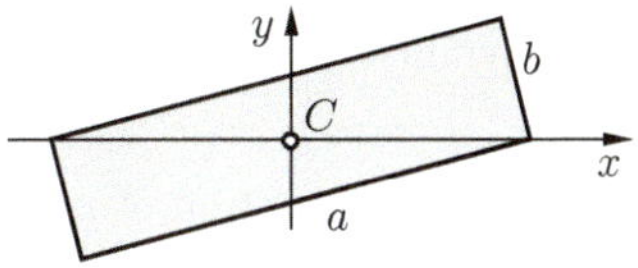

Solution For the *thin* plate exists a direct relation between the mass moments of inertia and the area moments of inertia (see vol. 1, chapter 9). With $\mathrm{d}m = \rho t \mathrm{d}A$ and $z \ll x, y$, it is given by

$$\Theta_x = \int (y^2 + z^2)\mathrm{d}m \cong \rho t \int y^2 \mathrm{d}A = \rho t I_x ,$$

$$\Theta_{xy} = -\int xy\mathrm{d}m = -\rho t \int xy\mathrm{d}A = \rho t I_{xy} .$$

Consequently, also the transformation relations (see vol. 1, chapter 9) can be applied. In this way, we find for the axes x and y which are inclined by $-\varphi$ to the pricipal axes 1 and 2:

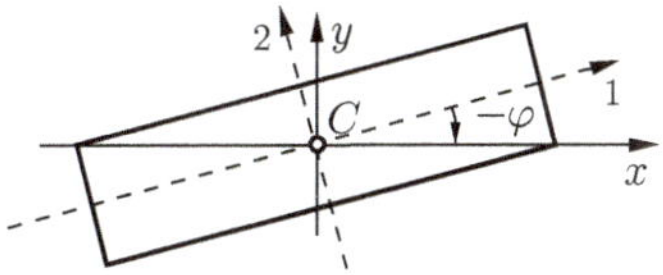

$$\Theta_x = \frac{\Theta_1 + \Theta_2}{2} + \frac{\Theta_1 - \Theta_2}{2} \cos 2\varphi ,$$

$$\Theta_{xy} = \frac{\Theta_1 - \Theta_2}{2} \sin 2\varphi .$$

With

$$\Theta_1 = \rho t \, \frac{ab^3}{12} = \frac{mb^2}{12} , \qquad \Theta_2 = \frac{ma^2}{12} ,$$

$$\cos\varphi = \frac{a}{\sqrt{a^2 + b^2}} , \qquad \sin\varphi = \frac{b}{\sqrt{a^2 + b^2}} ,$$

$$\cos 2\varphi = 2\cos^2\varphi - 1 = \frac{a^2 - b^2}{a^2 + b^2} , \quad \sin 2\varphi = 2\sin\varphi\cos\varphi = \frac{2ab}{a^2 + b^2} ,$$

we finally obtain

$$\underline{\underline{\Theta_x}} = \frac{m(a^2 + b^2)}{24} + \frac{m(b^2 - a^2)}{24} \, \frac{(a^2 - b^2)}{(a^2 + b^2)} = \underline{\underline{\frac{ma^2b^2}{6(a^2 + b^2)}}} ,$$

$$\underline{\underline{\Theta_{xy}}} = \frac{m(b^2 - a^2)}{24} \, \frac{2ab}{a^2 + b^2} = \underline{\underline{-\frac{m(a^2 - b^2)ab}{12(a^2 + b^2)}}} .$$

Remark: For $a = b$ follows $\Theta_x = ma^2/12$ and $\Theta_{xy} = 0$.

P5.9 **Problem 5.9** An angled homogeneous bar of weight $W = mg$ is pin-supported at A.

Formulate the equation of motion.

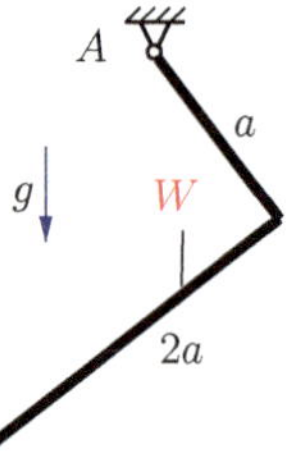

Solution The motion of the system is a pure rotation about the fixed point A due to the moment of the weight. The weight acts at the center of mass C. Its coordinates follow from (see volume 1, chapter 2)

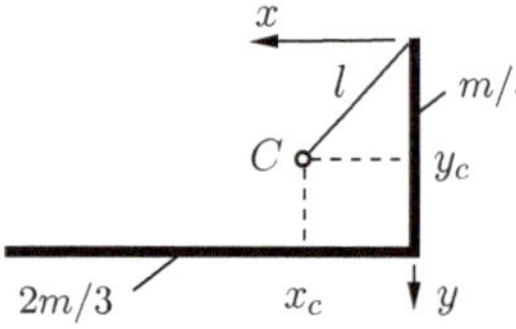

$$x_C = \frac{\frac{2m}{3}\,a}{m} = \frac{2}{3}\,a\,,$$

$$y_C = \frac{\frac{m}{3}\frac{a}{2} + \frac{2m}{3}\,a}{m} = \frac{5}{6}\,a$$

$$\rightsquigarrow \quad l = \sqrt{x_S^2 + y_S^2} = \frac{\sqrt{29}}{6}\,a\,.$$

To describe the motion, it is advantageous to introduce the angle φ, which characterizes the displaced position relative to the equilibrium position (where C is located below A). The principle of angular momentum then reads

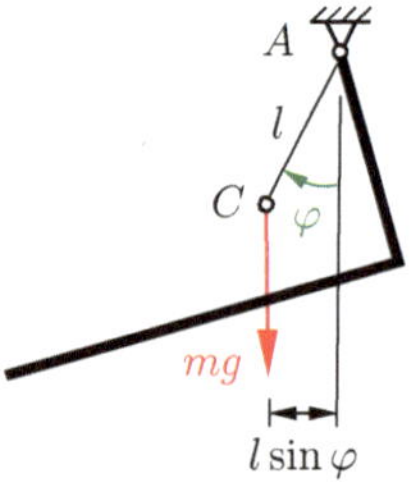

$$\overset{\curvearrowright}{A}: \quad \Theta_A\,\ddot{\varphi} = M_A\,,$$

where

$$\Theta_A = \frac{m}{3}\frac{a^2}{3} + \left[\frac{(2a)^2}{12} + (a^2 + a^2)\frac{2}{3}\,m\right] = \frac{5}{3}\,ma^2$$

$$M_A = -mgl\sin\varphi = -\frac{\sqrt{29}}{6}\,mga\sin\varphi\,.$$

This leads to the equation of motion

$$\frac{5}{3}\,ma^2\ddot{\varphi} = -\frac{\sqrt{29}}{6}\,mga\sin\varphi \quad \rightsquigarrow \quad \underline{\underline{\ddot{\varphi} + \frac{\sqrt{29}}{10}\,\frac{g}{a}\sin\varphi = 0}}\,.$$

Remarks: From $\ddot{\varphi}(\varphi)$ the angular velocity $\dot{\varphi}(\varphi)$ can be determined by integration (see page 4). For small displacements ($\varphi \ll 1$, $\sin\varphi \approx \varphi$), the equation of motion describes a harmonic vibration (see chapter 7).

P5.10

Problem 5.10 A circular disk (weight $W = mg$), which rotates with the angular velocity ω_0 about the vertical axis, is put on a rough horizontal plane (coefficient of kinetic friction μ).

After what time T and after how many turns u the disk comes to rest? Assume that the contact pressure is constant.

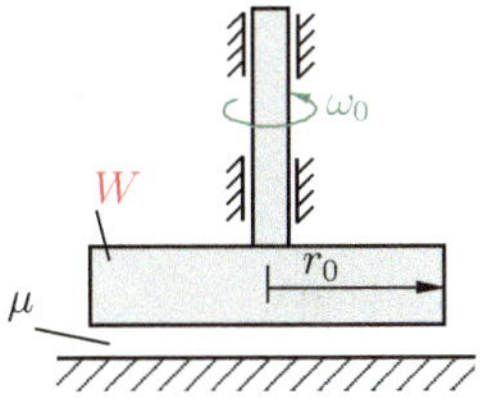

Solution The rotation about the fixed axis is described by

$$\Theta\,\dot{\omega} = M\,.$$

Here, the moment M is determined from the friction forces distributed over the circular area. From

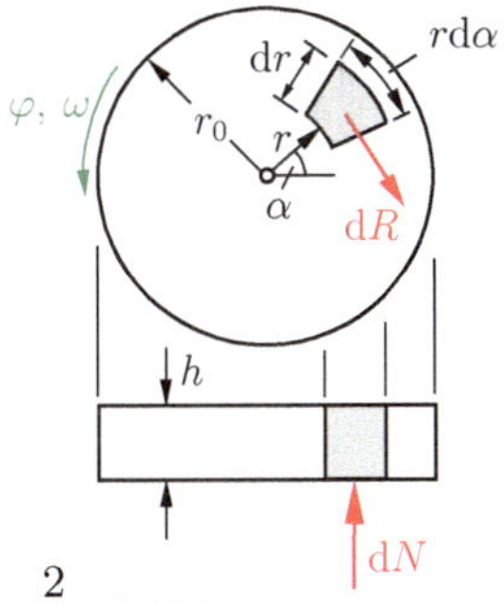

$$\mathrm{d}A = r\mathrm{d}\alpha\,\mathrm{d}r\,, \qquad m = \rho\pi r_0^2 h$$

$$\mathrm{d}N = \rho g h \mathrm{d}A\,, \qquad \mathrm{d}R = \mu \mathrm{d}N$$

follows (the friction moment is directed opposite to the angular velocity)

$$M = -\int r\,\mathrm{d}R = -\mu\rho g h \int_0^{r_0}\int_0^{2\pi} r^2 \mathrm{d}\alpha\,\mathrm{d}r = -\frac{2}{3}\,\mu r_0 m g\,.$$

Thus, with $\Theta = m r_0^2/2$ and $\dot{\omega} = \ddot{\varphi}$, the equation of motion leads to

$$\ddot{\varphi} = -\frac{4}{3}\,\frac{\mu g}{r_0}\,.$$

Twice integration and considering the initial conditions $\varphi(0) = 0$, $\dot{\varphi}(0) = \omega_0$ yields

$$\dot{\varphi} = \omega_0 - \frac{4}{3}\,\frac{\mu g}{r_0}\,t\,, \qquad \varphi = \omega_0 t - \frac{2}{3}\,\frac{\mu g}{r_0}\,t^2\,.$$

From the condition $\dot{\varphi}(T) = 0$, we finally obtain

$$\underline{\underline{T = \frac{3}{4}\,\frac{\omega_0 r_0}{\mu g}}}\,,$$

$$\varphi(T) = \frac{3}{8}\,\frac{\omega_0^2 r_0}{\mu g} \quad \leadsto \quad \underline{\underline{u}} = \frac{\varphi(T)}{2\pi} = \underline{\underline{\frac{3}{16\pi}\,\frac{\omega_0^2 r_0}{\mu g}}}\,.$$

P5.11 **Problem 5.11** A flywheel which initially rotates with speed n (rpm) about a fixed axis is brought to standstill in time t_b by a constant braking moment M_b.

Determine the mass moment of inertia Θ and the number of revolutions u of the flywheel during braking.

Solution During braking, the motion is described by

$$\Theta\ddot{\varphi} = -M_b \quad \rightsquigarrow \quad \ddot{\varphi} = -\frac{M_b}{\Theta} \, .$$

(Note that the braking moment acts opposite to the positive direction of rotation.) Twice integration and considering the initial conditions

$$\dot{\varphi}(0) = \omega_0 = 2\pi n \, , \qquad \varphi(0) = 0$$

leads to

$$\dot{\varphi} = -\frac{M_B}{\Theta}\, t + 2\pi n \, ,$$

$$\varphi = -\frac{M_B}{2\Theta}\, t^2 + 2\pi n t \, .$$

From the condition $\dot{\varphi}(t_b) = 0$ (standstill) we obtain

$$0 = -\frac{M_b}{\Theta}\, t_b + 2\pi n \quad \rightsquigarrow \quad \underline{\underline{\Theta = \frac{M_b t_b}{2\pi n}}} \, ,$$

$$\varphi_b = \varphi(t_b) = -\frac{M_b t_b^2}{2\Theta} + 2\pi n t_b = \pi n t_b \quad \rightsquigarrow \quad \underline{\underline{u}} = \frac{\varphi_b}{2\pi} = \underline{\underline{\frac{1}{2}\, n t_b}} \, .$$

The problem can be solved easier by using the principle of angular impulse and momentum and the work-energy theorem. With $\omega(t_b) = 0$ and $\omega_0 = 2\pi n$, they lead to

$$\int_0^{t_b} M \mathrm{d}\tau = \Theta\omega(t_b) - \Theta\omega_0 \; : \quad -M_b t_b = -\Theta\omega_0 \quad \rightsquigarrow \quad \underline{\underline{\Theta = \frac{M_b t_b}{2\pi n}}},$$

$$T - T_0 = W \; : \quad -\frac{1}{2}\Theta\omega_0^2 = -M_b\varphi_b \quad \rightsquigarrow \quad \underline{\underline{u}} = \frac{\varphi_b}{2\pi} = \underline{\underline{\frac{1}{2}\, n t_b}}.$$

Problem 5.12 On a homogeneous cylindrical roll of mass m_1, an inextensible rope is wound up, which is connected with a body ② of weight $W_2 = m_2 g$.

P5.12

Determine the acceleration of body ② and the force S in the rope when the roll can rotate about A without friction. The mass of the rope can be disregarded.

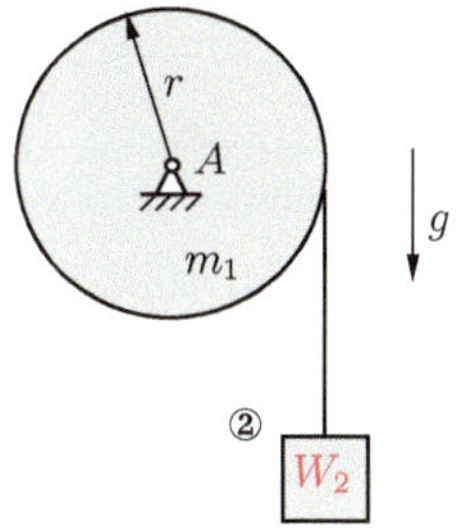

Solution We separate the roll and body ② by cutting the rope, choose positive directions of motion and formulate the equation of motion for the roll

$$\overset{\curvearrowright}{A}: \quad \Theta\ddot{\varphi} = r\,S$$

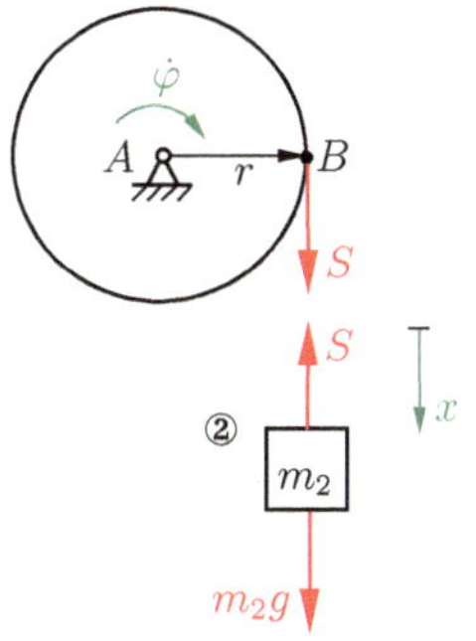

and the body ②

$$\downarrow: \quad m_2\ddot{x} = m_2 g - S\,.$$

The 2 equations of motion containing 3 unknowns ($\ddot{x}$, $\ddot{\varphi}$ and S) must be suplemented by a 3rd equation describing the kinematic constraint. The velocity $r\dot{\varphi}$ of point B of the roll and the velocity $\dot{x}$ of body ② and the rope, respectively, must be equal:

$$\dot{x} = r\dot{\varphi} \qquad \rightsquigarrow \qquad \ddot{x} = r\ddot{\varphi}\,.$$

This leads with $\Theta = m_1 r^2/2$ to

$$\underline{\underline{\ddot{x} = g\,\frac{2m_2}{m_1 + 2m_2}}}\,, \qquad \underline{\underline{S = m_2 g\,\frac{m_1}{m_1 + 2m_2}}}\,.$$

Remark: In the static case (roll is blocked) the force in the rope is $S_{St} = m_2 g$, while in dynamics $S < S_{St}$ holds.

P5.13

Problem 5.13 Two homogeneous drums (weights $m_1 g$, $m_2 g$), initially rotating with angular velocity ω_0 in the same direction, are placed on top of each other, such that they slide against each other (coefficient of dynamic friction μ).

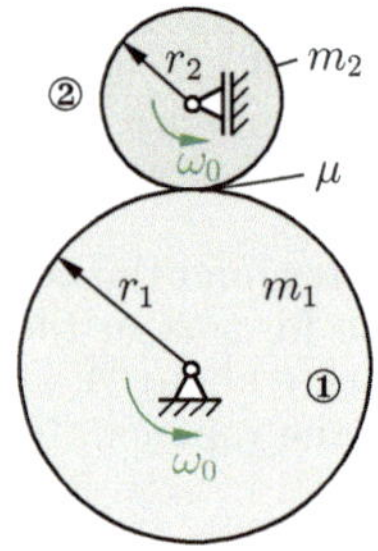

After which time t_R, the drums roll from one another and what are then their angular velocities?

Solution During sliding the equations of motion read

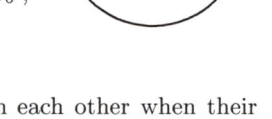

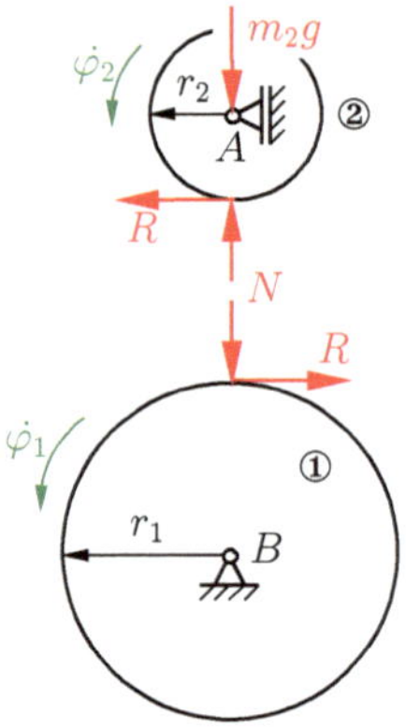

$$① \quad \overset{\curvearrowleft}{B}: \quad \Theta_1 \ddot{\varphi}_1 = -r_1 R\,,$$

$$② \quad \overset{\curvearrowleft}{A}: \quad \Theta_2 \ddot{\varphi}_2 = -r_2 R\,,$$

$$\uparrow: \quad 0 = N - m_2 g\,.$$

With the friction law $R = \mu N$ and $\Theta_1 = m_1 r_1^2/2$, $\Theta_2 = m_2 r_2^2/2$ and considering the initial conditions, it follows

$$\ddot{\varphi}_1 = -\frac{2\mu g}{r_1}\frac{m_2}{m_1} \quad \rightsquigarrow \quad \dot{\varphi}_1 = -\frac{2\mu g}{r_1}\frac{m_2}{m_1}t + \omega_0\,,$$

$$\ddot{\varphi}_2 = -\frac{2\mu g}{r_2} \quad \rightsquigarrow \quad \dot{\varphi}_2 = -\frac{2\mu g}{r_2}t + \omega_0\,.$$

The drums stop sliding and start rolling from each other when their velocities at the contact point are the same at time $t = t_R$:

$$r_1 \dot{\varphi}_1 = -r_2 \dot{\varphi}_2\,.$$

Note that positive angular velocities of the drums ① and ② lead to opposite directed velocities at the contact point! Thus, we obtain

$$\underline{\underline{t_R = \frac{\omega_0 (r_1 + r_2)}{2\mu g\left(1 + \dfrac{m_2}{1}\right)}}}$$

and for the angular velocities

$$\underline{\underline{\dot{\varphi}_1(t_R) = \omega_0 \frac{\dfrac{m_1}{m_2} - \dfrac{r_2}{r_1}}{1 + \dfrac{m_1}{m_2}}}}\,, \qquad \underline{\underline{\dot{\varphi}_2(t_R) = -\omega_0 \frac{\dfrac{r_1 m_1}{r_2 m_2} - 1}{1 + \dfrac{m_1}{m_2}}}}\,.$$

Remark: For $m_1/m_2 = r_2/r_1$ both drums come to standstill.

Problem 5.14 At an elevator, the cable drum ② is driven by the wheel ① without slip. P5.14

Determine the acceleration a of the elevator ③ (weight $W = mg$) and the cable force S, when the wheel is driven by a constant moment M_0. The weight of the cable can be neglected.

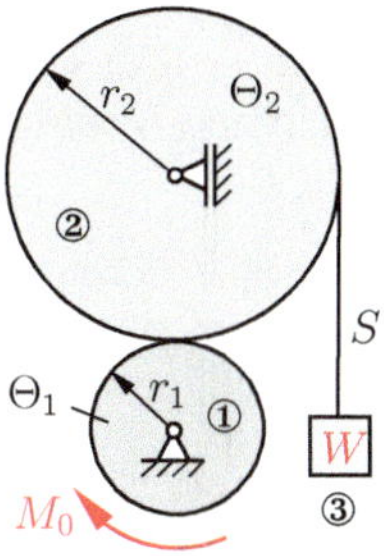

Solution We separate the system and obtain for the different parts

$$① \quad \overset{\curvearrowright}{A}: \quad \Theta_1\dot{\omega}_1 = M_0 - r_1 H ,$$

$$② \quad \overset{\curvearrowleft}{B}: \quad \Theta_2\dot{\omega}_2 = r_2 H - r_2 S ,$$

$$③ \quad \uparrow: \quad ma = m\dot{v} = S - mg .$$

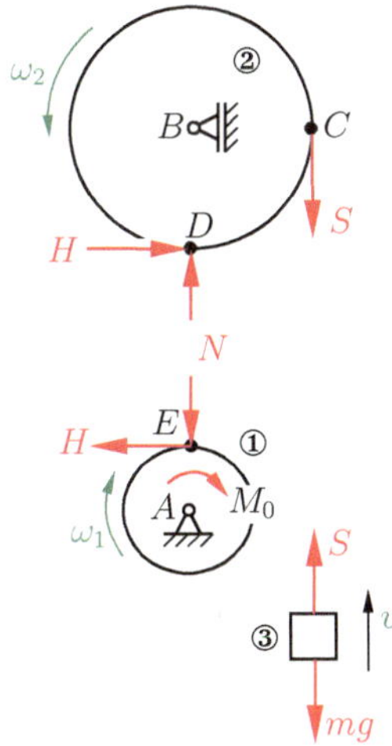

Since the velocity of point C of the drum ② and the velocity of the elevator (inextensible cable) as well as the velocities of D and E (no slip) must be equal, the kinematic relations are given by

$$r_2\omega_2 = v \quad \rightsquigarrow \quad r_2\dot{\omega}_2 = \dot{v} = a ,$$

$$r_2\omega_2 = r_1\omega_1 \quad \rightsquigarrow \quad r_2\dot{\omega}_2 = r_1\dot{\omega}_1 .$$

Thus, we have 5 equations for the 5 unknowns ($\dot{\omega}_1$, $\dot{\omega}_2$, a, H, S). Solving for a and S leads to

$$\underline{\underline{a = \frac{\dfrac{M_0}{r_1} - mg}{m + \dfrac{\Theta_1}{r_1^2} + \dfrac{\Theta_2}{r_2^2}}}} , \qquad \underline{\underline{S = m\,\frac{\dfrac{M_0}{r_1} + \left(\dfrac{\Theta_1}{r_1^2} + \dfrac{\Theta_2}{r_2^2}\right) g}{m + \dfrac{\Theta_1}{r_1^2} + \dfrac{\Theta_2}{r_2^2}}}} .$$

Remark: For $M_0 = r_1 mg$ we obtain $a = 0$ (statics) and $S = H = mg$.

P5.15 **Problem 5.15** A homogeneous drum (weight $W = mg$) rolls downwards a rough inclined plane (coefficient of static friction μ_0).

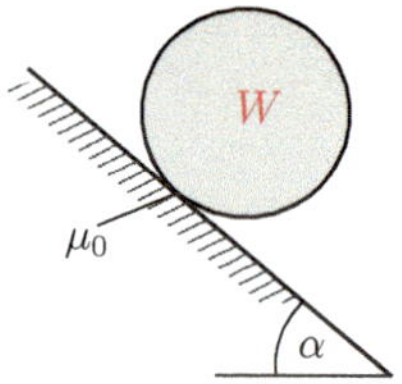

Determine its acceleration a. Under which circumstances pure rolling is possible?

Solution With $\ddot{y}_C = 0$ and $\ddot{x}_C = a$, we obtain from the principle of linear momentum

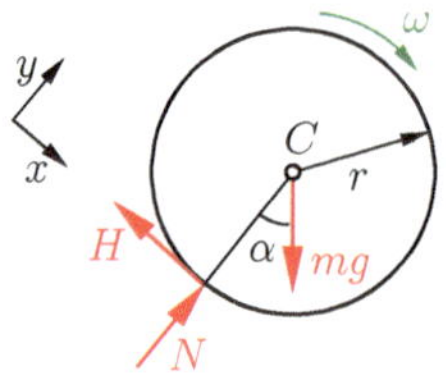

$$\searrow: \quad ma = mg\sin\alpha - H\,,$$

$$\nearrow: \quad 0 = N - mg\cos\alpha$$

and from the principle of angular momentum (with respect to the center of mass C)

$$\overset{\curvearrowright}{C}: \quad \Theta_C\,\dot{\omega} = rH\,.$$

After introducing $\Theta_C = mr^2/2$ and the kinematic constraint (rolling without slip)

$$r\omega = \dot{x}_C \quad \rightsquigarrow \quad r\dot{\omega} = \ddot{x}_C = a \quad \rightsquigarrow \quad \dot{\omega} = \frac{a}{r}\,,$$

solving for the acceleration yields

$$\underline{\underline{a = \frac{2}{3}\,g\sin\alpha\,.}}$$

The condition for non-slipping at the contact point of drum and plane (pure rolling) is given by

$$H \leq H_0 = \mu_0 N\,.$$

With

$$H = mg\sin\alpha - ma = \frac{1}{3}\,mg\sin\alpha\,, \qquad N = mg\cos\alpha\,,$$

it leads to

$$\underline{\underline{\mu_0 \geq \frac{1}{3}\,\tan\alpha\,.}}$$

Problem 5.16 A homogeneous bowling ball ($r = 0.11\,\text{m}$) is placed with an initial velocity $v_0 = 7\,\text{m/s}$ and an angular velocity $\omega_0 = 10\,\text{s}^{-1}$ in a rough bowling alley (kinetic friction coefficient $\mu = 0.15$).

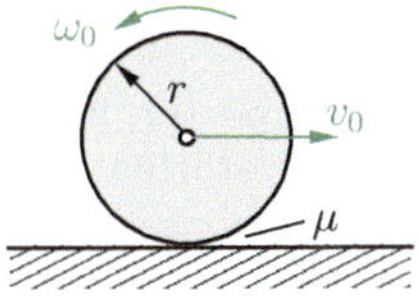

Determine the covered distance x_r until the ball rolls without slipping and its final speed v_r.

Solution We draw the free-body diagram and introduce positive directions for distance and angular velocity. After contact, the ball initially is slipping. Thus, the principles of linear and angular momentum yield

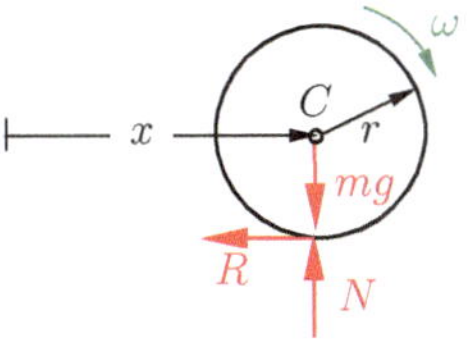

$$\rightarrow: \quad m\ddot{x} = -R\,,$$

$$\uparrow: \quad 0 = N - mg \quad \rightsquigarrow \quad N = mg\,,$$

$$\overset{\curvearrowright}{C}: \quad \Theta_C\,\dot{\omega} = rR\,.$$

We now introduce $\Theta_C = 2mr^2/5$ and the friction law $R = \mu N = \mu mg$ and subsequently integrate the 1st and the 2nd equation. Taking into account the initial conditions $v(t=0) = v_0$, $x(t=0) = 0$, $\omega(t=0) = -\omega_0$ leads to

$$v = \dot{x} = v_0 - \mu g t\,, \qquad x = v_0 t - \frac{1}{2}\,\mu g\, t^2\,, \qquad \omega = \frac{5\mu g}{2r}\,t - \omega_0\,.$$

When the ball is rolling, v and ω are related by

$$v = r\omega\,,$$

from which the time t_r follows for onset of rolling:

$$v_0 - \mu g t_r = \frac{5}{2}\,\mu\, g\, t_r - \omega_0 \qquad \rightsquigarrow \qquad t_r = \frac{2(v_0 + \omega_0 r)}{7\mu g} = 1.57\,\text{s}\,.$$

Finally, this leads to

$$\underline{\underline{x_r}} = x(t_r) = v_0 t_r - \frac{1}{2}\,\mu g\, t_r^2 = \underline{\underline{9.17\,\text{m}}}\,,$$

$$\underline{\underline{v_r}} = v(t_r) = v_0 - \mu g t_r = \underline{\underline{4.69\,\text{m/s}}}\,.$$

P5.17 **Problem 5.17** A cable drum on a rough surface is set into rolling motion (no slipping) by pulling with a force F.

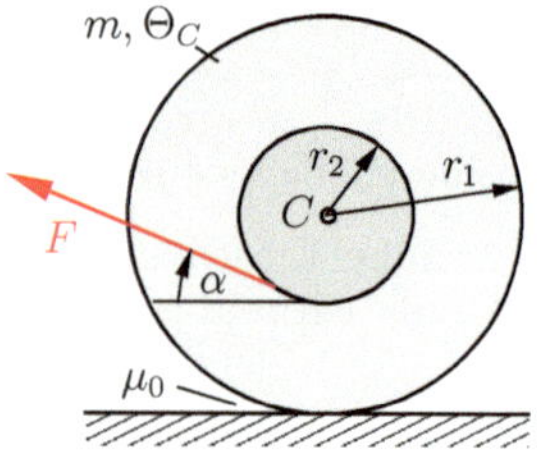

Determine the pulling angles α, for which the acceleration a_C takes extreme values.

What are the associated maximum forces, such that no slipping occurs?

Given: $W = mg$, $\Theta_C = 3r_1^2 m$, $r_1 = 2r_2$, μ_0.

Solution The principles of linear and angular momentum yield

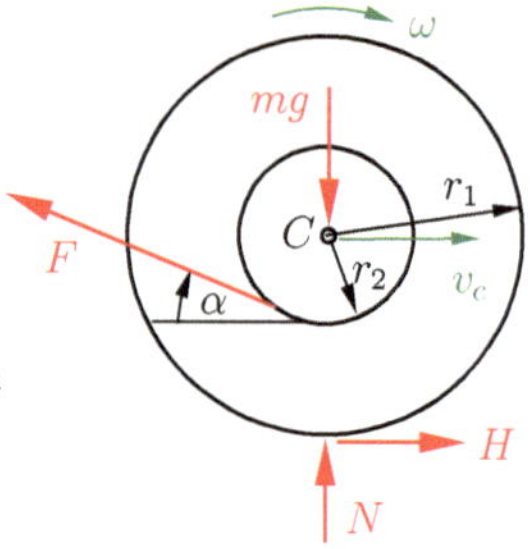

$$\rightarrow: \quad m a_C = -F\cos\alpha + H\,,$$

$$\uparrow: \quad 0 = N - mg + F\sin\alpha\,,$$

$$\overset{\curvearrowright}{C}: \quad \Theta_C\,\dot{\omega} = -r_1 H + r_2 F\,.$$

In connection with the kinematic condition for rolling,

$$v_C = r_1\omega \quad \rightsquigarrow \quad \dot{v}_C = a_C = r_1\dot{\omega}\,,$$

we obtain the acceleration and forces

$$a_C = -\frac{F}{m}\,\frac{\cos\alpha - r_2/r_1}{1 + \dfrac{\Theta_S}{r_1^2 m}} = -\frac{F}{8m}\,(2\cos\alpha - 1)\,,$$

$$H = F\,\frac{\dfrac{\Theta_C\cos\alpha}{r_1^2 m} + \dfrac{r_2}{r_1}}{1 + \dfrac{\Theta_C}{r_1^2 m}} = \frac{F}{8}\,(6\cos\alpha + 1), \qquad N = mg - F\sin\alpha\,.$$

The extreme values of a_c follow from $\mathrm{d}a_c/\mathrm{d}\alpha = 0$, i.e. $\sin\alpha = 0$, as

$$\underline{\underline{\alpha_1 = 0}} \quad \rightsquigarrow \quad \underline{\underline{a_{c1} = -F/8m}}\,,\,, \qquad \underline{\underline{\alpha_2 = \pi}} \quad \rightsquigarrow \quad \underline{\underline{a_{c2} = 3F/8m}}\,.$$

The associated maximum forces F_i are calculated via the maximum force of static friction:

$$|H|(\alpha_i) = H_0 = \mu_0 N(\alpha_i) \quad \rightsquigarrow \quad \underline{\underline{F_1 = \frac{8}{7}\,\mu_0 mg}}\,, \qquad \underline{\underline{F_2 = \frac{8}{5}\,\mu_0 mg}}\,.$$

Remark: For $\alpha_1 = 0$, the drum moves to the left and for $\alpha_2 = \pi$ to the right.

Problem 5.18 A homogeneous plate of weight $W = mg$ is symmetrically supported. P5.18

a) Determine the support reactions at A at motion initiation when the support B suddenly is removed.

b) What distance d must be chosen, such that the vertical force at A does not change in comparison to the static case?

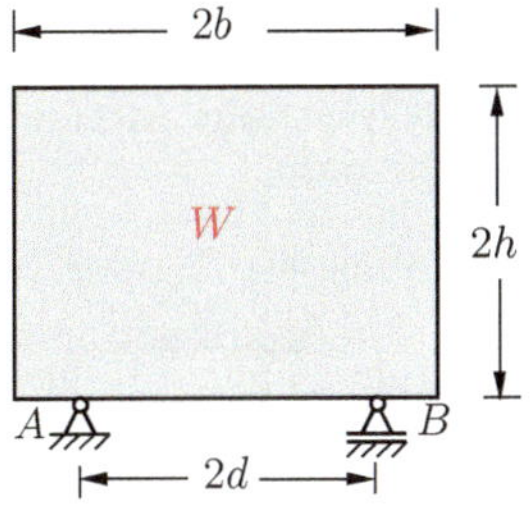

Solution **a)** Immediately after removing support B the principles of linear and angular momentum (with respect to the fixed point A) yield

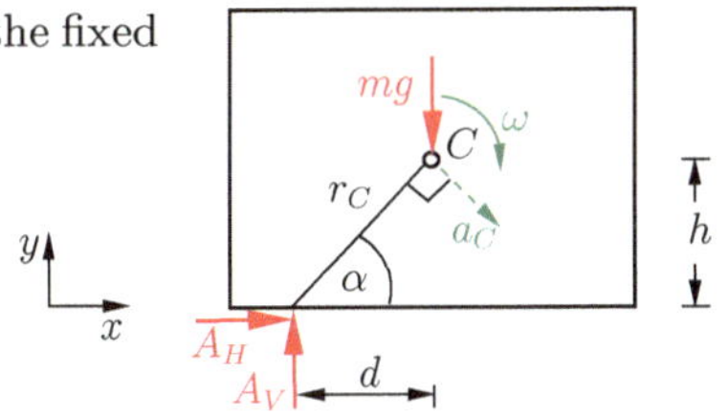

$$\rightarrow: \quad m\ddot{x}_C = A_H \, ,$$

$$\uparrow: \quad m\ddot{y}_C = A_V - mg \, ,$$

$$\overset{\curvearrowright}{A}: \quad \Theta_A \dot{\omega} = d\, mg \, ,$$

where

$$\Theta_A = \Theta_C + m r_C^2 = \frac{m}{12}(4b^2+4h^2) + m(d^2+h^2) = \frac{m}{3}(b^2+4h^2+3d^2) \, .$$

Since the center of mass C is purely rotating about A, its acceleration components at initiation of motion ($v_C = 0$) are given by

$$a_n = \frac{v_C^2}{r_C} = 0 \, , \qquad a_C = a_t = r_C \dot{\omega} \, ,$$

and we obtain with $\sin\alpha = h/r_C$, $\cos\alpha = d/r_C$

$$\ddot{x}_C = a_C \sin\alpha = h\,\dot{\omega} \, , \qquad \ddot{y}_C = -a_C \cos\alpha = -d\,\dot{\omega} \, .$$

Solving for the support reaction leads to

$$\underline{\underline{A_H = W\,\frac{3dh}{b^2+4h^2+3d^2}}} \, , \qquad \underline{\underline{A_V = W\,\frac{b^2+4h^2}{b^2+4h^2+3d^2}}} \, .$$

b) Before removing support B (statics), the vertical force at A is $A_V = W/2$. Thus, from the condition that there is not change, it follows

$$\frac{W}{2} = W\,\frac{b^2+4h^2}{b^2+4h^2+3d^2} \quad \rightsquigarrow \quad \underline{\underline{d = \frac{1}{\sqrt{3}}\sqrt{b^2+4h^2}}} \, .$$

Remark: Because of $d \leq b$, the force A_V remains unchanged only for $b \geq \sqrt{2}\,h$.

P5.19 **Problem 5.19** An initially upright bar (weight $W = mg$) starts to move, where it slides without friction along the wall and the base.

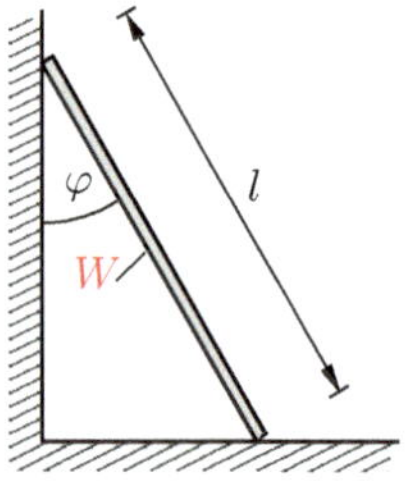

Determine the angular velocity $\ddot{\varphi}(\varphi)$.

What forces are acting on the wall and the base?

Solution As long as the bar slides along the wall and the base, the principles of linear and angular momentum (with respect to C) yield

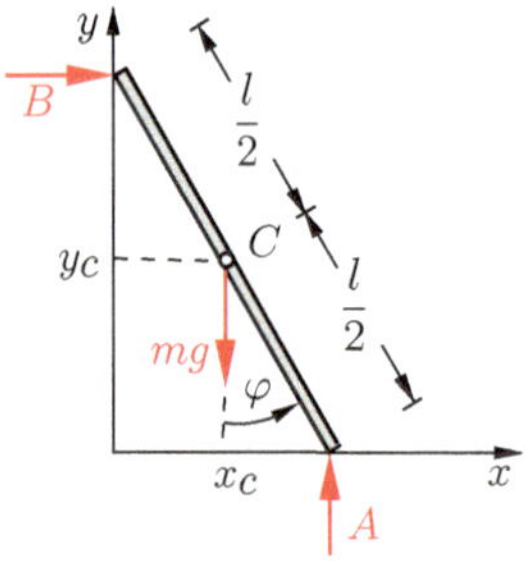

$$\rightarrow: \quad m\ddot{x}_C = B\,,$$

$$\uparrow: \quad m\ddot{y}_C = A - mg\,, \qquad ,$$

$$\overset{\curvearrowleft}{C}: \quad \Theta_C\ddot{\varphi} = A\,\frac{l}{2}\sin\varphi - B\,\frac{l}{2}\cos\varphi$$

where $\Theta_C = ml^2/12$.

The kinematic quantities x_C, y_C and φ are related by

$$x_C = \frac{l}{2}\sin\varphi\,, \qquad y_C = \frac{l}{2}\cos\varphi\,.$$

After differentiation one obtains

$$\dot{x}_C = \frac{l}{2}\,\dot{\varphi}\cos\varphi\,, \qquad \dot{y}_C = -\frac{l}{2}\,\dot{\varphi}\sin\varphi\,,$$

$$\ddot{x}_C = \frac{l}{2}\,(\ddot{\varphi}\cos\varphi - \dot{\varphi}^2\sin\varphi)\,, \qquad \ddot{y}_C = -\frac{l}{2}\,(\ddot{\varphi}\sin\varphi + \dot{\varphi}^2\cos\varphi)\,.$$

Thus, we have 5 equations for the 5 unknowns ($\ddot{x}_C$, $\ddot{y}_C$, $\ddot{\varphi}$, A, B). Solving for $\ddot{\varphi}$ leads to

$$\underline{\underline{\ddot{\varphi} = \frac{3g}{2l}\sin\varphi\,.}}$$

The forces are determined by

$$B = m\ddot{x}_C = m\,\frac{l}{2}\,(\ddot{\varphi}\cos\varphi - \dot{\varphi}^2\sin\varphi)\,,$$

$$A = mg + m\ddot{y}_C = mg - m\,\frac{l}{2}\,(\ddot{\varphi}\sin\varphi + \dot{\varphi}^2\cos\varphi)\,.$$

Here $\dot{\varphi}$ is required. We can calculate it from

$$\ddot{\varphi} = \frac{\mathrm{d}\dot{\varphi}}{\mathrm{d}\varphi}\frac{\mathrm{d}\varphi}{\mathrm{d}t} = \frac{\mathrm{d}\dot{\varphi}}{\mathrm{d}\varphi}\dot{\varphi} \quad \rightsquigarrow \quad \dot{\varphi}\,\mathrm{d}\dot{\varphi} = \ddot{\varphi}\,\mathrm{d}\varphi$$

by integration:

$$\frac{1}{2}\dot{\varphi}^2 = \int \ddot{\varphi}\,\mathrm{d}\varphi + C = \frac{3g}{2l}\int \sin\varphi\,\mathrm{d}\varphi + C = -\frac{3g}{2l}\cos\varphi + C\,.$$

The integration constant C follows from the initial condition $\dot{\varphi}(\varphi{=}0) = 0$ as $C = 3g/2l$, which leads to

$$\dot{\varphi}^2 = \frac{3g}{l}(1 - \cos\varphi)\,.$$

Thus, the froces are given by

$$\underline{\underline{B = \frac{3}{4}\,W(3\cos\varphi - 2)\sin\varphi}}\,,$$

$$\underline{\underline{A = \frac{1}{4}\,W(9\cos^2\varphi - 6\cos\varphi + 1)}}\,.$$

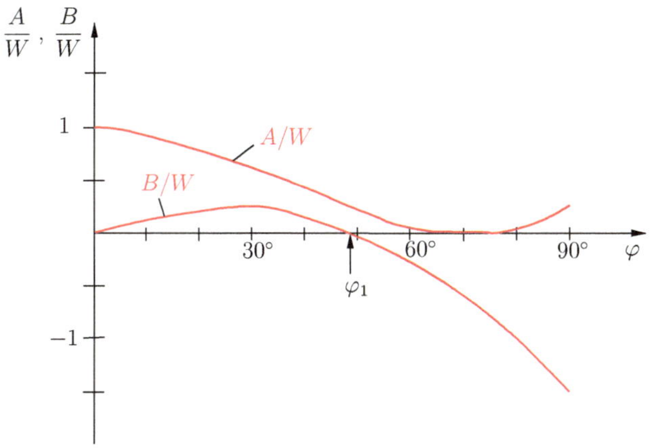

Remarks:

- The force B is zero for $\varphi_0 = 0$ and $3\cos\varphi_1 - 2 = 0 \rightsquigarrow \varphi_1 = 48.2°$.
- For $\varphi > \varphi_1$, the bar would loose contact with the wall at B. Therefore, the results remain valid only if the support in B is such that it can transmit a tension force.
- The force A is zero for $\cos\varphi = \dfrac{1}{3} \rightsquigarrow \varphi \approx 70.5°$.

P5.20 **Problem 5.20** A homogeneous bar (weight $W = mg$), initially at rest in the sketched position, is released and slides down along a frictionless semi-circular path.

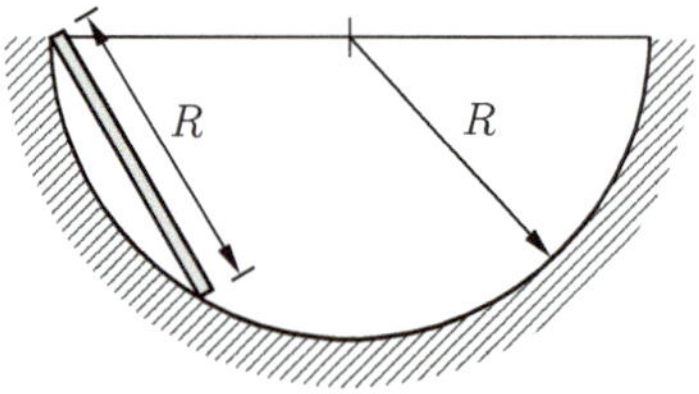

a) What are the velocity and acceleration of the center of mass in dependence on the position of the bar?

b) Determine the forces at the contact points.

Solution **a)** The center of mass C moves along a circular path with radius

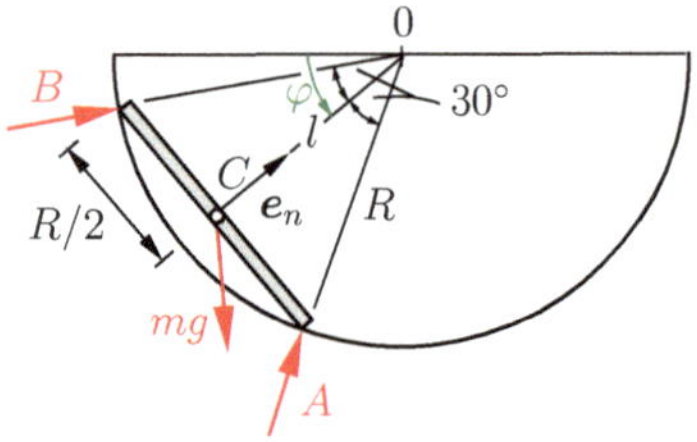

$$l = \sqrt{R^2 - \frac{R^2}{4}} = \frac{\sqrt{3}}{2}\,R$$

about 0. Thus, its velocity and acceleration are uniquely described by $\dot{\varphi}$ and $\ddot{\varphi}$.

Note that $\dot{\varphi}$ and $\ddot{\varphi}$ are identical to the angular velocity and angular acceleration of the bar!

The principle of angular momentum (pure rotation about 0)

$$\overset{\curvearrowleft}{0}: \quad \Theta_0 \ddot{\varphi} = mgl\cos\varphi$$

yields with

$$\Theta_0 = \frac{mR^2}{12} + ml^2 = \frac{5}{6}mR^2$$

the angular acceleration

$$\underline{\underline{\ddot{\varphi} = \frac{3\sqrt{3}}{5}\,\frac{g}{R}\cos\varphi}}\,.$$

From this result, using $\dot{\varphi}\mathrm{d}\dot{\varphi} = \ddot{\varphi}\mathrm{d}\varphi$ and the initial condition $\dot{\varphi}(\varphi = 30^\circ) = 0$, the angular velocity is determined by integration:

$$\frac{1}{2}\dot{\varphi}^2 = \int\limits_{30^o}^{\varphi} \ddot{\varphi}(\bar{\varphi})\mathrm{d}\bar{\varphi} = \frac{3\sqrt{3}}{5}\,\frac{g}{R}\left(\sin\varphi - \frac{1}{2}\right)$$

$$\leadsto \quad \underline{\underline{\dot{\varphi} = \sqrt{\frac{3\sqrt{3}}{5}\,\frac{g}{R}(2\sin\varphi - 1)}}}\,.$$

b) To determine the forces, we apply the principle of linear momentum in direction of $\boldsymbol{e}_n$ and the principle of angular momentum with respect to C:

$$\nearrow: \quad m a_n = A\cos 30^\circ + B\cos 30^\circ - mg\sin\varphi\,,$$

$$\overset{\curvearrowleft}{C}: \quad \Theta_C\ddot{\varphi} = \frac{R}{2}A\cos 30^\circ - \frac{R}{2}B\cos 30^\circ\,.$$

With $a_n = l\dot{\varphi}^2 = \frac{9}{10}\,g(2\sin\varphi - 1)$, $\Theta_C = mR^2/12$ and the already known $\ddot{\varphi}$ we obtain

$$\underline{\underline{A = \frac{mg}{10}\left[\frac{\sqrt{3}}{3}(28\sin\varphi - 9) + \cos\varphi\right]}}, \quad \underline{\underline{B = \frac{mg}{10}\left[\frac{\sqrt{3}}{3}(28\sin\varphi - 9) - \cos\varphi\right]}}.$$

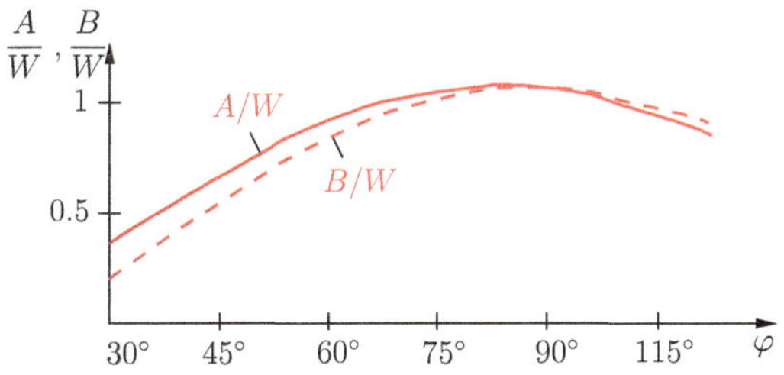

The quantities $\dot{\varphi}$ and $\ddot{\varphi}$ can alternatively be determined from the energy conservation law $T + V = T_0 + V_0$. Choosing the zero level of potential energy at the lowermost point of the circular path and using $v_C = l\dot{\varphi} = \sqrt{3}R\dot{\varphi}/2$, the different energy terms are given by

$$T_0 = 0\,,$$

$$V_0 = mg(R - l\sin 30^o) = mgR\left(1 - \frac{\sqrt{3}}{4}\right),$$

$$T = \frac{1}{2}mv_S^2 + \frac{1}{2}\Theta_C\dot{\varphi}^2 = \frac{5}{12}mR^2\dot{\varphi}^2\,,$$

$$V = mg(R - l\sin\varphi) = mgR\left(1 - \frac{\sqrt{3}}{2}\sin\varphi\right),$$

and we obtain

$$\underline{\underline{\dot{\varphi}^2 = \frac{3\sqrt{3}}{5}\frac{g}{R}(2\sin\varphi - 1)\,.}}$$

Differentiation with respect to time finally leads to

$$2\dot{\varphi}\ddot{\varphi} = \frac{3\sqrt{3}}{5}\frac{g}{R}2\dot{\varphi}\cos\varphi \qquad \rightsquigarrow \qquad \underline{\underline{\ddot{\varphi} = \frac{3\sqrt{3}}{5}\frac{g}{R}\cos\varphi\,.}}$$

P5.21 **Problem 5.21** The displayed system consists of two homogeneous wheels of mass m that are connected by an inextensible cable with a body of weight $W_1 = 5mg$.

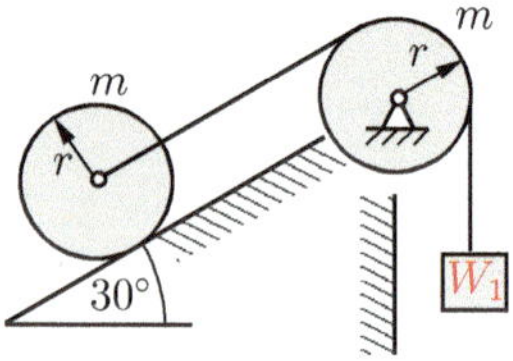

Determine the acceleration of the body and the cable forces, when the system freely moves and sliding occurs nowhere. The mass of the cable can be disregarded.

Solution We separate the system and write down the equations of motion for the different parts:

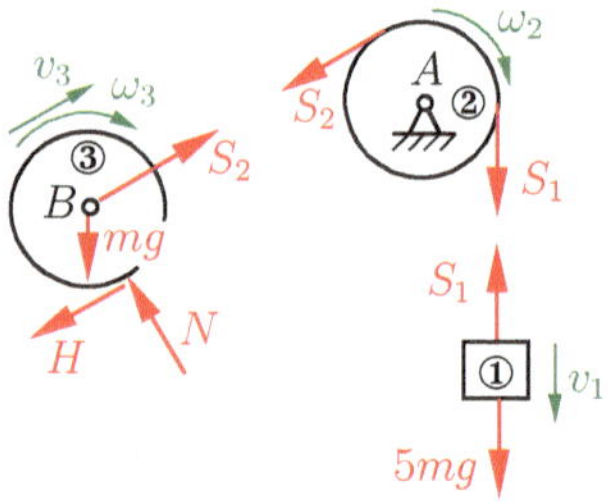

$$① \quad \downarrow: \quad 5m\dot{v}_1 = 5mg - S_1 ,$$

$$② \quad \overset{\curvearrowright}{A}: \quad \Theta_2\dot{\omega}_2 = rS_1 - rS_2 ,$$

$$③ \quad \nearrow: \quad m\dot{v}_3 = S_2 - H - mg\sin 30° ,$$

$$\overset{\curvearrowright}{B}: \quad \Theta_3\dot{\omega}_3 = rH ,$$

where

$$\Theta_2 = \Theta_3 = \frac{mr^2}{2} .$$

With the kinematic relations (constraint by the cable)

$$v_1 = r\omega_2 = v_3 = r\omega_3$$

$$\leadsto \quad \dot{\omega}_2 = \frac{\dot{v}_1}{r} , \qquad \dot{v}_3 = \dot{v}_1 , \qquad \dot{\omega}_3 = \frac{\dot{v}_1}{r}$$

we have 7 equations for the 7 unknowns ($\dot{v}_1$, $\dot{v}_3$, $\dot{\omega}_2$, $\dot{\omega}_3$, S_1, S_2, H). Solving yields for the acceleration

$$\underline{\underline{\dot{v}_1 = \frac{9}{14}\, g}}$$

and for the forces in the cable

$$\underline{\underline{S_1 = \frac{50}{28}\, mg}} , \qquad \underline{\underline{S_2 = \frac{41}{28}\, mg}} .$$

Problem 5.22 Two bodies (weights $m_1 g > m_2 g$) are connected by a cable which is passed over a wheel (moment of inertia Θ_0). The coefficient of kinetic friction between the body and the inclined plane is μ. **P5.22**

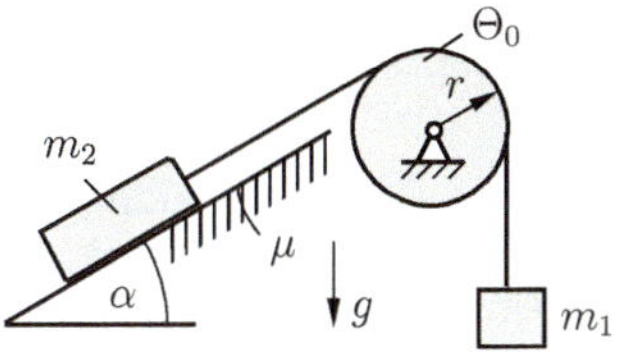

Determine the velocity of the bodies in dependence of their position. Assume that they are initially at rest and that there is no slip between cable and wheel.

Solution The velocity can be determined easiest by using the work-energy theorem

$$T - T_0 = U\,.$$

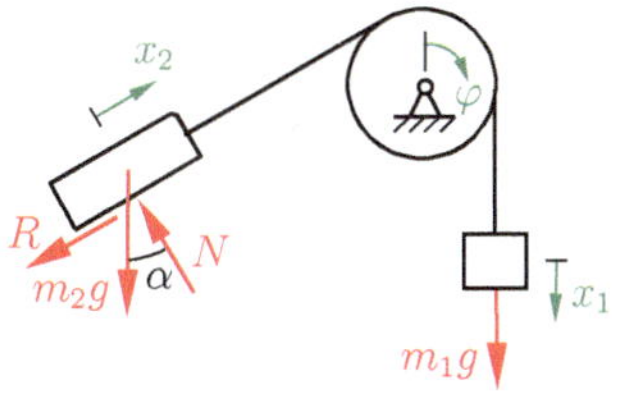

With the kinematic relations (constraint by the cable, no slipping)

$$x_1 = x_2 = r\varphi = x\,,$$

$$\rightsquigarrow \quad \dot{x}_1 = \dot{x}_2 = r\dot{\varphi} = \dot{x} = v\,,$$

the kinetic energies follow as

$$E_{k0} = 0 \qquad \text{(initial position = rest)}\,,$$

$$E_k = \frac{1}{2}\,m_1\dot{x}_1^2 + \frac{1}{2}\,m_2\dot{x}_2^2 + \frac{1}{2}\,\Theta_0\dot{\varphi}^2 = \frac{1}{2}\,m_1 v^2\left(1 + \frac{m_2}{m_1} + \frac{\Theta_0}{m_1 r^2}\right)\,.$$

Using the friction law $R = \mu N = \mu m_2 g\cos\alpha$, the work of the external forces is given by

$$U = m_1 g x_1 - R x_2 - (m_2 g\sin\alpha)x_2 = m_1 g x\left(1 - \mu\frac{m_2}{m_1}\cos\alpha - \frac{m_2}{m_1}\sin\alpha\right)\,.$$

Thus, the work-energy theorem leads to

$$\frac{1}{2}\,m_1 v^2\left(1 + \frac{m_2}{m_1} + \frac{\Theta_0}{m_1 r^2}\right) = m_1 g x\left(1 - \mu\,\frac{m_2}{m_1}\cos\alpha - \frac{m_2}{m_1}\sin\alpha\right)$$

$$\rightsquigarrow \quad \underline{\underline{v(x) = \sqrt{2gx}\sqrt{\frac{1 - \mu\dfrac{m_2}{m_1}\cos\alpha - \dfrac{m_2}{m_1}\sin\alpha}{1 + \dfrac{m_2}{m_1} + \dfrac{\Theta_0}{m_1 r^2}}}}}\,.$$

P5.23 **Problem 5.23** The cabin of an elevator (weight $W = mg$) hangs at a cable drum, which is braked by a band brake (kinetic friction coefficient μ).

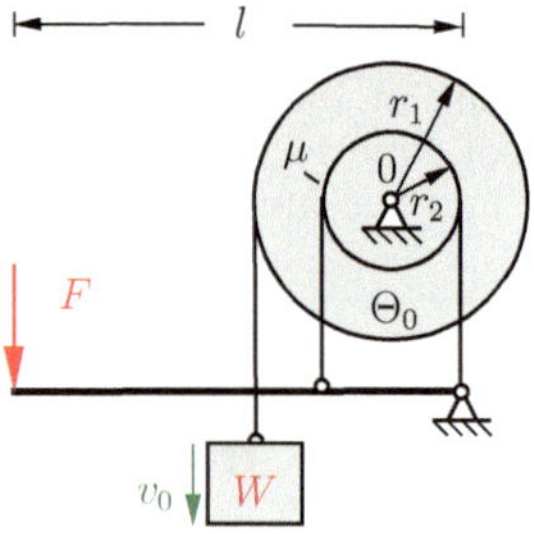

a) Determine the braking force F_0, such that the cabin moves with constant velocity v_0.

b) After which distance d, the cabin stops for a braking force $F > F_0$?

Solution **a)** The braking moment M_B, acting on the drum, is determined from the equilibrium condition at the lever

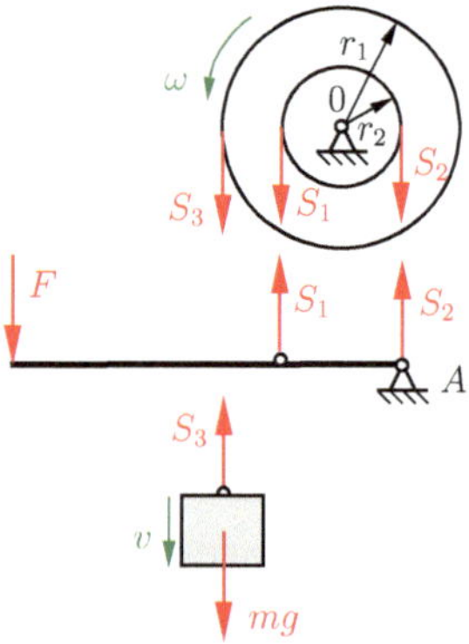

$$\overset{\curvearrowleft}{A}: \quad 0 = -2r_2 S_1 + lF \quad \rightsquigarrow \quad S_1 = \frac{Fl}{2r_2}$$

and the belt friction law $S_1 = S_2 e^{-\mu\pi}$:

$$M_B = r_2 S_2 - r_2 S_1 = (1 - e^{-\mu\pi})\frac{Fl}{2} .$$

The velocity of the cabin is constant, if the accelerations of the drum and the cabin are zero, i.e., if

$$\overset{\curvearrowleft}{0}: \quad r_1 S_3 = M_B , \quad S_3 = mg \quad \rightsquigarrow \quad \underline{\underline{F_0 = \frac{2r_1 mg}{(1 - e^{-\mu\pi})l}}} .$$

b) Now, we apply the work-energy theorem $T_1 - T_0 = U$ between initial state (0) and end state (stop) (1). With the kinematic relations $v_0 = r_1\omega_0$ and $d = r_1\varphi_d$ follows

$$E_{k1} = 0 , \qquad E_{k0} = \frac{1}{2}mv_0^2 + \frac{1}{2}\Theta_0\omega_0^2 = \frac{v_0^2}{2}\left(m + \frac{\Theta_0}{r_1^2}\right) ,$$

$$W = mgh - \int_0^{\varphi_d} M_B \mathrm{d}\varphi = mgd - (1 - e^{-\mu\pi})\frac{Fld}{2r_1} .$$

Thus, from

$$-\frac{v_0^2}{2}\left(m + \frac{\Theta_0}{r_1^2}\right) = mgd - (1 - e^{-\mu\pi})\frac{Fld}{2r_1} ,$$

we obtain by solving for d

$$\underline{\underline{d = \frac{v_0^2(m + \Theta_0/r_1^2)}{(1 - e^{-\mu\pi})Fl/2r_1 - mg}}} .$$

Problem 5.24 A bowling ball (radius r) has to overcome a hight difference on the return path which consists of two one-eighth circle arcs (radius $R = 5r$). **P5.24**

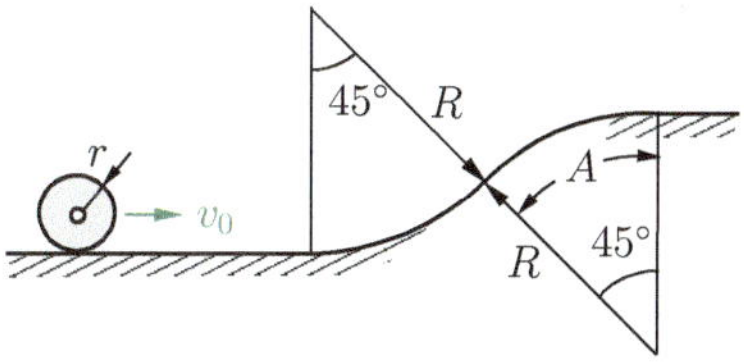

What speed v_0 may the ball have, so that lift-off from the path in the upper sector A is prevented?

Solution We first determine the velocity in sector A from the the energy conservation law

$$T + V = T_0 + V_0 \, .$$

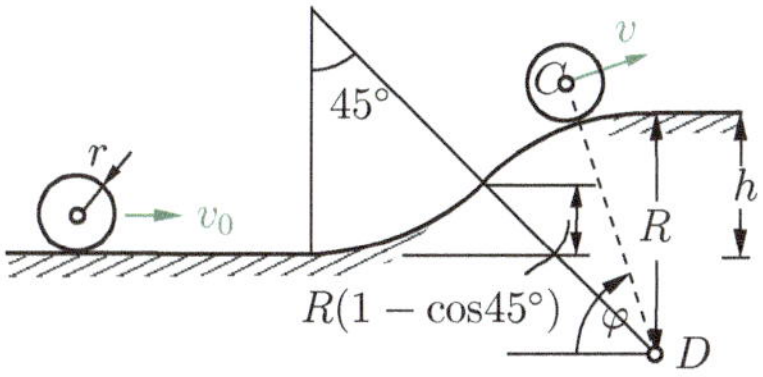

With $\Theta_C = 2mr^2/5$, $v = r\omega$ and $h = 2R(1 - \cos 45°) = 5r(2 - \sqrt{2})$ follow (zero level of potential at depth of D)

$$T_0 = \frac{1}{2}\, mv_0^2 + \frac{1}{2}\Theta_C \omega_0^2 = \frac{7}{10}\, mv_0^2 \, , \qquad T = \frac{7}{10}\, mv^2 \, ,$$

$$V_0 = mg(R - h + r) = (5\sqrt{2} - 4)mgr \, , \quad V = mg\, 6r \sin\varphi \, .$$

Solving for v leads to

$$v^2 = v_0^2 - \frac{60 \sin\varphi + 40 - 50\sqrt{2}}{7}\, gr \, .$$

In sector A the centripetal acceleration of the center of mass C is $a_n = v^2/6r$. Thus, from the equation of motion

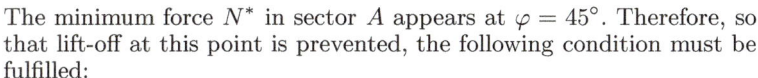

$$\searrow: \quad ma_n = -N + mg \sin\varphi \, ,$$

we obtain

$$N(\varphi) = -\frac{mv_0^2}{6r} + \frac{mg}{42}(102 \sin\varphi - 50\sqrt{2} + 40) \, .$$

The minimum force N^* in sector A appears at $\varphi = 45°$. Therefore, so that lift-off at this point is prevented, the following condition must be fulfilled:

$$N^* = N(45°) \geq 0 \quad \rightsquigarrow \quad \underline{\underline{v_0 \leq \sqrt{gr\, \frac{\sqrt{2} + 40}{7}} = 2.43\sqrt{gr}}} \, .$$

P5.25 **Problem 5.25** A homogeneous bar (weight $W = mg$) is hinge-connected with a body of mass m that can frictionless slide horizontally.

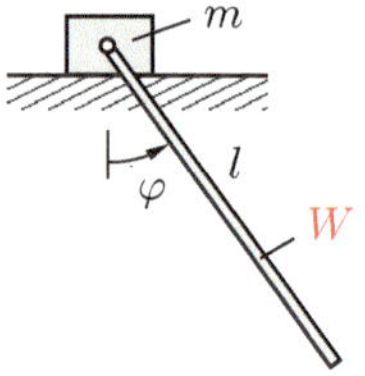

a) Determine the angular velocity $\dot{\varphi}(\varphi)$ of the bar, if it is released from rest from a horizontal position?

b) Determine the equation of motion.

Solution a) The position of the center of mass C^* of the total system is calculated as

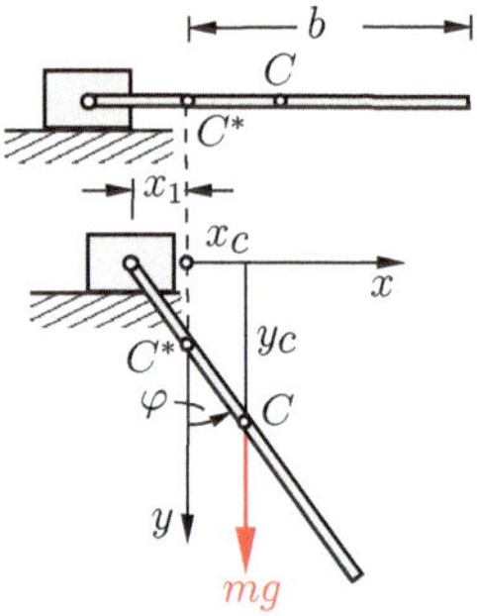

$$b = \frac{\frac{1}{2}\,lm + lm}{2m} = \frac{3}{4}\,l\,.$$

Since the system is not subjected to external forces in horizontal direction, there is no horizontal displacement of C^*. Thus, we have

$$x_1 = -\tfrac{1}{4} l \sin\varphi\,, \qquad \dot{x}_1 = -\tfrac{1}{4}\, l\dot{\varphi}\cos\varphi\,,$$

$$x_C = \tfrac{1}{4}\, l \sin\varphi\,, \qquad \dot{x}_C = \tfrac{1}{4}\, l\dot{\varphi}\cos\varphi\,,$$

$$y_C = \tfrac{1}{2}\, l\cos\varphi\,, \qquad \dot{y}_C = -\tfrac{1}{2}\, l\dot{\varphi}\sin\varphi\,.$$

We now determine the angular acceleration $\dot{\varphi}$ by using the energy conservation law $T + V = T_0 + V_0$. With $\Theta_C = ml^2/12$ follow

$$V_0 = 0\,, \qquad V = -mg\tfrac{1}{2}\, l\cos\varphi\,,$$

$$T_0 = 0\,, \qquad T = \tfrac{1}{2}\, m\dot{x}_1^2 + \left[\tfrac{1}{2}\, m(\dot{x}_S^2 + \dot{y}_S^2) + \tfrac{1}{2}\,\Theta_S\dot{\varphi}^2\right]$$

$$= \tfrac{1}{8}\, ml^2\dot{\varphi}^2 \left(\tfrac{1}{4}\cos^2\varphi + \tfrac{1}{4}\cos^2\varphi + \sin^2\varphi + \tfrac{1}{3}\right)$$

$$= \tfrac{1}{48} ml^2\dot{\varphi}^2\,(8 - 3\cos^2\varphi)$$

and we obtain after substitution

$$\frac{1}{24}\, l\dot{\varphi}^2(8 - 3\cos^2\varphi) - g\cos\varphi = 0 \quad \rightsquigarrow \quad \underline{\underline{\dot{\varphi} = \sqrt{\frac{24g\cos\varphi}{l(8 - 3\cos^2\varphi)}}}}\,.$$

b) Differentiating the energy conservation law (line above) with respect to time leads to the equation of motion:

$$\frac{l}{12}\dot{\varphi}\ddot{\varphi}(8 - 3\cos^2\varphi) + \frac{l}{4}\dot{\varphi}^3\cos\varphi\sin\varphi + g\dot{\varphi}\sin\varphi = 0$$

$$\rightsquigarrow \quad \underline{\underline{\ddot{\varphi}(8 - 3\cos^2\varphi) + 3\dot{\varphi}^2\cos\varphi\sin\varphi + 12\frac{g}{l}\sin\varphi = 0}}\,.$$

Problem 5.26 A symmetric disk with a half-circular boundary rolls without slipping on the flat ground. **P5.26**

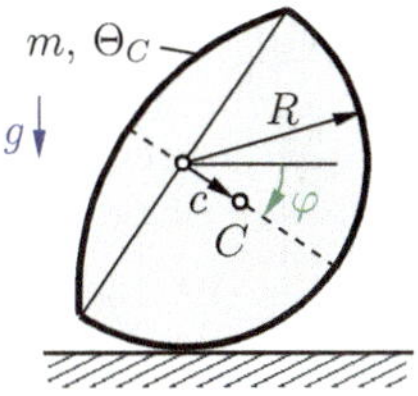

Determine the angular velocity in dependence of φ, if the body is released from rest at $\varphi = 0$. Calculate its maximum.

Given: R, $c = \kappa R$, m, $\Theta_C = \alpha\, mR^2$

Solution To solve the problem, it is advantageous to apply the energy conservation law. To formulate the energies, we introduce a coordinate system and find

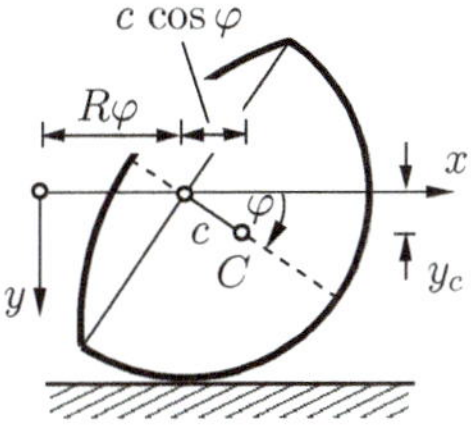

$$\begin{aligned}
x_C &= R\varphi + c\cos\varphi = R(\varphi + \kappa\cos\varphi)\ ,\\
y_C &= c\sin\varphi = \kappa R\sin\varphi\ ,\\
\dot{x}_C &= R\,\dot{\varphi} - c\dot{\varphi}\sin\varphi = R\,\dot{\varphi}(1 - \kappa\sin\varphi)\ ,\\
\dot{y}_C &= c\dot{\varphi}\cos\varphi = \kappa R\dot{\varphi}\cos\varphi\ .
\end{aligned}$$

Thus, the energies T_0, V_0 in the initial position and in an arbitrary displaced position are given by

$$\begin{aligned}
T_0 &= 0\ , \qquad V_0 = 0\ ,\\
T &= \frac{1}{2}\, m(\dot{x}_C^2 + \dot{y}_C^2) + \frac{1}{2}\Theta_C\dot{\varphi}^2 = \frac{1}{2}mR^2\dot{\varphi}^2(1 - 2\kappa\sin\varphi + \kappa^2\sin^2\varphi\\
&\quad + \kappa^2\cos^2\varphi) + \frac{1}{2}\,\alpha\, mR^2\dot{\varphi}^2 = \frac{1}{2}\, mR^2\dot{\varphi}^2(1 + \kappa^2 - 2\kappa\sin\varphi + \alpha)\ ,\\
V &= -mgy_C = -mgR\kappa\sin\varphi\ ,
\end{aligned}$$

and the energy conservation law $T + V = T_0 + V_0$ leads to

$$\frac{1}{2}\, mR^2\dot{\varphi}^2(1 + \kappa^2 - 2\kappa\sin\varphi + \alpha) - mgR\kappa\sin\varphi = 0$$

$$\rightsquigarrow \quad \underline{\underline{\dot{\varphi}(\varphi) = \sqrt{\frac{2g\kappa\sin\varphi}{R(1 + \kappa^2 - 2\kappa\sin\varphi + \alpha)}}}}\ .$$

The angular velocity takes its maximum at $\varphi = 90^\circ$ (lowest level of C):

$$\underline{\underline{\dot{\varphi}_{\max} = \sqrt{\frac{2g\kappa}{R[(1-\kappa)^2 + \alpha]}}}}\ .$$

P5.27 **Problem 5.27** A homogeneous circular disk (mass m, radius R) contains an excentric circular hole (radius $r = R/3$). The disk is released from rest from the displayed position.

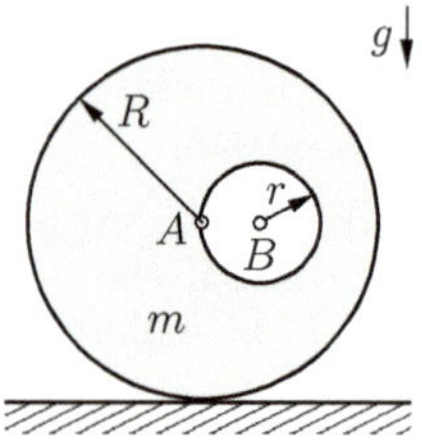

Determine the angular velocity for the instant, when B reaches its highest position

a) if the base is ideally frictionless,

b) if the disk rolls on the rough base.

Solution The solution is found by using the energy conservation law. For that purpose, we need the position of the center of mass and and the mass moment of inertia. It is useful to determine them by calculating the difference of two circular disks:

$$①: \quad A_1 = \pi R^2 , \quad m_1 = \frac{R^2\pi}{R^2\pi - r^2\pi}\, m = \frac{9}{8}\, m , \quad \Theta_1 = \frac{1}{2}\, m_1 R^2 ,$$

$$②: \quad A_2 = \pi r^2 , \quad m_2 = \frac{r^2\pi}{R^2\pi - r^2\pi}\, m = \frac{1}{8}\, m , \quad \Theta_2 = \frac{1}{2}\, m_2 r^2 .$$

This leads to

$$x_C = \frac{0 \cdot \pi R^2 - r \cdot \pi r^2}{\pi R^2 - \pi r^2} = -\frac{R}{24} , \qquad z_C = 0 ,$$

$$\begin{aligned}
\Theta_C &= [\Theta_1 + m_1 x_C^2] - [\Theta_2 + m_2 (x_C + r)^2] \\
&= \frac{9}{16}\, mR^2 + \frac{9}{8(24)^2}\, mR^2 - \frac{1}{16 \cdot 9}\, mR^2 \\
&\quad - \frac{m}{8}\left(\frac{R}{24} + \frac{R}{3}\right)^2 \\
&= \frac{311}{576}\, mR^2 .
\end{aligned}$$

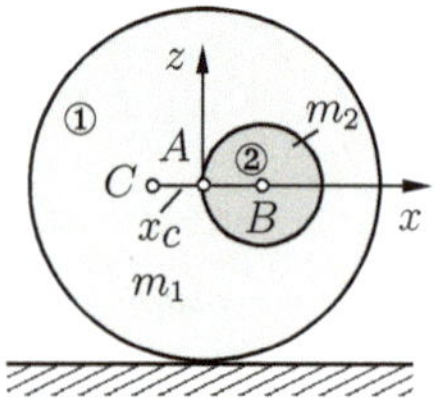

a) In the initial state, the disk is at rest and, if the zero level of potential energy is chosen at level of A, we have

$$V_0 = 0 , \qquad T_0 = 0 .$$

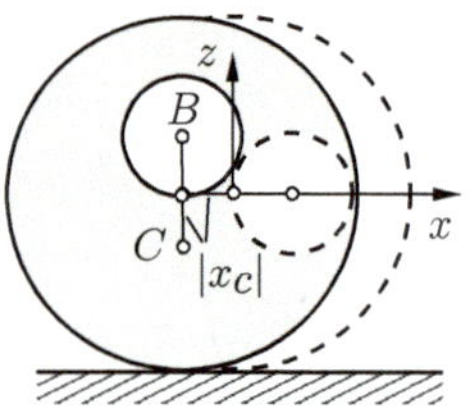

Because the base is frictionless, there act no forces in horizontal direction. Therefore, the center of mass C experiences no velocity change in this direction, i.e. $\dot{x}_C = 0$. When B reaches its highest position, C just reaches its lowest position. In vertical direction, the velocities of these points change at that instant their sign, and consequently,

$\dot{z}_C = 0$ holds. Thus, the energies at that instant are

$$V_1 = -mg|x_C|\,, \qquad T_1 = \frac{1}{2}\,\Theta_C\omega^2$$

and from the energy conservation law

$$V_1 + T_1 = V_0 + T_0\,,$$

it follows

$$\frac{1}{2}\,\Theta_C\omega^2 = mg|x_S| \quad \rightsquigarrow \quad \underline{\underline{\omega = \pm\sqrt{\frac{2mg|x_C|}{\Theta_C}} = \pm\sqrt{\frac{48\,g}{311\,R}}}}\,.$$

b) At the initial state, we have as before

$$V_0 = 0\,, \qquad T_0 = 0\,.$$

When the disk is rolling, the instantaneous center of rotation Π is located at the contact point between disk and base. When the center of mass reaches its lowest position (and B its highest position), its velocity has the horizontal direction and the magnitude $\dot{x}_S = \omega(R - |x_S|)$. Therewith, the energies at this moment are given by

$$V_1 = -mg|x_C|\,, \qquad T_1 = \frac{1}{2}\,m\dot{x}_C^2 + \frac{1}{2}\,\Theta_C\omega^2\,,$$

and the energy conservation law leads to

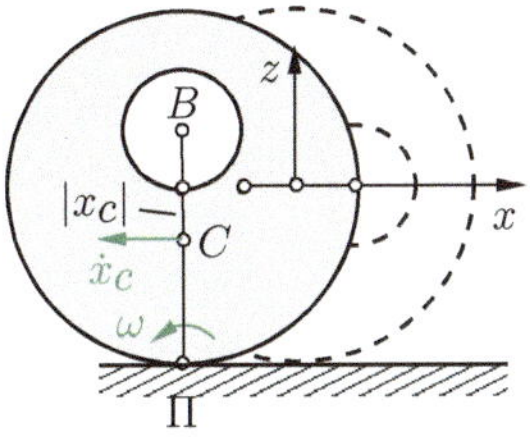

$$\frac{1}{2}\,\omega^2[m\,(R - |x_C|)^2 + \Theta_C] = mg|x_C|$$

and finally to

$$\underline{\underline{\omega = \pm\sqrt{\frac{2mg|x_C|}{m\,(R - |x_C|)^2 + \Theta_C}} = \pm\sqrt{\frac{2\,g}{35\,R}}}}\,.$$

Remark: The angular velocity in case a) is higher than in case b). The reason for that lies in the fact that the kinetic energy in case b) is split into translational and the rotational parts, while in case a), there exists only a rotational part.

P5.28 **Problem 5.28** At a pivoted massless arm, a motor is attached (point mass m_1). Its motor shaft is rigidly connected with the center of mass C of a disk (mass m_2, moment of inertia Θ_2).

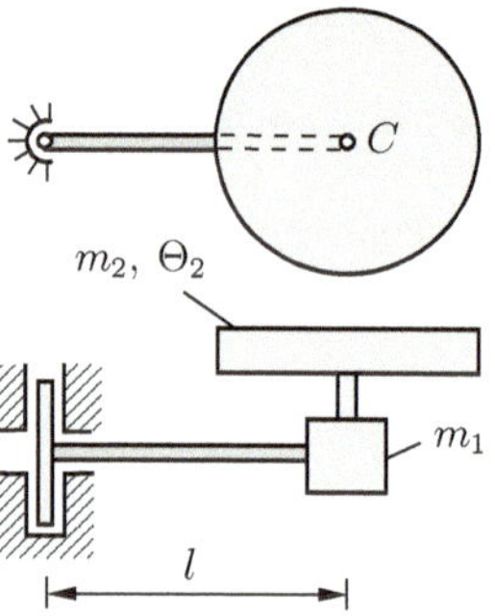

a) What are the angular velocities of the arm and the disk, if the motor delivers during the time interval Δt a constant torque M_0 and the system initially was at rest?

b) Determine the work done by the motor.

Solution **a)** We separate the system to make the *internal moment* M_0 visible. The principles of linear and angular impulse and momentum then lead to

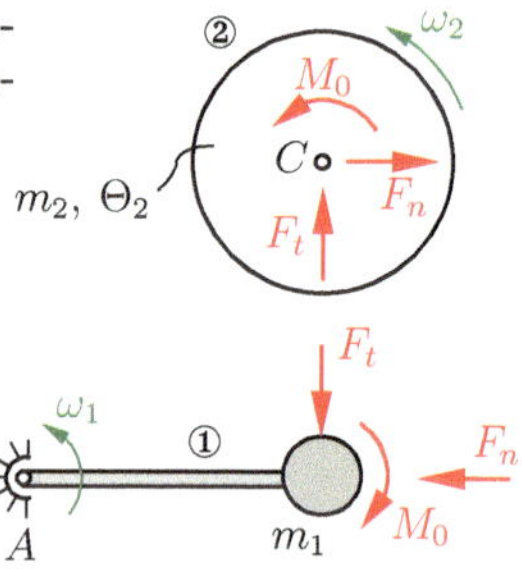

$$① \quad \stackrel{\curvearrowleft}{A}: \quad \Theta_A\omega_1 = -\widehat{M}_0 - l\widehat{F}_t\,,$$

$$② \quad \uparrow: \quad m_2 v_C = \widehat{F}_t\,,$$

$$\stackrel{\curvearrowleft}{C}: \quad \Theta_2\omega_2 = \widehat{M}_0\,,$$

where

$$\widehat{M}_0 = \int_0^{\Delta t} M_0 \mathrm{d}\bar{t} = M_0\Delta t\,, \qquad \widehat{F}_t = \int_0^{\Delta t} F_t(\bar{t})\mathrm{d}\bar{t}\,, \qquad \Theta_A = m_1 l^2\,.$$

With the kinematic relation $v_C = l\omega_1$ follows by eliminating $\widehat{F}_t$

$$\underline{\underline{\omega_1 = -\frac{M_0\Delta t}{(m_1+m_2)l^2}}}\,, \qquad \underline{\underline{\omega_2 = +\frac{M_0\Delta t}{\Theta_2}}}\,.$$

b) The work U done by the motor is determined from the kinetic energy of the system after time Δt:

$$\underline{\underline{U}} = \frac{1}{2}[(m_1+m_2)l^2]\omega_1^2 + \frac{1}{2}\Theta_2\omega_2^2 = \underline{\underline{\frac{1}{2}(M_0\Delta t)^2\left[\frac{1}{(m_1+m_2)l^2} + \frac{1}{\Theta_2}\right]}}\,.$$

Remark: Since no *external moment* is present, the angular momentum of the total system remains zero:

$$\Theta_A\omega_1 + m_2 l^2\omega_1 + \Theta_2\omega_2 = 0\,.$$

Problem 5.29 On a disk ① (moment of inertia Θ_1), which rotates with an angular velocity ω_1, a second disk ② (mass m_2, moment of inertia Θ_2) rotates with ω_2 about the shaft B (= center of mass of ②). When ② slides down along B and touches ①, both disks rub on each other such that ② comes to rest relatively to disk ① after time t_0. P5.29

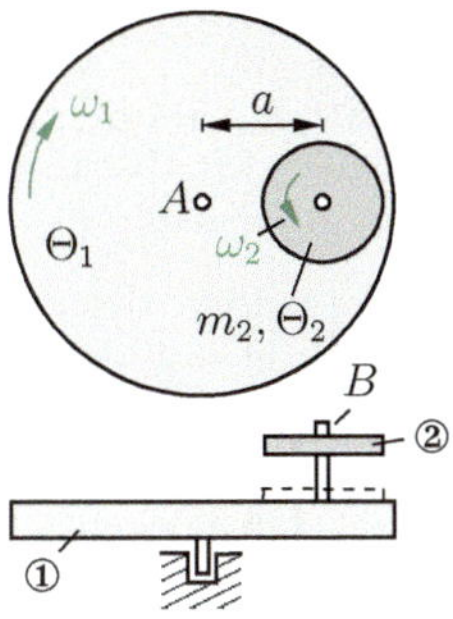

Determine the common angular velocity ω_0 after time t_0.

Solution We separate the systen and formulate the principles of angular impulse and momentum for ① (with respect to fixed axis A) and ② (with respect to its center of mass B):

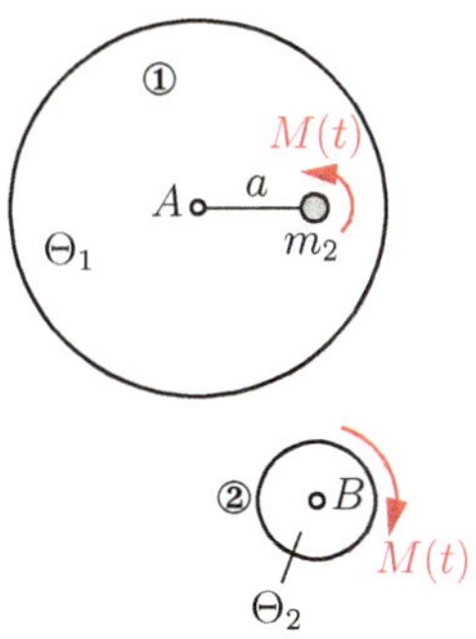

$$\overset{\curvearrowright}{A}: \quad (\Theta_1 + m_2 a^2)[\omega_0 - \omega_1] = -\widehat{M}\,,$$

$$\overset{\curvearrowright}{B}: \quad \Theta_2[\omega_0 - (-\omega_2)] = +\widehat{M}\,,$$

where

$$\widehat{M} = \int_0^{t_0} M(\bar{t})\mathrm{d}\bar{t}\,.$$

Notice that the mass m_2 of disk ② is considered in the moment of inertia of ①.

Eliminating $\widehat{M}$ leads to

$$\underline{\underline{\omega_0 = \frac{(\Theta_1 + m_2 a^2)\omega_1 - \Theta_2\omega_2}{\Theta_1 + \Theta_2 + m_2 a^2}}}\,.$$

The problem can be solved easier by applying the principle of angular momentum to the total system. Because there acts no *external moment*, conservation of angular momentum about the fixed axis A yields

$$\overset{\curvearrowright}{A}: \quad (\Theta_1 + m_2 a^2)\omega_1 - \Theta_2\omega_2 = [\Theta_1 + (\Theta_2 + m_2 a^2)]\omega_0\,,$$

i.e. we find the same result as above.

Remark: For $\Theta_2\omega_2 = (\Theta_1 + m_2 a^2)\omega_1$, the system comes to rest.

P5.30 **Problem 5.30** A homogeneous circular disk (weight $W_1 = mg$) rotates with angular velocity ω_0 about pin B of an initially resting bar (weight $W_2 = 2mg$, $l = 2r$), which is pin-supportet at A.

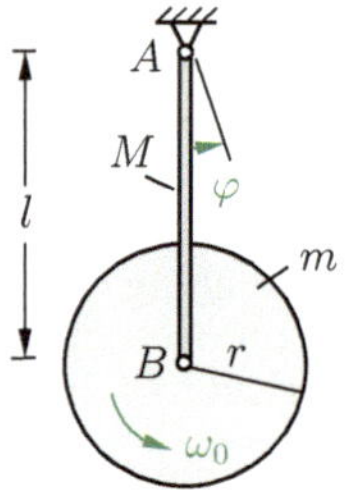

a) Determine the amplitude φ_2 of the bar, if the disk is suddenly blocked at the bar.

b) Determine the energy loss due to the blockade.

Solution The angular velocity ω_1 of the system immediately after blocking is determined from the conservation of angular momentum. With the moments of inertia of the disk and the blocked system

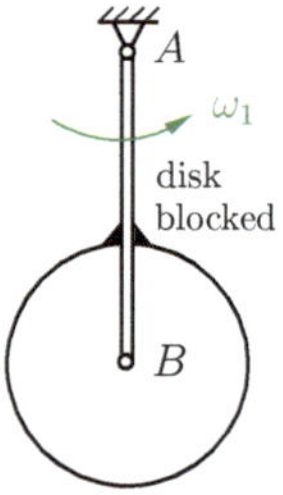

$$\Theta_B = \tfrac{1}{2}mr^2 ,$$
$$\Theta_A = (\tfrac{1}{2}mr^2 + ml^2) + Ml^2/3 = \tfrac{43}{6}mr^2 ,$$

it follows

$$\Theta_B\omega_0 = \Theta_A\omega_1 \quad \rightsquigarrow \quad \omega_1 = \frac{\Theta_B}{\Theta_A}\,\omega_0 = \frac{3}{43}\,\omega_0 \, .$$

The amplitude φ_2 of the blocked system is calculated from the energy conservation after blocking: $T_2 + V_2 = T_1 + V_1$. Choosing zero potential at the level of A, we obtain

$$T_2 = 0 ,$$
$$V_2 = -mgl\sin\varphi_2 - Mg(l/2)\sin\varphi_2 ,$$
$$T_1 = \frac{1}{2}\Theta_A\omega_1^2 = \frac{mr^2\omega_0^2}{4}\,\frac{3}{43} ,$$
$$V_1 = -lmg - lMg/2 = -4rmg ,$$

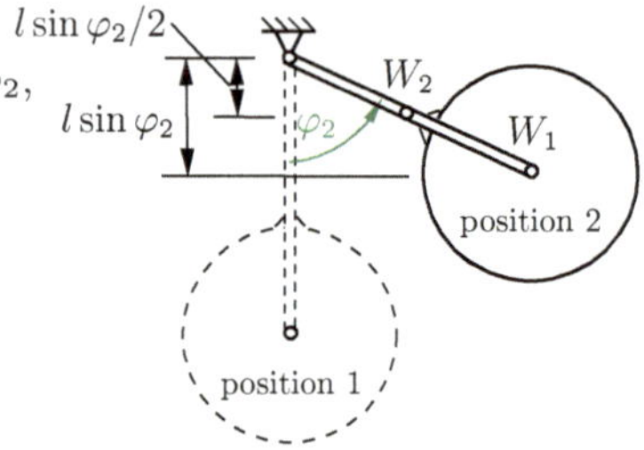

what leads to

$$\underline{\underline{\sin\varphi_2 = 1 - \frac{3}{688}\,\frac{r\omega_0^2}{g}}} \, .$$

b) The energy loss is determined from the difference ΔT of kinetic energies immediately before and after blocking:

$$\underline{\underline{\Delta T}} = \frac{\Theta_B\omega_0^2}{2} - \frac{\Theta_A\omega_1^2}{2} = \frac{mr^2\omega_0^2}{4} - \frac{mr^2\omega_0^2}{4}\,\frac{3}{43} = \underline{\underline{\frac{10}{43}\,mr^2\omega_0^2}} \, .$$

Problem 5.31 A yo-yo (weight mg, moment of inertia Θ_A) moves at time $t = 0$ with speed v_0 and angular velocity ω_0. **P5.31**

Determine the velocities v_1 and ω_1 at time t_1 when pulling the string with a force $S(t) = S_0\, t/t_1$.

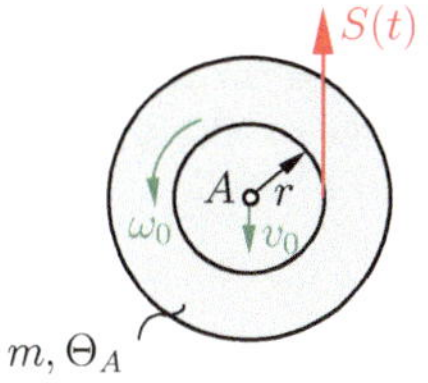

Solution The equations of motion read

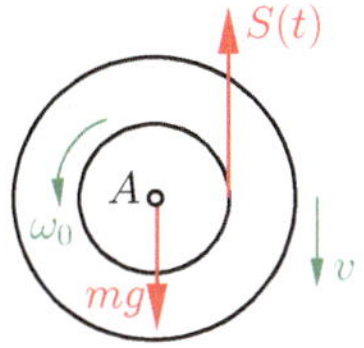

$$\downarrow: \qquad mv_1 - mv_0 = \widehat{F}_x \ ,$$

$$\overset{\curvearrowleft}{A}: \quad \Theta_A\omega_1 - \Theta_A\omega_0 = \widehat{M}_A \ ,$$

where

$$\widehat{F}_x = \int_0^{t_1} [mg - S(\bar{t})]\mathrm{d}\bar{t} = mgt_1 - \frac{1}{2}S_0t_1 \ ,$$

$$\widehat{M}_A = \int_0^{t_1} rS(\bar{t})\mathrm{d}\bar{t} = \frac{1}{2}rS_0t_1 \ .$$

Solving for v_1 and ω_1 yields

$$\underline{\underline{v_1 = v_0 + gt_1 - \frac{S_0t_1}{2m}}} \ , \qquad \underline{\underline{\omega_1 = \omega_0 + \frac{rS_0t_1}{2\Theta_A}}} \ .$$

Problem 5.32 A homogeneous bar of mass m moves on a horizontal plane purely translationally with the velocity v_0. **P5.32**

Determine its angular velocity ω_1, when its end suddenly latches into the fixed bearing A.

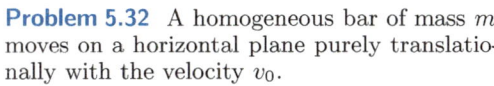

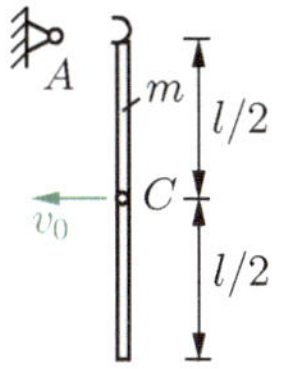

Solution Initially, the angular momentum of the bar with respect to A is $L_0 = \frac{l}{2}(mv_0)$. After latching, the angular momentum is $L_1 = \Theta_A\omega_1 = \frac{1}{3}(ml^2\omega_1)$. Since there acts no external moment with respect to A, the angular momentum is conserved:

$$\frac{1}{3}ml^2\omega_1 = \frac{1}{2}lmv_0 \qquad \leadsto \qquad \underline{\underline{\omega_1 = \frac{3v_0}{2l}}} \ .$$

Remark: Latching leads to an energy loss ΔT:

$$\Delta T = T_0 - T_1 = \frac{1}{2}mv_0^2 - \frac{1}{2}\Theta_A\omega_1^2 = \frac{1}{8}mv_0^2 = \frac{1}{4}T_0 \ .$$

P5.33

Problem 5.33 Two homogeneous bars of masses m and $2m$ are attached at a massless shaft which is driven by a torque M_0.

Determine the equation of motion and the support reactions at the bearings.

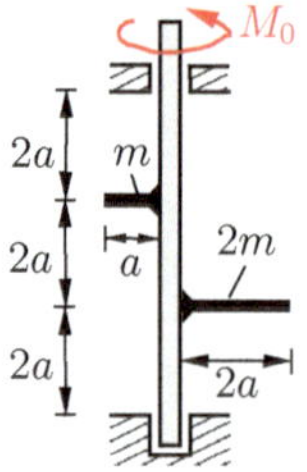

Solution We first determine the axial moment of inertia and products of inertia for the *body-fixed* system ξ, η, ζ

$$\Theta_\zeta = \frac{2m}{3}\,(2a)^2 + \frac{m}{3}\,a^2 = 3\,ma^2\,,$$

$$\Theta_{\xi\zeta} = -m\,2a\,\frac{-a}{2} = ma^2\,,$$

$$\Theta_{\eta\zeta} = 0\,.$$

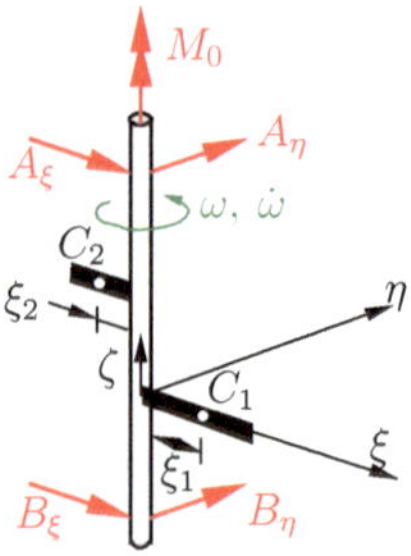

With the external moments

$$M_\xi = 2aB_\eta - 4aA_\eta\,,\quad M_\eta = 4aA_\xi - 2aB_\xi\,,\quad M_\zeta = M_0\,,$$

it follows from the priciple of angular momentum with respect to the body-fixed system

$$\overset{\curvearrowright}{\xi}:\quad 2a(B_\eta - 2A_\eta) = -\dot\omega\,ma^2 \quad\rightsquigarrow\quad B_\eta - 2A_\eta = -\frac{ma\dot\omega}{2}\,,$$

$$\overset{\curvearrowright}{\eta}:\quad 2a(2A_\xi - B_\xi) = \omega^2 ma^2 \quad\rightsquigarrow\quad 2A_\xi - B_\xi = \frac{ma\omega^2}{2}\,,$$

$$\overset{\curvearrowright}{\zeta}:\quad M_0 = 3\,ma^2\dot\omega \quad\rightsquigarrow\quad \underline{\underline{\dot\omega = \frac{M_0}{3\,ma^2}}}\,.$$

The last equation is the equation of motion of the shaft. To determine the support reactions, the principle of linear momentum must be applied. Since the motion of the mass centers C_1 and C_2 is circular, the acceleration components of e.g. C_1 are given by $\xi_1\dot\omega$ in η-direction and $-\xi_1\omega^2$ in ξ-direction. Thus, we obtain in ξ- and in η-direction

$$-2m\,a\,\omega^2 + m(a/2)\omega^2 = A_\xi + B_\xi\,,\qquad 2m\,a\,\dot\omega - m(a/2)\dot\omega = A_\eta + B_\eta$$

and finally

$$\underline{\underline{A_\xi = -\frac{1}{3}\,ma\omega^2}},\quad \underline{\underline{A_\eta = \frac{2}{9}\,\frac{M_0}{a}}},\quad \underline{\underline{B_\xi = -\frac{5}{6}\,ma\omega^2}},\quad \underline{\underline{B_\eta = \frac{5}{18}\,\frac{M_0}{a}}}.$$

Problem 5.34 A homogeneous plate (mass m) rotates with the constant angular velocity ω about a fixed axis. P5.34

a) Calculate the support reactions at A and B.

b) Determine the additional masses m_1 that must be attached at D and E, such that the support reactions are zero.

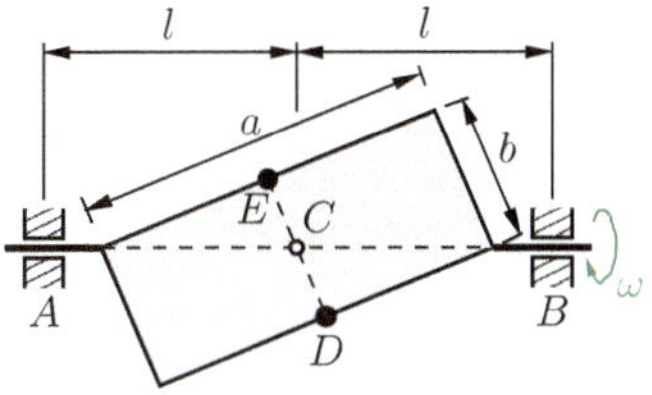

Solution **a)** Because the center of mass is located on the axis of rotation (acceleration of C is zero), the equation of motion reduces to

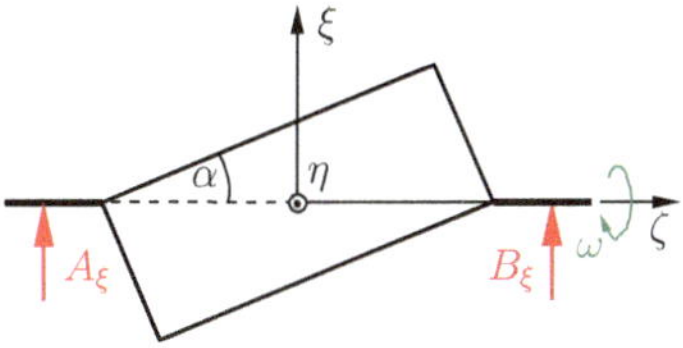

$$A_\xi + B_\xi = 0\,.$$

With $\dot\omega = 0$ and

$$\Theta_{\xi\zeta} = -\frac{m}{12}\,\frac{(a^2-b^2)ab}{a^2+b^2}$$

(see Problem 5.8), it follows from the principle of moment of momentum $M_\eta = \dot\omega\Theta_{\eta\zeta} + \omega^2\Theta_{\xi\zeta}$ (notice the positive sense of rotation about the η-axis!)

$$\overset{\curvearrowleft}{\eta}: \quad -lA_\xi + lB_\xi = \omega^2\Theta_{\xi\zeta}\,.$$

Therewith we obtain

$$\underline{\underline{A_\xi = -B_\xi = -\frac{\omega^2\Theta_{\xi\zeta}}{2l} = \frac{m\omega^2}{24l}\,\frac{(a^2-b^2)ab}{a^2+b^2}}}\,.$$

b) Such that A_ξ and B_ξ are zero, the total product of inertia $\Theta^*_{\xi\zeta}$ (including m_1) must vanish. Thus, with

$$\sin\alpha = \frac{b}{\sqrt{a^2+b^2}}\,, \qquad \cos\alpha = \frac{a}{\sqrt{a^2+b^2}}$$

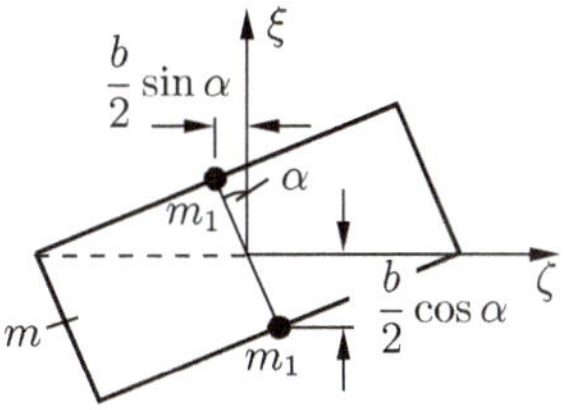

the following condition must be fulfilled:

$$\Theta^*_{\xi\zeta} = -\frac{m}{12}\,\frac{(a^2-b^2)ab}{a^2+b^2} + 2m_1\Big(\frac{b}{2}\sin\alpha\Big)\Big(\frac{b}{2}\cos\alpha\Big) = 0$$

$$\leadsto \quad \underline{\underline{m_1 = \frac{m}{6}\Big(\frac{a^2}{b^2} - 1\Big)}}\,.$$

Remark: The masses m_1 must be attached at both sides, such that the center of mass remains still located on the axis of rotation.

P5.35 **Problem 5.35** A homogeneous circular disk (radius r, mass m) is mounted obliquely (angle α) and with an eccentricity e to a rigid thin shaft. The system rotates with constant angular velocity ω_0.

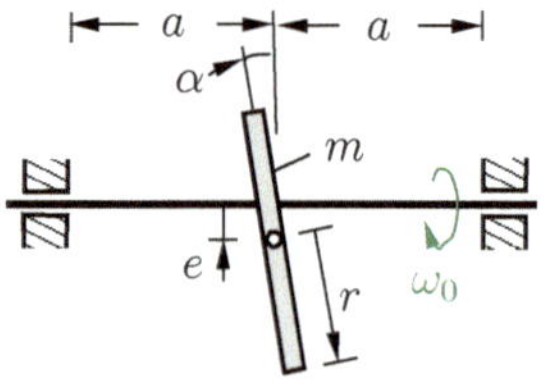

Determine the forces in the bearings.

Solution To fully describe the motion of the rigid body, the principles of linear and angular momentum must be applied. The latter is advantageously formulated with the aid of Euler's equations:

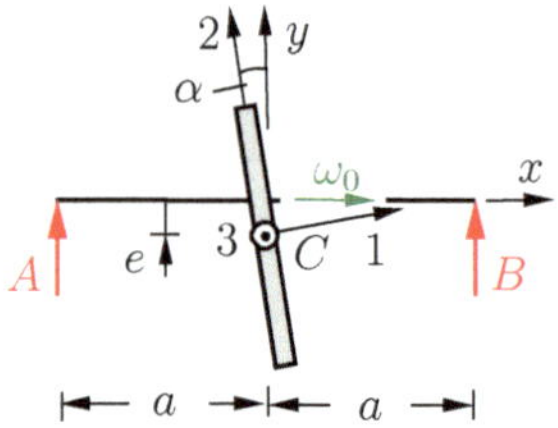

$$\Theta_1\dot{\omega}_1 - (\Theta_2 - \Theta_3)\omega_2\omega_3 = M_1\,,$$
$$\Theta_2\dot{\omega}_2 - (\Theta_3 - \Theta_1)\omega_3\omega_1 = M_2\,,$$
$$\Theta_3\dot{\omega}_3 - (\Theta_1 - \Theta_2)\omega_1\omega_2 = M_3\,.$$

With

$$\omega_1 = \omega_0 \cos\alpha, \quad \dot{\omega}_1 = 0, \quad \omega_2 = -\omega_0 \sin\alpha, \quad \dot{\omega}_2 = 0, \quad \omega_3 = \dot{\omega}_3 = 0$$

and

$$\Theta_1 = \frac{mr^2}{2}\,, \qquad \Theta_2 = \Theta_3 = \frac{mr^2}{4}\,,$$

it follows

$$M_1 = 0\,, \quad M_2 = 0\,, \quad M_3 = \frac{mr^2}{4}\,\omega_0^2 \sin\alpha\cos\alpha = \frac{mr^2}{8}\,\omega_0^2 \sin 2\alpha\,.$$

Here M_3 is the moment of the external forces about the principal axis 3. Its relation with the forces in the bearings is given by

$$\overset{\curvearrowleft}{C}: \quad M_3 = aB - aA \quad \leadsto \quad aB - aA = \frac{mr^2}{8}\,\omega_0^2 \sin 2\alpha\,.$$

Now, we use the principle of linear momentum. Since C moves circularly with constant angular velocity, its acceleration $|a| = e\omega_0^2$ is directed to the ratation axis. Thus,

$$\uparrow: \quad m\,e\omega^2 = A + B\,.$$

Solving these two equations leads to

$$\underline{\underline{A = \frac{m\omega_0^2}{16}\left(8e - \frac{r^2}{a}\,\sin 2\alpha\right)}}\,, \qquad \underline{\underline{B = \frac{m\omega_0^2}{16}\left(8e + \frac{r^2}{a}\,\sin 2\alpha\right)}}\,.$$

Problem 5.36 A homogeneous bar (mass m) on a frictionless horizontal base is accelerated by a force F acting at point A in length direction of the bar. P5.36

Determine the distribution of the normal force N along the bar.

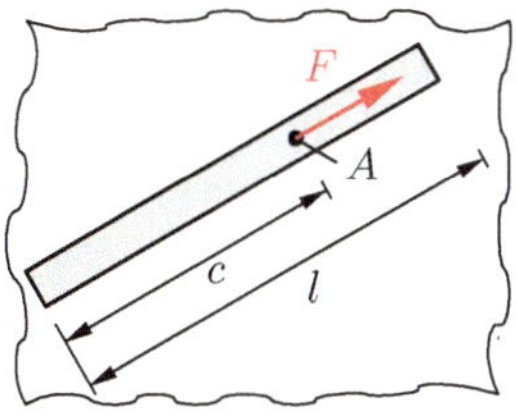

Solution The acceleration of the bar (in length direction) is $a = F\,/\,m$. To calculate the normal force, we first introduce the body-fixed coordinate x and then cut the bar at an arbitrary location x above and/or below the point A. For the lower cut ($x < c$), the principle of linear momentum yields with $\bar{m} = m\,x/l$

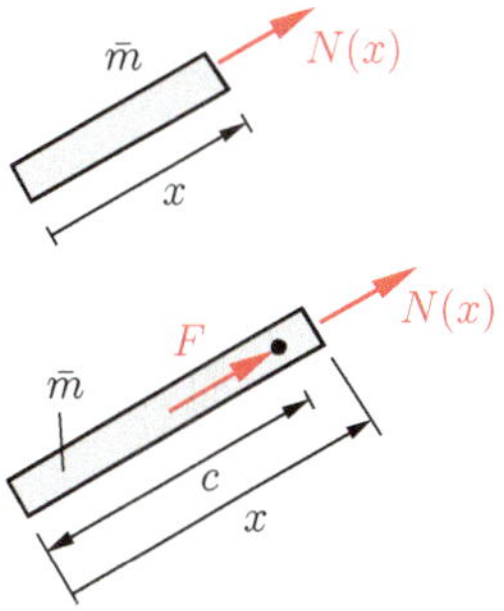

$$\underline{\underline{N(x)}} = \bar{m}\,a = \frac{\bar{m}}{m}\,F = \underline{\underline{\frac{x}{l}\,F}}\,.$$

For the upper cut ($x > c$), we obtain

$$N(x) + F = \bar{m}\,a = \frac{\bar{m}}{m}\,F$$

$$\rightsquigarrow \quad \underline{\underline{N(x) = -F\,(1 - \frac{x}{l})}}\,.$$

The normal force is linearly distributed and has a jump of magnitude F at $x = c$. This leads to the below displayed graph of $N(x)$.

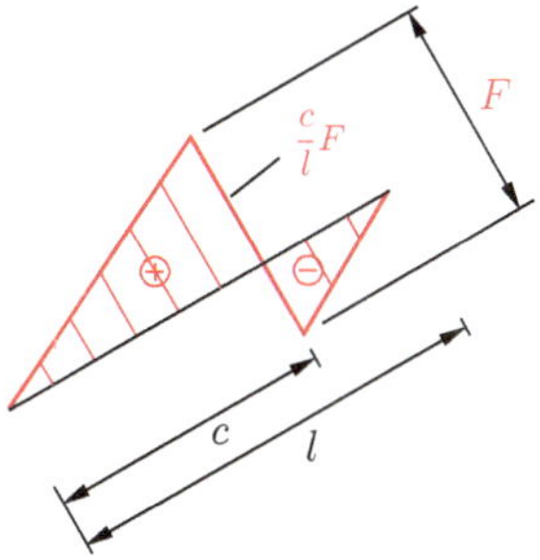

P5.37

Problem 5.37 A pin-supported homogeneous beam (mass m, length l) is initially at rest. Starting at time $t = 0$ it is subjected to a constant force F, acting perpendicularly to its longitudinal axis.

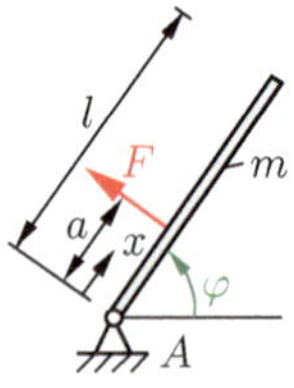

Determine the stress resultants (M, V, N) in the beam domain $a \leq x \leq l$ at time $t > t_0$. Neglect the weight of the beam.

Solution Since the motion of the beam is a pure rotation, its acceleration and velocity are uniquely described by $\ddot{\varphi}$ and $\dot{\varphi}$. The principle of angular momentum with respect to A yields

$$\overset{\curvearrowleft}{A}: \quad \Theta_A \ddot{\varphi} = a F \quad \rightsquigarrow \quad \ddot{\varphi} = \frac{a F}{\Theta_A}.$$

Integration in conjunction with the initial condition $\dot{\varphi}(0) = 0$ leads to the angular velocity

$$\dot{\varphi}(t) = \frac{a F}{\Theta_A} t.$$

Now, we introduce body-fixed coordinates, cut the system at a position $x > a$ and introduce the stress resultants. Here, we first restrict ourselves to the shear force V and the bending moment M. For the free part of the beam the principle of angular momentum with respect to the center of mass $\bar{C}$ and the principle of linear momentum in z-direction yields

$$\overset{\curvearrowleft}{\bar{C}}: \quad \bar{\Theta}_{\bar{C}} \ddot{\varphi} = -M(x) - V(x) \frac{l - x}{2},$$

$$\searrow: \quad \bar{m} \ddot{z}_{\bar{C}} = -V(x).$$

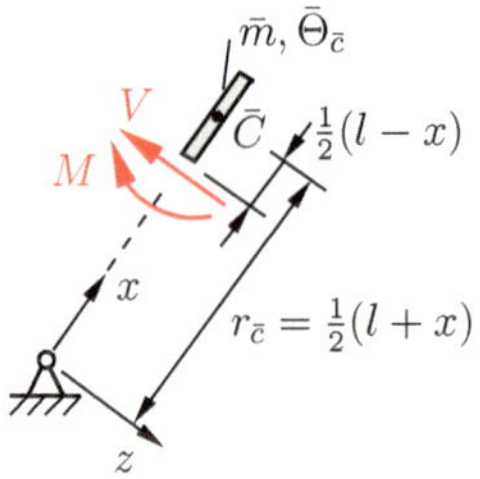

From the kinematics (circular motion) follows the acceleration $\ddot{z}_{\bar{C}}$:

$$\ddot{z}_{\bar{C}} = -r_{\bar{C}} \ddot{\varphi} = -\frac{l + x}{2} \ddot{\varphi}.$$

Therewith and with $\bar{m} = (1 - \frac{x}{l}) m$ and $\Theta_A = \dfrac{m l^2}{3}$, we obtain the shear force:

$$\underline{\underline{V(x)}} = \bar{m} \frac{l + x}{2} \ddot{\varphi} = \bar{m} \frac{l + x}{2} \frac{a F}{\Theta_A} = \underline{\underline{\frac{3}{2} \frac{a F}{l} \left[1 - \left(\frac{x}{l} \right)^2 \right]}}.$$

Introducing this result into the 1st equation of motion leads with $\bar{\Theta}_{\bar{c}} = \dfrac{1}{12}\,\bar{m}\,(l-x)^2$ to the bending moment

$$
\begin{aligned}
\underline{\underline{M(x)}} &= -Q\frac{l-x}{2} - \bar{\Theta}_{\bar{c}}\,\ddot{\varphi} \\
&= -\frac{3}{4}\,a\,F\left(1-\frac{x}{l}\right)^2\left(1+\frac{x}{l}\right) - \frac{m\,l^2}{12}\left(1-\frac{x}{l}\right)^3\frac{a\,F}{\Theta_A} \\
&= \underline{\underline{-a\,F\left(1-\frac{x}{l}\right)^2\left(1+\frac{1}{2}\frac{x}{l}\right)}}\,.
\end{aligned}
$$

The normal force N is obtained by using the principle of linear momentum in x-direction:

$$\nearrow: \quad \bar{m}\,\ddot{x}_{\bar{c}} = -N(x)\,.$$

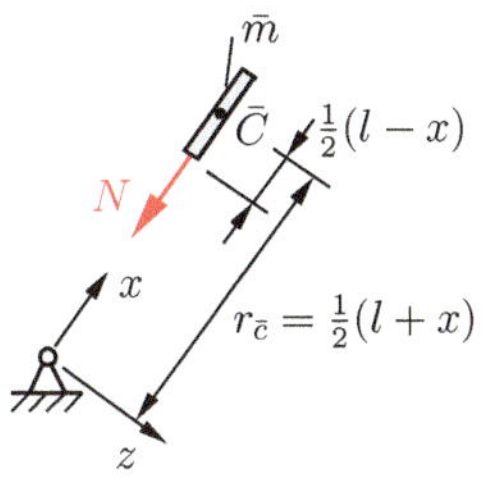

Here, $\ddot{x}_{\bar{c}} = -r_{\bar{c}}\,\dot{\varphi}^2$ is the centripetal acceleration. After introducing the already known quantities, this leads to

$$
\begin{aligned}
\underline{\underline{N(x)}} &= \bar{m}\,r_{\bar{C}}\,\dot{\varphi}^2 \\
&= m\,(1-\frac{x}{l})\,\frac{x+l}{2}\left(\frac{M_0}{\Theta_A}\,t\right)^2 \\
&= \underline{\underline{\frac{9}{2}\,\frac{M_0^2\,t^2}{m\,l^3}\left[1-\left(\frac{x}{l}\right)^2\right]}}\,.
\end{aligned}
$$

Contrary to the bending moment and the shear force, the normal force increases with the square of time t.
The graphs of normal force, shear force and bending moment are sketched below for the special case $a = l/3$.

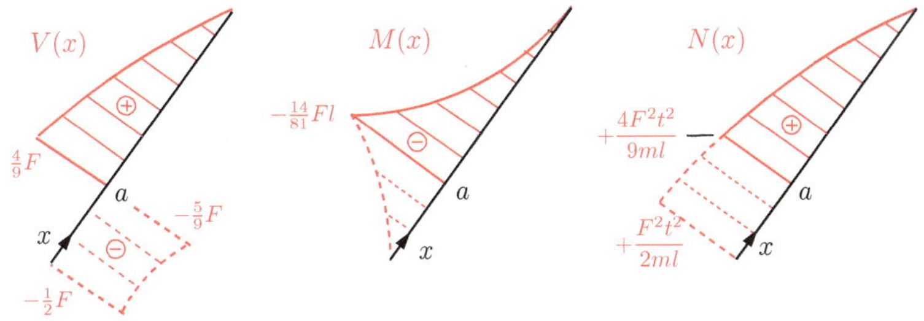

P5.38 **Problem 5.38** A homogeneous beam (mass m, length l) is lifted from a horizontal base by a force $F = 4\,mg$.

Determine the stress resultants immediately after lift-off.

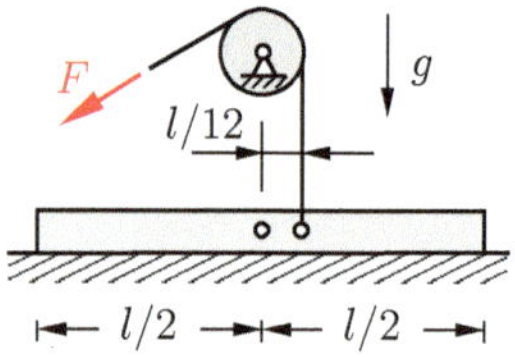

Solution Immediately after lift-off, the beam still has a horizontal position. The principles of linear and angular momentum then yield with $\Theta_C = ml^2/12$

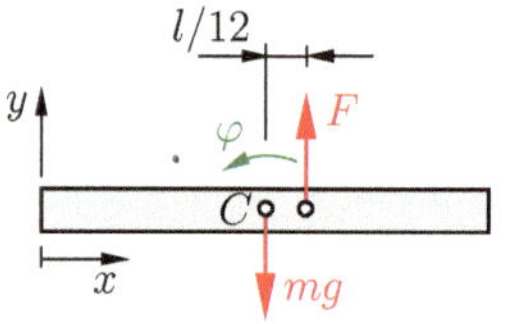

$$\uparrow: \quad m\ddot{y}_C = F - mg \quad \rightsquigarrow \quad \ddot{y}_C = 3g\,,$$

$$\stackrel{\curvearrowleft}{C}: \quad \Theta_C\,\ddot{\varphi} = F\frac{l}{12} \quad \rightsquigarrow \quad \ddot{\varphi} = \frac{Fl}{12\,\Theta_C} = 4\frac{g}{l}\,.$$

We now cut the beam at position x, introduce the stress resultants and consider first a cut left to the point where F applies ($x < \frac{7}{12}l$). In this case the principles of linear and angular momentum yield for part ①

$$\uparrow: \quad m_1\ddot{y}_{C_1} = -m_1\,g - V_1\,,$$

$$\stackrel{\curvearrowleft}{C_1}: \quad \Theta_{C_1}\ddot{\varphi} = M_1 - V_1\,\frac{x}{2}\,.$$

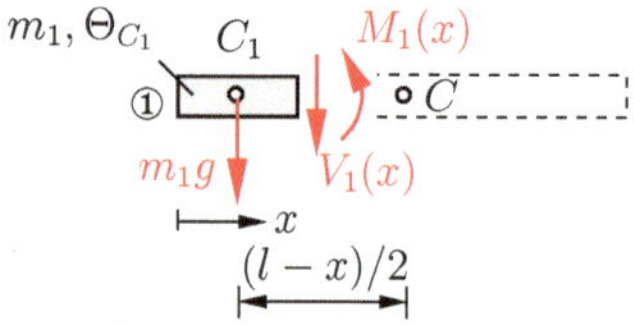

With

$$m_1(x) = \frac{x}{l}m\,, \qquad \Theta_{C_1} = \frac{1}{12}m_1x^2 = \frac{1}{12}\frac{x^3}{l}m$$

and the kinematic relation

$$\ddot{y}_{C_1} = \ddot{y}_C - \frac{1}{2}(l-x)\,\ddot{\varphi} = 3g - \frac{1}{2}(l-x)\,4\frac{g}{l} = g\left(1 + 2\frac{x}{l}\right),$$

we obtain the resultants

$$\underline{\underline{V_1(x)}} = -m_1\,(g + \ddot{y}_{C_1}) = \underline{\underline{-2\,mg\frac{x}{l}\left(1 + \frac{x}{l}\right)}},$$

$$\underline{\underline{M_1(x)}} = \frac{x}{2}V_1 + \Theta_{C_1}\ddot{\varphi} = -mgl\left(\frac{x}{l}\right)^2\left(1 + \frac{x}{l}\right) + \frac{1}{12}\frac{x^3}{l}m\,4\frac{g}{l}$$

$$= \underline{\underline{-\frac{mgl}{3}\left(\frac{x}{l}\right)^2\left(3 + 2\frac{x}{l}\right)}}\,.$$

For a cut right to the point where F applies ($x > \frac{7}{12}l$), we obtain in the same way now for part ②

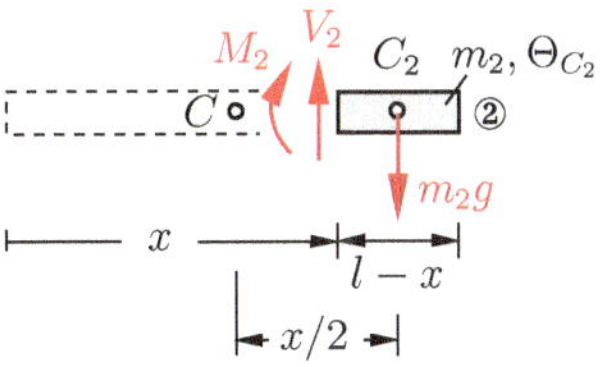

$$\uparrow: \quad m_2 \ddot{y}_{S_2} = -m_2\, g + V_2 \,,$$

$$\overset{\curvearrowleft}{C_2}: \quad \Theta_{C_2} \ddot{\varphi} = -M_2 - V_2 \frac{l-x}{2} \,.$$

With

$$m_2 = \frac{l-x}{l} m \,, \qquad \Theta_{C_2} = \frac{1}{12} m_2 (l-x)^2 = \frac{1}{12} \frac{(l-x)^3}{l} m \,,$$

$$\ddot{y}_{C_2} = \ddot{y}_C + \frac{x}{2}\ddot{\varphi} = g\left(3 + 2\frac{x}{l}\right) \,,$$

it follows

$$\underline{\underline{V_2(x)}} = m_2\,(g + \ddot{y}_{C_2}) = \underline{\underline{2mg\left(1 - \frac{x}{l}\right)\left(2 + \frac{x}{l}\right)}} \,,$$

$$\underline{\underline{M_2(x)}} = -\frac{l-x}{2} V_2 - \Theta_{C_2}\ddot{\varphi} = \underline{\underline{-\frac{mgl}{3}\left(1 - \frac{x}{l}\right)^2 \left(7 + 2\frac{x}{l}\right)}} \,.$$

The graphs of $V(x)$ and $M(x)$ are plotted below:

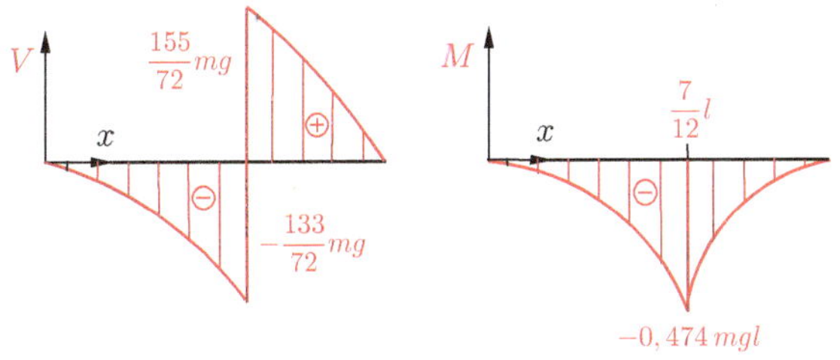

Remarks:

- At the point of application of F, the shear for has a jump: $\Delta V = F = 4\,mg$.
- Also in the dynamic case, the relation $V = \dfrac{\mathrm{d}M}{\mathrm{d}x}$ holds. This follows from the principle of angular momentum for a beam element of length $\mathrm{d}x$ by considering that the moment of inertia is small of higher order.
- The acceleration $\ddot{y}$ is positive for all points: $\ddot{y}(x) = \ddot{y}_C + (x - l/2)\ddot{\varphi} = g(1 + 4\,x/l) > 0$, i.e., the beam actually completely lifts off from the base.

P5.39

Problem 5.39 A chimney stack of mass m and length l is blown up at the base and falls over.

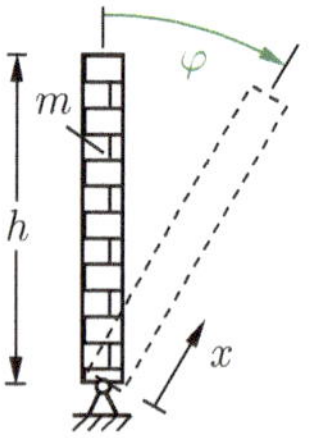

Assuming a hinged support at the base and a constant mass distribution, determine the stress resultants during the motion. At which point x_B and at which angle φ_B the maximum transmissible moment $M_B = mgh/100$ is exceeded? Calculate the normal force at this point.

Solution We first determine the angular acceleration $\ddot{\varphi}$ and angular velocity $\dot{\varphi}$ of the chimney. With $\Theta_A = mh^2/3$, the principle of angular momentum leads to

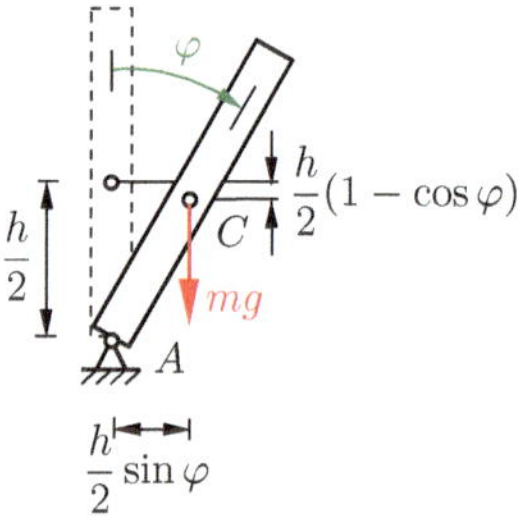

$$\overset{\curvearrowright}{A}: \ \Theta_A\,\ddot{\varphi} = mg\,\frac{h}{2}\sin\varphi \ \leadsto\ \ddot{\varphi} = \frac{3}{2}\frac{g}{h}\sin\varphi\,.$$

To calculate $\dot{\varphi}$ it is practical to use the energy conservation law:

$$-mg\frac{h}{2}(1-\cos\varphi) + \frac{1}{2}\,\Theta_A\dot{\varphi}^2 = 0 \quad\leadsto\quad \dot{\varphi}^2 = 3\,\frac{g}{h}\,(1-\cos\varphi)\,.$$

Now we cut the system, introduce the stress resultants and formulate the equations of motion for the cut part:

$$\nwarrow:\quad \bar{m}\ddot{y}_{\bar{C}} = V - \bar{m}g\sin\varphi\,,$$

$$\nearrow:\quad \bar{m}\ddot{x}_{\bar{C}} = -N - \bar{m}g\cos\varphi\,,$$

$$\overset{\curvearrowright}{\bar{C}}:\quad \Theta_{\bar{C}}\,\ddot{\varphi} = M + \frac{1}{2}(h-x)V\,.$$

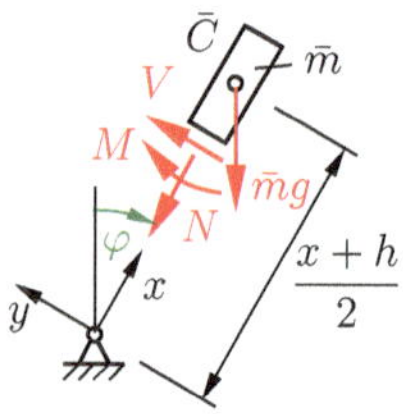

The unknown accelerations $\ddot{x}_{\bar{C}}, \ddot{y}_{\bar{C}}$ therein can be calculated from the known quantities $\ddot{\varphi}$ and $\dot{\varphi}^2$:

$$\ddot{x}_{\bar{C}} = -x_{\bar{C}}\,\dot{\varphi}^2 = -\frac{h+x}{2}\,3\,\frac{g}{h}(1-\cos\varphi) = -\frac{3}{2}\,g\left(1+\frac{x}{h}\right)(1-\cos\varphi)\,,$$

$$\ddot{y}_{\bar{C}} = -x_{\bar{C}}\,\ddot{\varphi} = -\frac{h+x}{2}\,\frac{3}{2}\,\frac{g}{h}\sin\varphi = -\frac{3}{4}\,g\left(1+\frac{x}{h}\right)\sin\varphi\,.$$

Thus, with

$$\bar{m} = \frac{h-x}{h}\,x = \left(1-\frac{x}{h}\right)m\,, \qquad \Theta_{\bar{C}} = \frac{\bar{m}(h-x)^2}{12} = \frac{mh^2}{12}\left(1-\frac{x}{h}\right)^3,$$

we obtain the stress resultants

$$\underline{\underline{V(x)}} = \bar{m}\,(\ddot{y}_{\bar{C}} + g\sin\varphi) = mg\left(1-\frac{x}{h}\right)\left[-\frac{3}{4}\left(1+\frac{x}{h}\right)\sin\varphi + \sin\varphi\right]$$

$$= \underline{\underline{\frac{mg}{4}\sin\varphi\left(1-\frac{x}{h}\right)\left(1-3\frac{x}{h}\right)}}\,,$$

$$\underline{\underline{N(x)}} = -\bar{m}\,(\ddot{x}_{\bar{C}} + g\cos\varphi)$$

$$= \underline{\underline{-mg\left(1-\frac{x}{h}\right)\left[-\frac{3}{2}\left(1+\frac{x}{h}\right)(1-\cos\varphi)+\cos\varphi\right]}}\,,$$

$$\underline{\underline{M(x)}} = \Theta_{\bar{C}}\,\ddot{\varphi} - \frac{h}{2}\left(1-\frac{x}{h}\right)[\,\bar{m}\ddot{y}_{\bar{C}} + \bar{m}g\sin\varphi\,]$$

$$= \frac{mh^2}{12}\left(1-\frac{x}{h}\right)^3\frac{3}{2}\frac{g}{h}\sin\varphi - \frac{h}{2}mg\left(1-\frac{x}{h}\right)^2\left[-\frac{3}{4}\left(1+\frac{x}{h}\right)+1\right]\sin\varphi$$

$$= \underline{\underline{\frac{mgh}{4}\sin\varphi\left(1-\frac{x}{h}\right)^2\frac{x}{h}}}\,.$$

The bending moment has its maximum, where the shear force is zero. From $V = 0$ follows

$$\left(1-\frac{x}{h}\right)\left(1-3\frac{x}{h}\right) = 0 \qquad \leadsto \qquad \underline{\underline{x_B = \frac{h}{3}}}$$

(The 2nd solution $x = h$ is not of interest because the bending moment is zero at this point). This leads to the maximum bending moment

$$M_{max} = M(x_B) = \frac{1}{27}\,mgh\sin\varphi\ .$$

From the condition $M_{max} = M_B$ the angle φ_B follows:

$$\frac{mgh}{27}\sin\varphi_B = \frac{mgh}{100} \quad \leadsto \quad \sin\varphi_B = \frac{27}{100} \quad \leadsto \quad \underline{\underline{\varphi_B = 15.66^\circ}}\,.$$

Finally, the normal force at x_B and at the angle φ_B is obtained as

$$\underline{\underline{N\left(\frac{h}{3},\varphi_B\right)}} = -\frac{2mg}{3}\left[-\frac{3}{2}\frac{4}{3}(1-\cos\varphi_B)+\cos\varphi_B\right]$$

$$= -\frac{2}{3}mg\,(-2+3\cos\varphi_B) = \underline{\underline{-0.59\,mg}}\ .$$

P5.40 **Problem 5.40** A homogeneous arc of mass m rotates with the angular velocity ω and angular acceleration $\dot{\omega}$ about the vertical axis.

Calculate the bending moments and shear forces at location A.

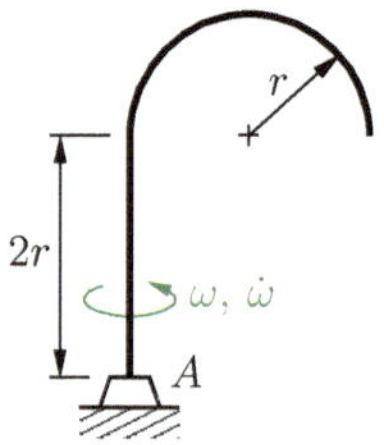

Solution We introduce body fixed coordinates ξ, η, ζ and cut the system at A. Since the center of mass C rotates along a circular path, its acceleration components are

$$a_\xi = -\eta_S \dot{\omega}\ , \qquad a_\eta = -\eta_S \omega^2\ .$$

Using the density ρ and the cross section as auxiliary quantities its distance η_C from the ζ-axis follows as

$$\eta_C = \frac{r(\pi r \rho A)}{2r\rho A + \pi r \rho A} = \frac{\pi}{2+\pi}\, r\ .$$

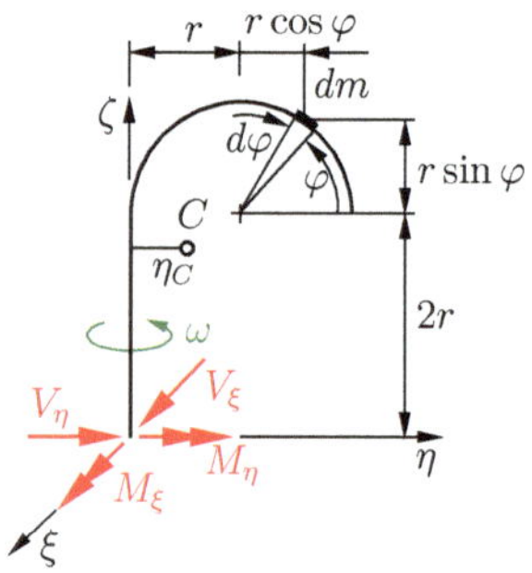

Therewith, the principle of linear momentum yields the shear forces

$$\underline{\underline{V_\xi}} = m a_\xi = \underline{\underline{-mr\dot{\omega}\,\frac{\pi}{2+\pi}}}\ , \qquad \underline{\underline{V_\eta}} = m a_\eta = \underline{\underline{-mr\omega^2\,\frac{\pi}{2+\pi}}}\ .$$

The bending moments are given by the principle of angular momentum:

$$M_\xi = \dot{\omega}\Theta_{\xi\zeta} - \omega^2\Theta_{\eta\zeta}\ , \qquad M_\eta = \dot{\omega}\Theta_{\eta\zeta} + \omega^2\Theta_{\xi\zeta}\ .$$

With $\mathrm{d}m = \rho A r \mathrm{d}\varphi = m\mathrm{d}\varphi/(2+\pi)$, we obtain $\Theta_{\xi\zeta} = 0$ and

$$\Theta_{\eta\zeta} = -\int \eta\zeta \mathrm{d}m = -\frac{mr^2}{2+\pi}\int_0^\pi (1+\cos\varphi)(2+\sin\varphi)\mathrm{d}\varphi$$
$$= -\frac{2(1+\pi)}{2+\pi}\, mr^2\ .$$

Thus, it follows

$$\underline{\underline{M_\xi = \frac{2(1+\pi)}{2+\pi}\, mr^2\omega^2}}\ , \qquad \underline{\underline{M_\eta = -\frac{2(1+\pi)}{2+\pi}\, mr^2\dot{\omega}}}\ .$$

Chapter 6

Impact

6

Impact: A sudden collision of two bodies leading to a sudden change of their motion is called *Impact.*

Notation:

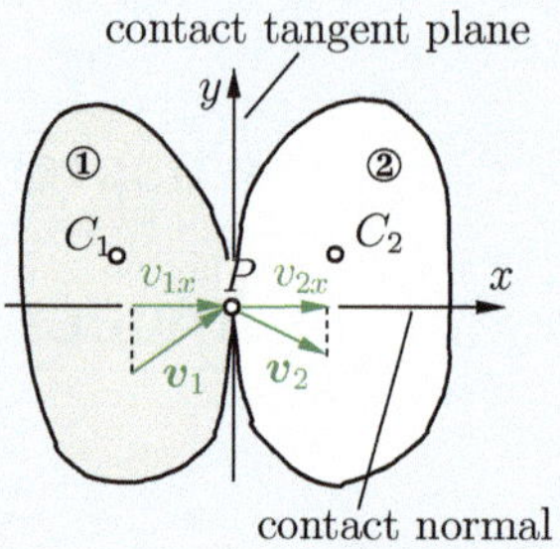

$\boldsymbol{v}_1$, $\boldsymbol{v}_2$ $\widehat{=}$ velocities at contact points P before impact,

v_{1x}, v_{2x} $\widehat{=}$ velocity components at points P in direction of the contact normal before impact,

$\overline{v}_{1x}$, $\overline{v}_{2x}$ $\widehat{=}$ velocity components at points P in direction of the contact normal after impact,

$\widehat{F}_x = \int\limits_0^{t^*} F_x(\bar{t})\,\mathrm{d}\bar{t}$ $\widehat{=}$ linear impulse.

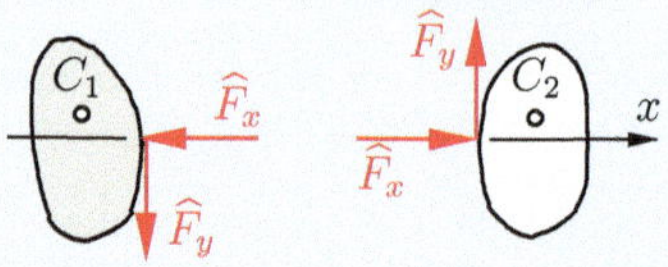

Assumptions:

- The impact duration t^* is neglegible small.
- Position changes of the bodies during impact are neglegible small.
- Other forces (e.g. weight) are small compared with the impulsive forces at the contact point and can be neglected during impact.

Impact Hypothesis:

$$e = -\frac{\overline{v}_{1x} - \overline{v}_{2x}}{v_{1x} - v_{2x}} = -\frac{\text{relative separation velocity}}{\text{relative approach velocity}}$$

where $e \;\widehat{=}$ coefficient of restitution: $0 \le e \le 1$

$e = 1$: *elastic impact* (no energy loss),

$e = 0$: *plastic impact* ($\overline{v}_{2x} = \overline{v}_{1x}$, no separation of bodies).

Solution of plane impact problems: Application of the principles of linear and angular impulse and momentum to each body (see p.102):

$$m(\overline{v}_{Cx} - v_{Cx}) = \sum \widehat{F}_x\,, \qquad m(\overline{v}_{Cy} - v_{Cy}) = \sum \widehat{F}_y\,,$$

$$\Theta_A(\overline{\omega} - \omega) = \sum \widehat{M}_A$$

$A \mathrel{\widehat{=}}$ fixed point or center of mass C. The equations are complemented by the impact hypothesis and the kinematic relations between the velocities at the contact points and the velocities of the centers of mass and the angular velocities.

Remarks:

- If the bodies are *smooth*, the direction of linear impulse (due to contact) coincides with that of the contact normal.
- If the bodies are sufficiently *rough* (no slip during contact), the velocity components in tangential direction at the contact points after impact are equal.

Impact		Remarks
direct	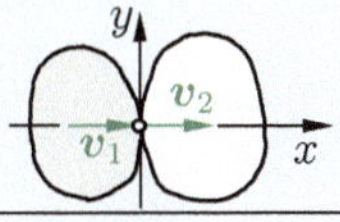	$\boldsymbol{v}_1$, $\boldsymbol{v}_2$ in direction of contact normal, impulse always in direction of contact normal ($\widehat{F}_y = 0$).
oblique	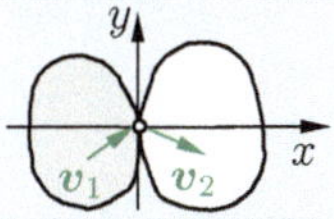	$\boldsymbol{v}_1$, $\boldsymbol{v}_2$ not in direction of contact normal.
central	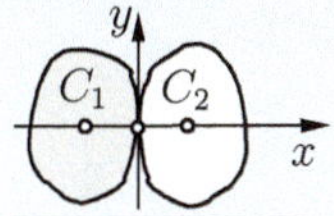	centers of mass located on contact normal.
eccentric	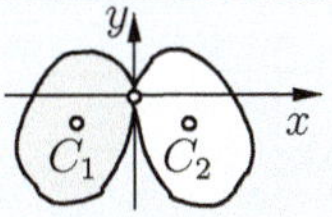	centers of mass not located on contact normal.

Direct central impact:

$$\bar{v}_1 = \frac{m_1 v_1 + m_2 v_2 - e\, m_2 (v_1 - v_2)}{m_1 + m_2},$$

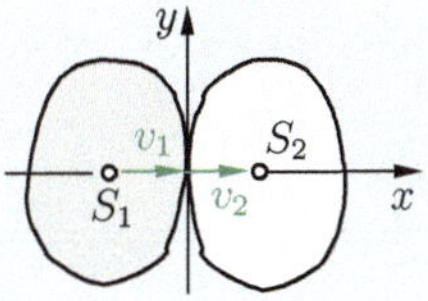

$$\bar{v}_2 = \frac{m_1 v_1 + m_2 v_2 + e\, m_1 (v_1 - v_2)}{m_1 + m_2},$$

$$\Delta T = \frac{1 - e^2}{2} \frac{m_1 m_2}{m_1 + m_2} (v_1 - v_2)^2 \;=\; \text{energy loss.}$$

P6.1 **Problem 6.1** A point mass m_1, suspended from a wire, is released from rest at height h_1 and collides at A with a mass point $m_2 = 2m_1$ which initially is at rest. After impact mass m_1 swings back to a height of $h_1/2$.

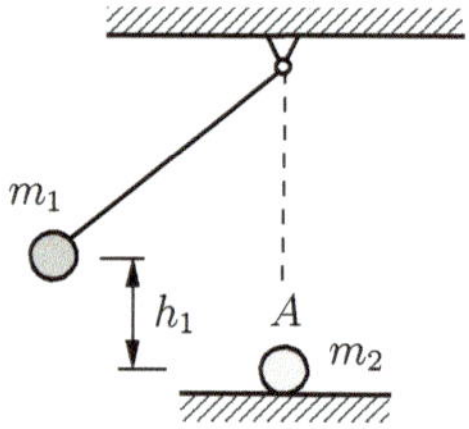

Determine the coefficient of restitution e and the velocity of m_2 immediately after impact.

Solution The velocities of m_1 and m_2 immediately before impact are

$$v_1 = \sqrt{2gh_1}\,, \qquad v_2 = 0\,.$$

Thus, denoting the velocities after impact by $\bar{v}_1$, $\bar{v}_2$, the principle of linear impulse yield

$$① \quad \rightarrow: \quad m_1(\bar{v}_1 - v_1) = -\widehat{F}\,,$$

$$② \quad \rightarrow: \quad m_2\bar{v}_2 = +\widehat{F}$$

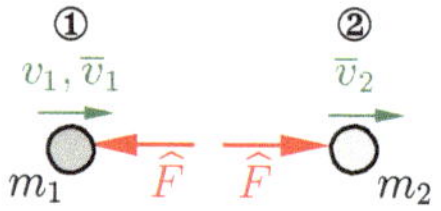

Eliminating $\widehat{F}$ and using the impact hypothesis

$$e = -\frac{\bar{v}_1 - \bar{v}_2}{v_1}$$

leads to

$$\bar{v}_1 = v_1\,\frac{m_1 - e\,m_2}{m_1 + m_2}\,, \qquad \bar{v}_2 = v_1(1+e)\,\frac{m_1}{m_1 + m_2}\,.$$

The velocity $\bar{v}_1$ can be determined from the energy conservation law for m_1 after impact:

$$\frac{1}{2}m_1\bar{v}_1^2 = m_1 g\frac{h_1}{2} \quad \rightsquigarrow \quad \bar{v}_1 = \sqrt{gh_1} = \frac{v_1}{\sqrt{2}}\,.$$

Thus, from the two equations above we obtain

$$\underline{\underline{e = \frac{m_1}{m_2} - \frac{m_1 + m_2}{\sqrt{2}\,m_2}}}\,,$$

$$\underline{\underline{\bar{v}_2 = v_1\,\frac{m_1}{m_2}\left(1 - \frac{1}{\sqrt{2}}\right)}}\,.$$

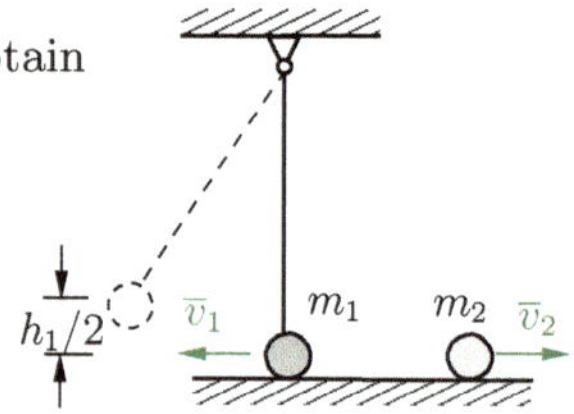

Problem 6.2 A ball is dropped from height h_0 onto a flat surface and bounces back five times (coefficient of restitution $e = 0.85$). **P6.2**

a) Determine the height h_5 the ball reaches after the last impact.

b) Determine the energy loss for each impact.

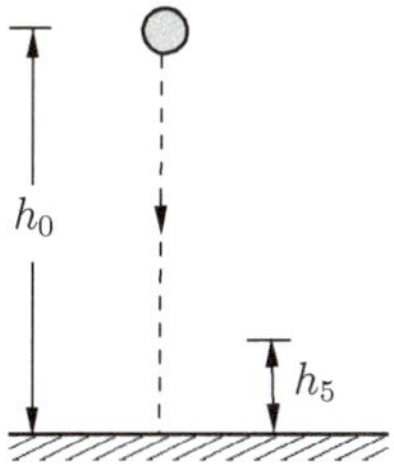

Solution **a)** If the positive direction of velocity is chosen upwards, the ball has immediately before the 1st hit the velocity

$$v_1 = -\sqrt{2gh_0} \ .$$

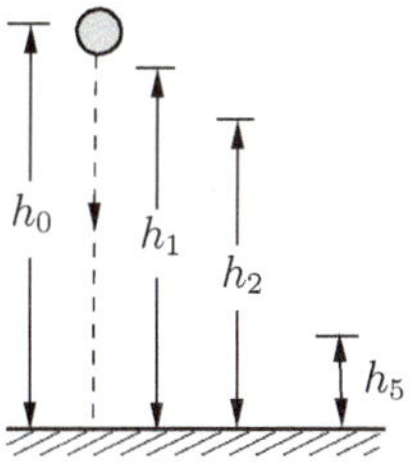

Denoting the velocity immediately after impact by $\overline{v}_1$, the impact hypothesis leads to

$$e = -\frac{\overline{v}}{v} \quad \rightsquigarrow \quad \overline{v}_1 = -e\,v_1 = e\sqrt{2gh_0} \,.$$

From the energy conservation law follows the new height, reached after the first hit:

$$\frac{1}{2}\,m\overline{v}_1^2 = mgh_1 \qquad \rightsquigarrow \qquad h_1 = \frac{\overline{v}_1^2}{2g} = e^2 h_0 \ .$$

In the same way we find for all further hits

$$h_i = e^2 h_{i-1} \,,$$

and therefore

$$\underline{\underline{h_5}} = e^2 h_4 = e^4 h_3 = \ldots = e^{10} h_0 = \underline{\underline{0.197\, h_0}} \ .$$

b) The energy loss during an impact can be determined from the difference in potential energy calculated from the heights before and after impact. Choosing zero potential at the surface we obtain with $V_{i-1} = mgh_{i-1}$ and $V_i = mgh_i$ (the kinetic energy is zero when reaching the heights)

$$\Delta V_i = V_{i-1} - V_i = mg(h_{i-1} - h_i) = (1 - e^2)\, mgh_{i-1} \,.$$

Remark: The energy loss can alternatively be determined from the difference in kinetic energies. Note that the energy loss decreases with the number of hits.

P6.3 **Problem 6.3** At a quality test, balls ① fall from height H on a *rigid smooth* plate ② which is inclined by $\alpha = 20°$ against the horizontal.

Which distance d must have a collecting container ③, so that only the balls with a coefficient of restitution $e \geq 0.8$ reach the container?

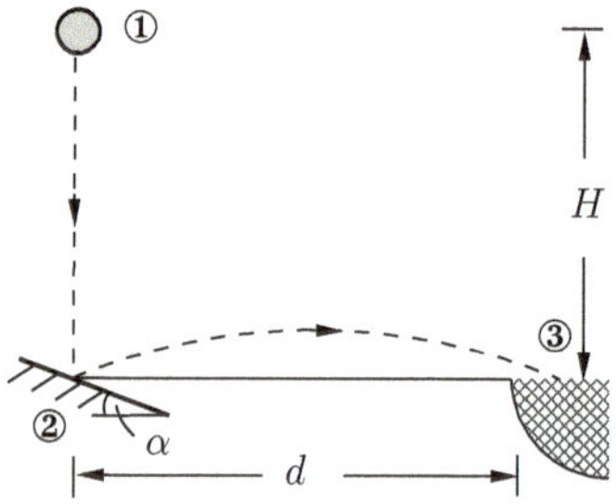

Solution That balls with $e \geq 0.8$ reach the container, the distance d must exactly be equal to the flight distance of balls with the coefficient of restitution $e = 0.8$. The balls ① hit plate ② with the velocity

$$v_1 = \sqrt{2gH}\,.$$

Using the displayed coordinate system ($x \widehat{=}$ contact normal), its components are

$$v_{1x} = -v_1 \cos\alpha\,,$$

$$v_{1y} = -v_1 \sin\alpha\,.$$

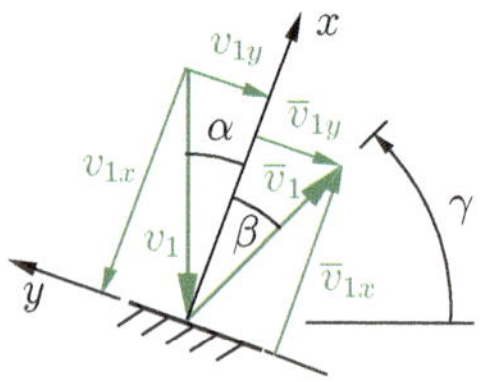

Since the impulse in y-direction is zero (smooth plate), $\overline{v}_{1y} = v_{1y}$ holds. The impact hypothesis leads then with $v_{2x} = \overline{v}_{2x} = 0$ (rigid plate) to

$$e = -\frac{\overline{v}_{1x}}{v_{1x}} \quad \leadsto \quad \overline{v}_{1x} = -e\,v_{1x} = e\,v_1 \cos\alpha\,.$$

Thus, for the limit case $e = 0,8$ follows for the angle β

$$\tan\beta = \frac{|\overline{v}_{1y}|}{|\overline{v}_{1x}|} = \frac{|v_{1y}|}{e\,v_{1x}} = \frac{1}{e}\tan\alpha \quad \leadsto \quad \beta = 24.46°$$

and for the launching angle γ to the horizontal

$$\gamma = 90° - \alpha - \beta = 45.54°\,.$$

With the equations for the projectile motion (see page 32) we obtain the flight distance d of balls with $e = 0.8$:

$$\underline{\underline{d}} = \frac{1}{g}\,\overline{v}_1^2 \sin 2\gamma = \frac{1}{g}(\overline{v}_{1x}^2 + \overline{v}_{1y}^2)\sin 2\gamma$$

$$= 2H(e^2\cos^2\alpha + \sin^2\alpha)\sin 2\gamma = \underline{\underline{1.364\,H}}\,.$$

Problem 6.4 Two old cars (no ABS) move along the highway with same speed v_0 in a short distance b when a traffic jam gets in sight in distance c. Both cars make a full brake and get sliding, car ① a shock moment Δt later than car ②. **P6.4**

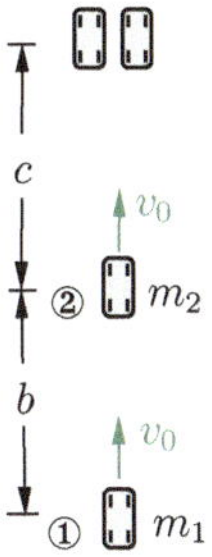

a) When and where both cars clash? Determine their velocities after the clash (coefficient of restitution e).
b) Does car ② come to stand ahead of the jam?

Given: $v_0 = 30\,\text{m/s}$ (=108 km/h), $m_1 = 1.5\,m$, $m_2 = m$, $\Delta t = 1\,\text{s}$, $\mu_1 = \mu_2 = 1/2$, $e = 0{,}8$, $b = 25\,\text{m}$, $c = 150\,\text{m}$. Neglect the lengths of the cars.

Solution **a)** We introduce the coordinate s and start counting time when car ② slames on brakes. During sliding the acceleration of both cars is the same: $a_1 = a_2 = -\mu g$. Considering the initial conditions and the shock delay, the velocities and covered distances are given by

$$v_2 = v_0 - \mu g t\,, \qquad s_2 = v_0 t - \tfrac{1}{2}\mu g t^2\,,$$
$$v_1 = v_0 - \mu g(t - \Delta t)\,, \qquad s_1 = v_0 t - \tfrac{1}{2}\mu g(t - \Delta t)^2 - b\,.$$

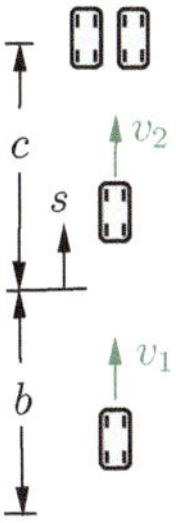

The collision time t^* and the associated position and velocities are found from the condition $s_1(t^*) = s_2(t^*)$:

$$\underline{\underline{t^*}} = \frac{b + \mu g(\Delta t)^2/2}{\mu g \Delta t} = \underline{\underline{5.6\,\text{s}}}\,,$$
$$\underline{\underline{s^*}} = s_2(t^*) = v_0 t^* - \tfrac{1}{2}\mu g t^{*2} = \underline{\underline{91.17\,\text{m}}}\,,$$
$$v_2^* = v_2(t^*) = v_0 - \mu g t^* = 2.56\,\text{m/s}\,,$$
$$v_1^* = v_1(t^*) = v_0 - \mu g(t^* - \Delta t) = 7.46\,\text{m/s}\,.$$

The velocities after collision are calculated by using the formulas on page 149 for central impact:

$$\underline{\underline{\overline{v}_2}} = \frac{m_1 v_1^* + m_2 v_2^* + e\, m_1(v_1^* - v_2^*)}{m_1 + m_2} = \underline{\underline{25.4\,\text{m/s}}}\,,$$
$$\underline{\underline{\overline{v}_1}} = \frac{m_1 v_1^* + m_2 v_2^* - e\, m_2(v_1^* - v_2^*)}{m_1 + m_2} = \underline{\underline{14.1\,\text{m/s}}}\,.$$

b) With $\overline{v}_2$, $\overline{v}_1$ and s^* the positions for standstill would be

$$s_2^{**} = s^* + \frac{\overline{v}_2^2}{2\mu g} = 65.8\,\text{m}\,, \qquad s_1^{**} = s^* + \frac{\overline{v}_1^2}{2\mu g} = 20.3\,\text{m}\,.$$

Because $s_2^{**} > c$, car ② would crash into the jam.

P6.5 **Problem 6.5** A bullet (mass m) with the velocity v hits centrally an initially resting frictionless supported board of mass M. It penetrates the board and has thereafter the velocity $\bar{v}$.

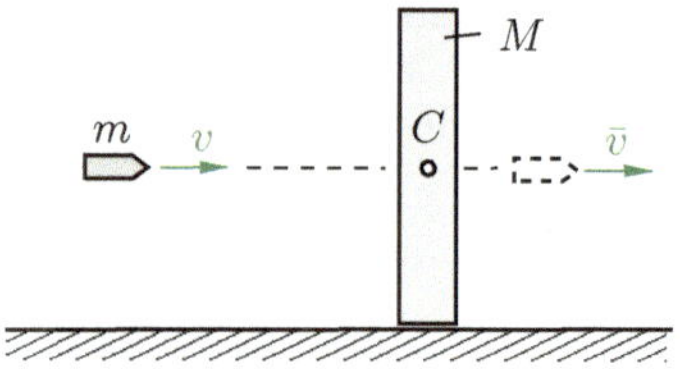

a) Determine the velocity $\bar{w}$ of the board after the penetrating shot.

b) How much energy is needed for producing the penetration hole?

Given: $M = 10\,m$, $\bar{v} = v/2$.

Solution **a)** The velocity $\bar{w}$ of the board is determined by using the conservation of linear momentum law with the known bullet speeds v and $\bar{v}$:

$$\rightarrow: \quad mv = m\bar{v} + M\,\bar{w}\,.$$

Solving for $\bar{w}$ leads to

$$\underline{\underline{\bar{w}}} = \frac{m}{M}(v - \bar{v}) = \frac{1}{10}\,\frac{v}{2} = \underline{\underline{\frac{v}{20}}}\,.$$

b) The energy needed for producing the penetration hole is calculated from the energy loss of the system. Since the potential energy does not change, only the kinetic energy before and after penetration must be considered:

$$T = \frac{1}{2}\,mv^2\,, \qquad \bar{T} = \frac{1}{2}\,m\bar{v}^2 + \frac{1}{2}\,M\bar{w}^2\,.$$

The difference yields the energy loss

$$\begin{aligned}
\underline{\underline{\Delta T}} = T - \bar{T} &= \frac{1}{2}\,m(v^2 - \bar{v}^2) - \frac{1}{2}\,M\bar{w}^2 \\
&= \frac{1}{2}\,m\Big[(v^2 - \bar{v}^2) - \frac{m}{M}(v - \bar{v})^2\Big] \\
&= \frac{1}{2}\,mv^2\Big[1 - \Big(\frac{\bar{v}}{v}\Big)^2 - \frac{m}{M}\Big(1 - \frac{\bar{v}}{v}\Big)^2\Big] = \frac{1}{2}\,mv^2\Big[1 - \frac{1}{4} - \frac{1}{10}\,\frac{1}{4}\Big] \\
&= \underline{\underline{\frac{29}{40}\Big(\frac{1}{2}\,mv^2\Big)}}\,.
\end{aligned}$$

Problem 6.6 Two cars (point masses m_1, m_2) collide at an intersection. From a radar measurement the velocity v_1 before collision and from the sliding track the velocity $\overline{v}_1$ and angle γ after collision are known. **P6.6**

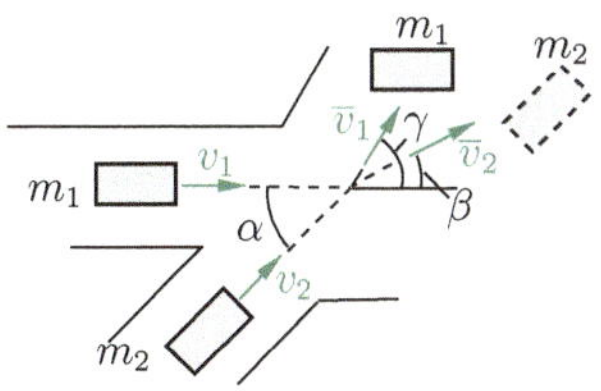

Determine the velocities v_2, $\overline{v}_2$ and the angle β of the 2nd car by assuming that no energy is absorbed during impact.

Given: $m_1 = m_2 = m$, v_1, $\overline{v}_1 = v_1/3$, $\alpha = 45°$, $\gamma = 60°$.

Solution Before and after collision the cars have the velocities v_1, v_2, $\overline{v}_1$ and $\overline{v}_2$. They are related by the conservation of linear momentum law

$$\rightarrow: \quad m_1 v_1 + m_2 v_2 \cos\alpha = m_1 \overline{v}_1 \cos\gamma + m_2 \overline{v}_2 \cos\beta \,,$$

$$\uparrow: \quad m_2 v_2 \sin\alpha = m_1 \overline{v}_1 \sin\gamma + m_2 \overline{v}_2 \sin\beta \,,$$

which, after inserting the given quantities, leads to

$$v_1 + v_2\sqrt{2}/2 = \overline{v}_1/2 + \overline{v}_2 \cos\beta \,,$$

$$v_2\sqrt{2}/2 = \overline{v}_1\sqrt{3}/2 + \overline{v}_2 \sin\beta \,.$$

Conservation of energy during impact requires

$$\frac{1}{2} m_1 v_1^2 + \frac{1}{2} m_2 v_2^2 = \frac{1}{2} m_1 \overline{v}_1^2 + \frac{1}{2} m_2 \overline{v}_2^2 \quad \rightsquigarrow \quad v_1^2 + v_2^2 = \overline{v}_1^2 + \overline{v}_2^2 \,.$$

Herewith we have three equations for the three unknowns v_2, $\overline{v}_2$ and β. To solve them, we first eliminate β:

$$\begin{aligned}
\overline{v}_2^2 &= \overline{v}_2^2(\cos^2\beta + \sin^2\beta) \\
&= (v_1^2 + v_2^2/2 + \overline{v}_1^2/4 + v_1 v_2\sqrt{2} - v_1\overline{v}_1 - v_2\overline{v}_1\sqrt{2}/2) \\
&\quad + (v_2^2/2 + \overline{v}_1^2 3/4 - v_2\overline{v}_1\sqrt{6}/2) \\
&= v_1^2 + v_2^2 + \overline{v}_1^2 + v_1 v_2\sqrt{2} - v_1\overline{v}_1 - v_2\overline{v}_1(1+\sqrt{3})\sqrt{2}/2
\end{aligned}$$

Introduction into energy conservation then leads to

$$\underline{\underline{v_2}} = \frac{\overline{v}_1(v_1 - 2\overline{v}_1)}{v_1\sqrt{2} - \overline{v}_1(1+\sqrt{3})\sqrt{2}/2} = \underline{\underline{0.144\, v_1}} \,.$$

$$\overline{v}_2^2 = v_1^2 + v_2^2 - \overline{v}_1^2 \quad \rightsquigarrow \quad \underline{\underline{\overline{v}_2 = 0.953\, v_1}} \,.$$

and subsequently to

$$\overline{v}_2 \sin\beta = v_2\sqrt{2}/2 - \overline{v}_1\sqrt{3}/2 \quad \rightsquigarrow \quad \sin\beta = -0.196 \quad \rightsquigarrow \quad \underline{\underline{\beta = -11.3°}} \,.$$

P6.7

Problem 6.7 A point mass (mass m_1) is shot upwards a rough inclined plane (coefficient of kinetic friction μ) with an initial velocity v_0. In point B it hits a resting point mass m_2, which after this 1st impact bounces against a spring (stiffness k) at the end of the smooth path $\overline{BC}$.

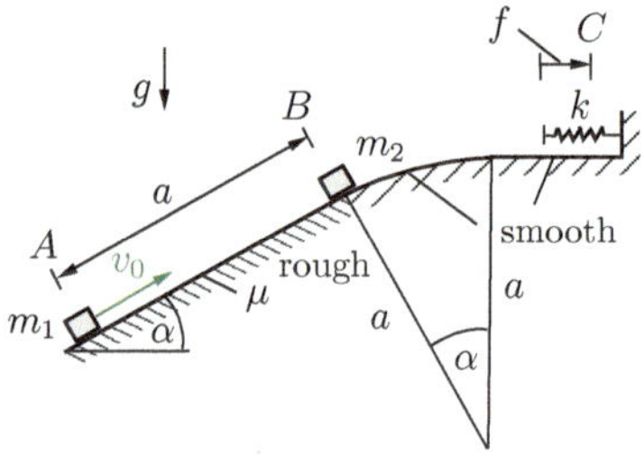

a) Determine the velocity v_1 of the point mass m_1 immediately before the impact.

b) Calculate the coefficient of restitution e if the maximum compression of the spring is f.

Given: $m_1 = m_2 = m = 0.1\,\text{kg}$, $g = 9.81\,\text{m/s}$, $\alpha = 30°$, $v_0 = 6\,\text{m/s}$, $\mu = 0.5$, $k = 400\,\text{N/m}$, $a = 1\,\text{m}$.

Solution **a)** Application of the work-energy theorem $V_A + T_A = V_B + T_B + U_{AB}$ to the mass m_1 between the points A and B yields with $R = \mu N = \mu m_1 g \cos\alpha$ and

$$V_A = 0, \quad T_A = \tfrac{1}{2} m_1 v_0^2, \quad U_{AB} = -R\,a = -\mu\, m_1 g a \cos\alpha$$
$$V_B = m_1 g a \sin\alpha, \quad T_B = \tfrac{1}{2} m_1 v_1^2$$

the velocity

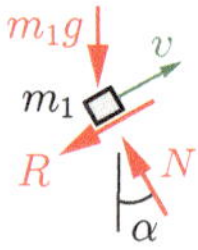

$$\underline{\underline{v_1 = \sqrt{v_0^2 - 2ga(\sin\alpha + \mu\cos\alpha)} = 4.21\,\text{m/s}}}\,.$$

b) The velocity of m_2 immediately after impact follows with $v_2 = 0$ and $m_1 = m_2$ as

$$\bar{v}_2 = \frac{m_1 v_1 + m_2 v_2 + e m_1 (v_1 - v_2)}{m_1 + m_2} = \frac{1+e}{2}\, v_1\,. \qquad (a)$$

Now we apply the energy conservation law $V_B + T_B = V_C + T_C$ to m_2 between the points B and C. With

$$V_B = 0, \quad T_B = \tfrac{1}{2} m_2 \bar{v}_2^2, \quad V_C = m_2 g a (1 - \cos\alpha) + \tfrac{1}{2} c f^2, \quad T_C = 0$$

follows

$$\bar{v}_2 = \sqrt{\frac{c}{m_2} f^2 + 2ga(1 - \cos\alpha)} = 3.55\,\text{m/s}\,.$$

Introduction into (a) finally leads to

$$\underline{\underline{e = 2\,\frac{\bar{v}_2}{v_1} - 1}} = 2\,\frac{3.55}{4.21} - 1 = \underline{\underline{0.69}}\,.$$

Problem 6.8 A bowling ball ① hits with speed v_1 the pin ②. Assume that all surfaces are smooth and that the impact is partially elastic (coefficient of restitution e). **P6.8**

Determine the velocities of pin and ball immediately after impact.

Given: $v_1 = 7\,\mathrm{m/s}$, $m_1 = 4.5\,\mathrm{kg}$, $m_2 = 1.5\,\mathrm{kg}$, $\Theta_c = 0.013\,m_2\,\mathrm{m}^2$, $\alpha = 10°$, $c = 0.02\,\mathrm{m}$, $e = 0.9$.

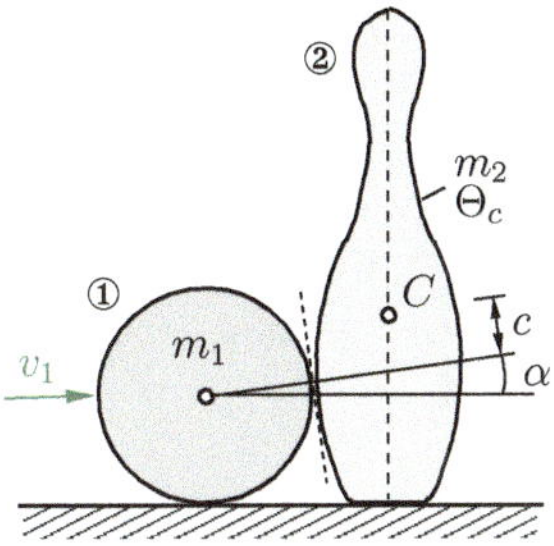

Solution The impact is eccentric. Since the surfaces are smooth, the linear impulse $\widehat{F}$ acts in the line of impact and thus, the impulse laws lead to

$$① \rightarrow: \quad m_1(\overline{v}_1 - v_1) = -\widehat{F}\cos\alpha\,,$$

$$② \nearrow: \quad m_2\overline{v}_2 = \widehat{F}\,,$$

$$\overset{\curvearrowleft}{C}: \quad \Theta_c\overline{\omega}_2 = c\,\widehat{F}\,.$$

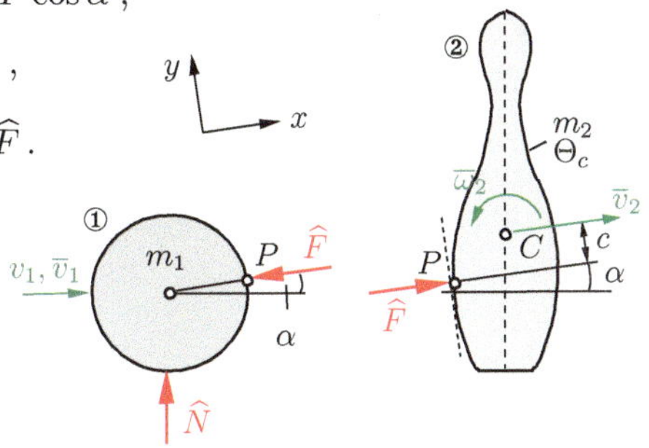

Note that the weights can be neglected during impact and that the ball cannot move vertically. Using the impact hypothesis

$$e = -\frac{\overline{v}^P_{1x} - \overline{v}^P_{2x}}{v^P_{1x} - v^P_{2x}}$$

and

$$v^P_{1x} = v_1\cos\alpha\,, \qquad v^P_{2x} = 0\,, \qquad \overline{v}^P_{1x} = \overline{v}_1\cos\alpha\,, \qquad \overline{v}^P_{2x} = \overline{v}_2 + c\,\overline{\omega}_2$$

we obtain by solving for $\widehat{F}$

$$\widehat{F} = (1+e)\,\frac{m_2 v_1\cos\alpha}{1 + m_2/m_1 + c^2 m_2/\Theta_c} = 14.41\,\mathrm{kgm/s^2}\,.$$

Thus, the velocities follow as

$$\underline{\underline{\overline{v}_1}} = v_1 - \frac{\widehat{F}}{m_1}\cos\alpha = \underline{\underline{3.8\,\mathrm{m/s}}}\,, \qquad \underline{\underline{\overline{v}_2}} = \frac{\widehat{F}}{m_2} = \underline{\underline{9.6\,\mathrm{m/s}}}\,.$$

Note that the pin after impact has an angular velocity:

$$\overline{\omega}_2 = c\,\widehat{F}/\Theta_c = 13.2\ 1/\mathrm{s}\,.$$

P6.9 **Problem 6.9** A point mass m_1 impinges with the velocity v_1 eccentrically a homogeneous bar (mass m_2), which initially rests on a *smooth* plane.

Determine the velocities of the point mass and the bar after an ideal elastic impact.

Given.: $m_2 = 2m_1$, $b = l/4$.

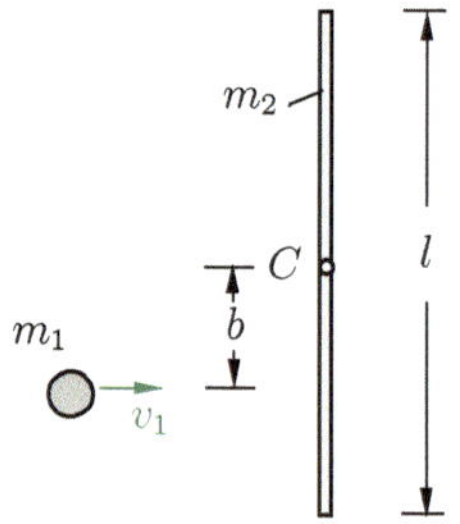

Solution The principle of impulse an momentum applied to the point mass ① and the bar ② yield (with $v_{2C} = 0$, $\omega_2 = 0$)

$$① \quad \rightarrow: \quad m_1(\overline{v}_1 - v_1) = -\widehat{F}\,,$$

$$② \quad \rightarrow: \quad m_2\,\overline{v}_{2C} = \widehat{F}\,,$$

$$\overset{\curvearrowleft}{C}: \quad \Theta_C\overline{\omega}_2 = b\widehat{F}\,.$$

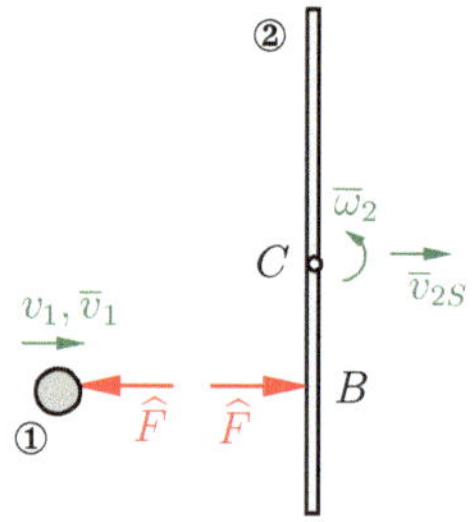

These equations are complemented by the impact hypothesis

$$e = -\frac{\overline{v}_1 - \overline{v}_{2B}}{v_1} = 1$$

and the kinematic relation

$$\overline{v}_{2B} = \overline{v}_{2C} + b\,\overline{\omega}_2\,.$$

Thus, we have five equations for the five unknowns ($\overline{v}_1$, $\overline{v}_{2C}$, $\overline{v}_{2B}$, $\overline{\omega}_2$, $\widehat{F}$). With $\Theta_C = m_2 l^2/12$ we obtain

$$\underline{\underline{\overline{v}_1 = v_1\,\frac{\dfrac{m_1}{m_2}\left(1+\dfrac{12b^2}{l^2}\right)-1}{\dfrac{m_1}{m_2}\left(1+\dfrac{12b^2}{l^2}\right)+1} = -\frac{v_1}{15}}}\,,$$

$$\underline{\underline{\overline{v}_{2C} = \frac{m_1}{m_2}(v_1 - \overline{v}_1) = \frac{8}{15}\,v_1}}\,, \qquad \underline{\underline{\overline{\omega}_2 = \frac{b}{\Theta_C}\,m_2\overline{v}_{2C} = \frac{8}{5}\,\frac{v_1}{l}}}\,.$$

Remark: If the impact is *purely elastic* the impact hypothesis can be replaced by the energy conservation law. In our case it reads

$$\frac{1}{2}\,m_1 v_1^2 = \frac{1}{2}\,m_1\overline{v}_1^2 + \frac{1}{2}\,m_2\overline{v}_{2C}^2 + \frac{1}{2}\,\Theta_C\overline{\omega}_2^2\,.$$

Problem 6.10 A homogeneous angle-shaped body (mass m, dimensions a, $t \ll a$) slides along a smooth plane and impinges with its edge *plastically* against an obstacle H. **P6.10**

Calculate the minimum impact speed v, so that the body tilts over.

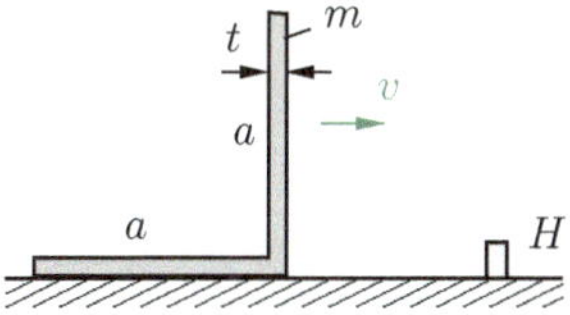

Solution Since the impact at H is *plastic*, the edge of the angle does not separate from H. Therefore, tilting can be regarded as a pure rotation about the fixed point H. With the distance $a/4$ of the center of mass, the moment of angular momentum L_H before impact (= moment of linear momentum with respect to H) and $\overline{L}_H$ after impact are given by

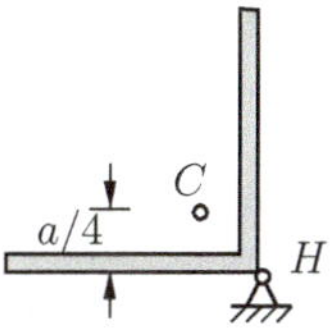

$$L_H = \frac{a}{4}(mv)\,, \qquad \overline{L}_H = \Theta_H \overline{\omega}$$

where

$$\Theta_H = 2\left(\frac{m}{2}\frac{a^2}{3}\right) = \frac{ma^2}{3}\,.$$

Since there acts no angular impulse with respect to H (weight can be disregarded during impact), the moment of momentum is conserved:

$$L_H = \overline{L}_H \quad \rightsquigarrow \quad \frac{a}{4}mv = \frac{ma^2}{3}\overline{\omega} \quad \rightsquigarrow \quad \overline{\omega} = \frac{3}{4}\frac{v}{a}\,.$$

The body can only tilt over, if the center of mass reaches the highest possible position. The minimum velocity required for that follows from energy conservation, applied to the motion after impact:

$$\frac{1}{2}\Theta_H \overline{\omega}^2_{\min} = mg\left(\sqrt{2}\frac{a}{4} - \frac{a}{4}\right)$$

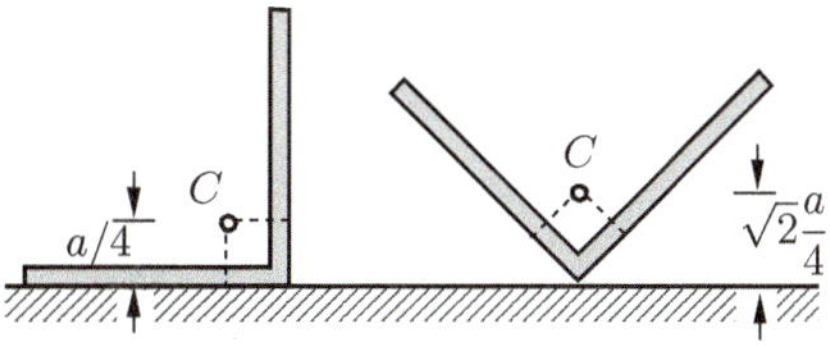

This finally leads to

$$\underline{\underline{v_{\min} = \sqrt{\frac{8}{3}(\sqrt{2}-1)ga} = 1.05\sqrt{ga}\,.}}$$

P6.11 **Problem 6.11** At a hammer mill, the thumb C of the rotating flywheel (moment of inertia Θ_A, angular velocity ω_1) hits the resting hammer (moment of inertia Θ_B) which is pivoted at B.

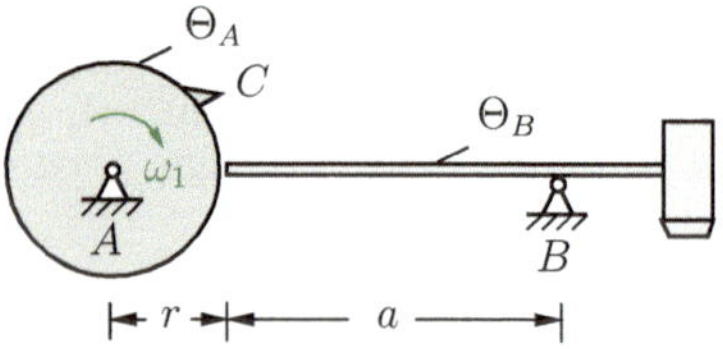

a) Determine the angular velocities $\overline{\omega}_1$ of the flywheel and $\overline{\omega}_2$ of the hammer shortly after impact (coefficient of restitution e)

b) Calculate the energy loss for the case $\Theta_B/\Theta_A = a^2/r^2$.

Solution **a)** The principles of angular impulse and momentum for ① and ② yield

$$\overset{\curvearrowright}{A}: \quad \Theta_A(\overline{\omega}_1 - \omega_1) = -r\widehat{F}\,,$$

$$\overset{\curvearrowleft}{B}: \quad \Theta_B\overline{\omega}_2 = a\widehat{F}\,.$$

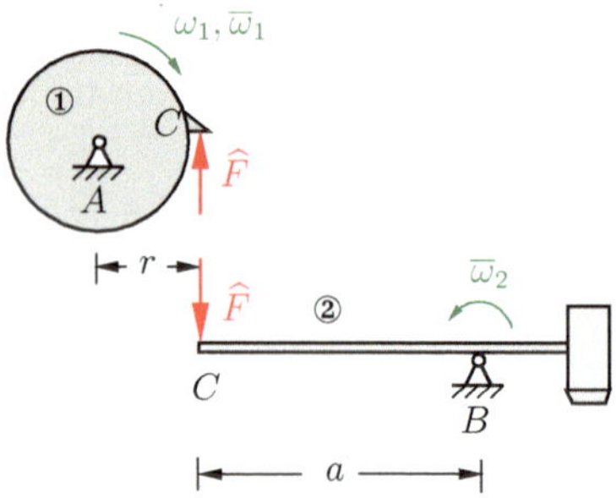

With the impact hypothesis

$$e = -\frac{\overline{v}_{1C} - \overline{v}_{2C}}{v_{1C} - v_{2C}}$$

and the kinematic relations

$$v_{1C} = r\,\omega_1\,, \qquad \overline{v}_{1C} = r\overline{\omega}_1\,, \qquad v_{2C} = 0\,, \qquad \overline{v}_{2C} = a\,\overline{\omega}_2$$

we obtain

$$\underline{\underline{\overline{\omega}_1 = \omega_1 \frac{1 - e\dfrac{r^2\Theta_B}{a^2\Theta_A}}{1 + \dfrac{r^2\Theta_B}{a^2\Theta_A}}}}\,, \qquad \underline{\underline{\overline{\omega}_2 = \omega_1 \frac{(1+e)\dfrac{r}{a}}{1 + \dfrac{r^2\Theta_B}{a^2\Theta_A}}}}\,.$$

b) The energy loss is calculated from the difference of kinetic energies before and after impact. With $r^2\Theta_B = a^2\Theta_A$ follows

$$\underline{\underline{\Delta T}} = \frac{1}{2}\Theta_A\omega_1^2 - \left[\frac{1}{2}\Theta_A\overline{\omega}_1^2 + \frac{1}{2}\Theta_B\overline{\omega}_2^2\right]$$

$$= \frac{1}{8}\Theta_A\omega_1^2[4 - (1-e)^2 - (1+e)^2] = \underline{\underline{\frac{1}{4}(1-e^2)\Theta_A\omega_1^2}}\,.$$

Remark: In case of an ideal elastic impact ($e = 1$) we would obtain for the given datai $\overline{\omega}_1 = 0$ and $\overline{\omega}_2 = \omega_1\,r/a$.

P6.12

Problem 6.12 A springy pivoted hammer (torsional spring constant c_T) can rotate in a horizontal plane. It consists of a homogeneous bar of mass m_2 and the hammer head (point mass m_3). The hammer is hit by a ball of mass m_1.

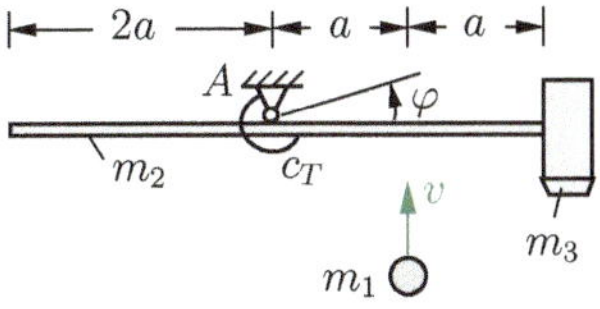

Determine the required speed v of the ball so that the hammer for a given coefficient of restitution e just reaches a maximum angle φ_0.

Given: $m_1 = m_2 = m_3 = m$, a, c_T, $e = \frac{1}{2}$.

Solution Since the hammer initially is at rest, the angular momentum of the system with respect to the fixed point A before impact is given by the moment of linear momentum $L_A = a\,m_1 v$. Immediately after impact the hammer (moment of inertia Θ_A) has the angular velocity $\overline{\omega}$ and the ball (mass m_1) the velocity $\overline{v}$. Thus, conservation of angular momentum leads to

$$\stackrel{\curvearrowleft}{A}: \quad a\,m_1 v = \Theta_A\,\overline{\omega} + a\,m_1\overline{v}\,.$$

With the impact hypothesis

$$e = \frac{\overline{\omega} a - \overline{v}}{v}$$

follows

$$\overline{\omega} = \frac{m_1(1+e)a}{\Theta_A + m_1 a^2}\,v \quad \text{where} \quad \Theta_A = \frac{m_2(4a)^2}{12} + (2a)^2 m_3 = \frac{16}{3}\,ma^2\,.$$

Immediately after impact (no spring deformation) the energy of the hammer-spring system is solely given by its kinetic energy $\frac{1}{2}\Theta_A\overline{\omega}^2$. At maximum angle φ_0 the potential energy is just $\frac{1}{2}c_T\varphi_0^2$ while the kinetic energy is zero. Thus, from the energy conservation law after introducing $\overline{\omega}$ and the given data we obtain the required velocity:

$$\frac{1}{2}\,\Theta_A\overline{\omega}^2 = \frac{1}{2}\,c_T\varphi_0^2 \qquad \rightsquigarrow \qquad \overline{\omega} = \sqrt{\frac{c_T}{\Theta_A}}\,\varphi_0$$

$$\rightsquigarrow \quad \underline{\underline{v = \sqrt{\frac{c_T}{\Theta_A}}\,\frac{\Theta_A + m_1 a^2}{m_1(1+e)a}\,\varphi_0 = \frac{19}{18}\sqrt{\frac{3c_T}{m}}\,\varphi_0\,.}}$$

P6.13 **Problem 6.13** A soccer ball (mass m, moment of inertia Θ_C) hits with speed v_0 horizontally the rough post of the goal. The impact (coefficient of restitution e) is *central* under the angle α to the goal line.

Determine the required spin (angular velocity ω_0) such that the ball crosses after impact the goal line if during impact there is *no slip* (static friction!).

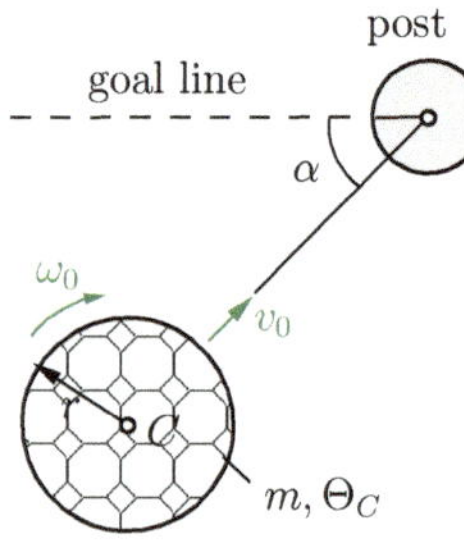

Solution The principles of impulse and momentum, applied to the ball, yield

$$\nearrow: \quad m(\overline{v}_x - v_0) = -\widehat{F}\,,$$

$$\nwarrow: \quad m\overline{v}_y = \widehat{H}\,,$$

$$\overset{\curvearrowright}{C}: \quad \Theta_C(\overline{\omega} - \omega_0) = -r\widehat{H}\,.$$

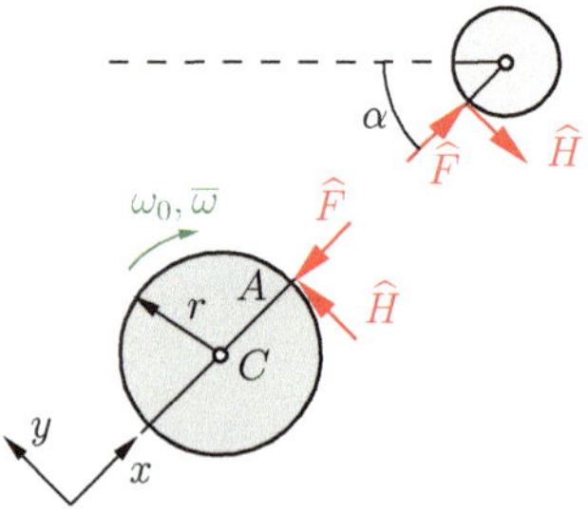

With the impact hypothesis (rigid post)

$$e = -\frac{\overline{v}_x}{v_0}$$

and the condition for static friction (velocity of contact point A in y-direction is zero after impact)

$$\overline{v}_{Ay} = \overline{v}_y - r\,\overline{\omega} = 0$$

we obtain for $\overline{v}_x$ and $\overline{v}_y$

$$\overline{v}_x = -e\,v_0\,, \qquad \overline{v}_y = \frac{r\omega_0}{1 + \dfrac{r^2 m}{\Theta_C}}\,.$$

The ball only crosses the goal line if

$$\overline{v}_\xi = \overline{v}_x \sin\alpha + \overline{v}_y \cos\alpha > 0\,.$$

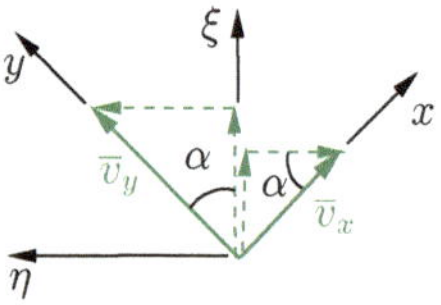

This leads for the require spin to

$$\underline{\underline{\omega_0 > \frac{e v_0}{r}\left(1 + \frac{r^2 m}{\Theta_C}\right)\tan\alpha\,.}}$$

Problem 6.14 The motion of a homogeneous door (mass m, width b, thickness $t \ll b$) is limited by a stopper B. **P6.14**

In which distance c from the door hinge A the stopper must be fixed, that the impulse $\widehat{A}$ at the hinge is zero during impact?

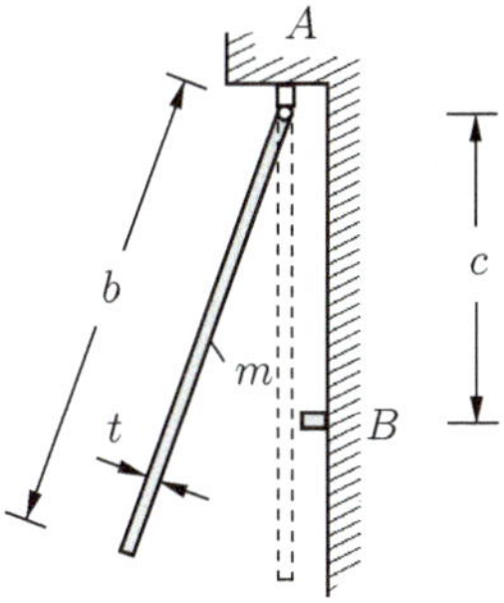

Solution We assume that before impact the door has the angular velocity ω. Then the principles of impulse and momentum read

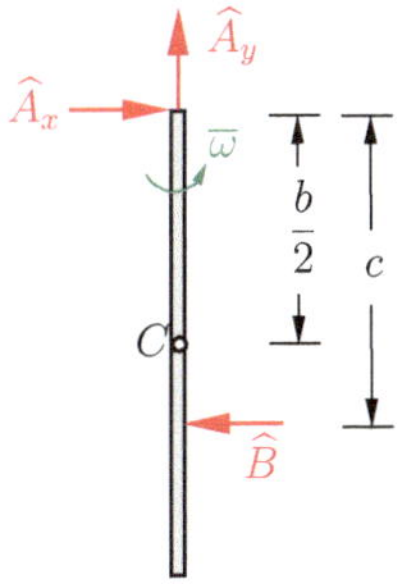

$$\rightarrow: \quad m(\overline{v}_{Cx} - v_{Cx}) = \widehat{A}_x - \widehat{B} \,,$$

$$\uparrow: \quad m(\overline{v}_{Cy} - v_{Cy}) = \widehat{A}_y \,,$$

$$\overset{\curvearrowleft}{A}: \quad \Theta_A(\overline{\omega} - \omega) = -c\widehat{B} \,,$$

where

$$v_{Cx} = \frac{b}{2}\,\omega \,, \qquad \overline{v}_{Cx} = \frac{b}{2}\,\overline{\omega} \,, \qquad v_{Cy} = \overline{v}_{Cy} = 0 \,.$$

Using the impact hypothesis

$$e = -\frac{\overline{v}_{Bx}}{v_{Bx}} = -\frac{c\overline{\omega}}{c\omega} = -\frac{\overline{\omega}}{\omega}$$

and solving the equations for the hinge impulses leads to

$$\widehat{A}_y = 0 \,, \qquad \widehat{A}_x = (1+e)\omega\left[\frac{\Theta_A}{c} - \frac{mb}{2}\right] .$$

From the condition $\widehat{A}_x = 0$ with $\Theta_A = mb^2/3$ finally follows

$$\underline{\underline{c = \frac{2\Theta_A}{mb} = \frac{2}{3}\,b}} \,.$$

P6.15 **Problem 6.15** The double pendulum, consisting of two homogeneous bars, is struck horizontally in point D by a linear impulse $\widehat{F}$.

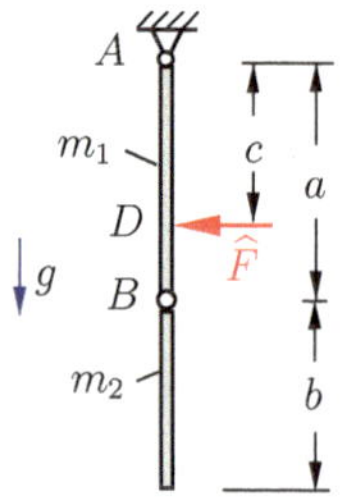

a) Determine the impulsive forces in A and B, the state of motion of the lower bar immediately after impact.

b) Under which circumstances the magnitudes of angular velocities after impact are related as $|\overline{\omega}_2| = 2|\overline{\omega}_1|$?

Solution **a)** We separate the two bars, draw the free-body diagram and formulate the priciples of linear and angular momentum (note that the weights can be neglected during impact):

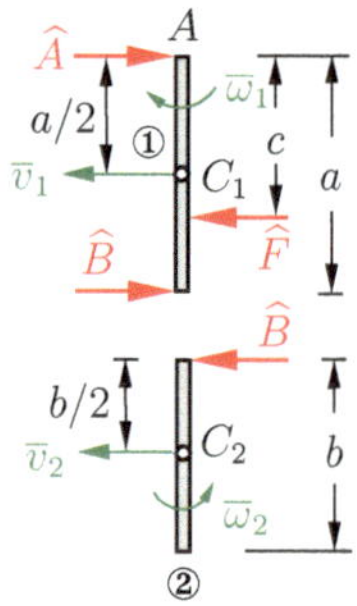

$$
\begin{aligned}
&① \quad \leftarrow: \quad && m_1\overline{v}_1 = \widehat{F} - \widehat{A} - \widehat{B}\,,\\
&\quad\ \ \overset{\curvearrowright}{A}: && \Theta_A\,\overline{\omega}_1 = c\,\widehat{F} - a\,\widehat{B}\,,\\
&② \quad \leftarrow: && m_2\,\overline{v}_2 = \widehat{B}\,,\\
&\quad\ \ \overset{\curvearrowright}{C}_2: && \Theta_{C_2}\,\overline{\omega}_2 = (b/2)\,\widehat{B}
\end{aligned}
$$

where

$$\Theta_A = \tfrac{1}{3}m_1a^2\,, \qquad \Theta_{C_2} = \tfrac{1}{12}m_2b^2\,.$$

Since the velocities of both bars at point B must be equal after impact, and $\overline{v}_1$ and $\overline{\omega}_1$ are simply related, the kinematic relations read

$$a\,\overline{\omega}_1 = \overline{v}_2 + (b/2)\,\overline{\omega}_2\,, \qquad \overline{v}_1 = (a/2)\,\overline{\omega}_1\,.$$

Solving these six equations yields

$$\underline{\underline{\widehat{A} = \Big(1 - \frac{c}{a}\,\frac{1}{1+m_1/m_2}\Big)\widehat{F}}}\,, \qquad \underline{\underline{\widehat{B} = \frac{c}{a}\,\frac{\widehat{F}}{1+m_1/m_2}}}\,,$$

$$\overline{v}_1 = 4\,\frac{\widehat{B}}{m_2}\,, \qquad \overline{\omega}_1 = 4\,\frac{\widehat{B}}{am_2}\,, \qquad \underline{\underline{\overline{v}_2 = \frac{\widehat{B}}{m_2}}}\,, \qquad \underline{\underline{\overline{\omega}_2 = \frac{b/2}{\Theta_{C_2}}\,\widehat{B}}}\,.$$

b) Inserting the quantities into the condition $|\overline{\omega}_2| = 2|\overline{\omega}_1|$ leads to

$$\frac{b/2}{\Theta_{C_2}}\,\widehat{B} = 8\,\frac{\widehat{B}}{am_2} \quad \rightsquigarrow \quad bam_2 = 16\,\Theta_{C_2} \quad \rightsquigarrow \quad \underline{\underline{a = \frac{4}{3}\,b}}\,.$$

Remark: Note that $\widehat{A} = 0$ if c is chosen such that the bracket is zero.

Problem 6.16 A bar (mass m_1) hits under the angle $\alpha = 45°$ with speed v a resting homogeneous quadratic plate of mass m_2.

Determine the velocities of the bar and the plate after impact (coefficient of restitution e). Assume that all surfaces are *smooth*.

P6.16

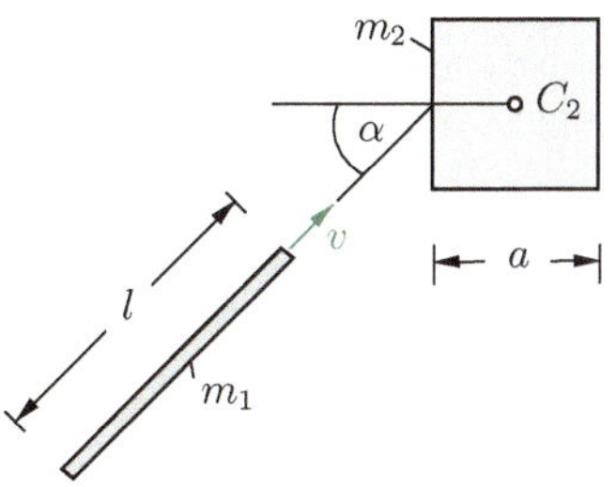

Solution If we denote the velocities of bar ① with v and of the plate ② with w, the principles of impulse and momentum lead to

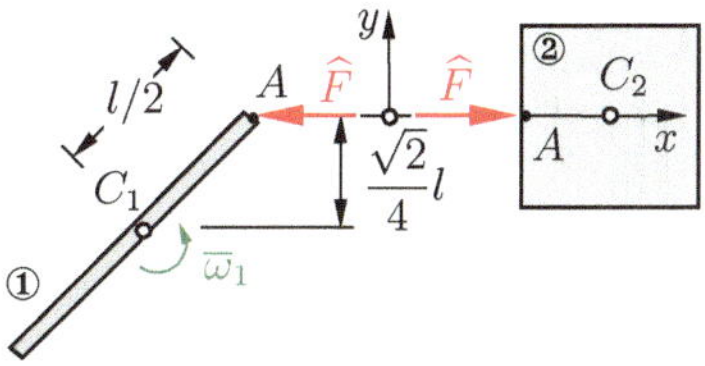

$$① \quad \rightarrow: \quad m_1(\bar{v}_{Cx} - v_{Cx}) = -\widehat{F}\,,$$

$$\uparrow: \quad m_1(\bar{v}_{Cy} - v_{Cy}) = 0\,, \qquad \overset{\curvearrowleft}{C_1}: \quad \Theta_1\bar{\omega}_1 = \frac{\sqrt{2}}{4}\, l\widehat{F}\,,$$

$$② \quad \rightarrow: \quad m_2\bar{w}_{Cx} = \widehat{F}\,, \qquad \uparrow: \quad m_2\bar{w}_{Cy} = 0\,, \qquad \overset{\curvearrowright}{C_2}: \quad \Theta_2\bar{\omega}_2 = 0\,,$$

where $\Theta_1 = m_1 l^2/12$ and $v_{Cx} = v_{Cy} = v/\sqrt{2}$. Hence, using the impact hypothesis

$$e = -\frac{\bar{v}_{Ax} - \bar{w}_{Ax}}{v_{Ax}}$$

where

$$v_{Ax} = \frac{v}{\sqrt{2}}\,, \qquad \bar{v}_{Ax} = \bar{v}_{Cx} - \frac{\sqrt{2}}{4}\, l\bar{\omega}_1\,, \qquad \bar{w}_{Ax} = \bar{w}_{Cx}$$

we obtain

$$\underline{\underline{\bar{v}_{Cx} = \frac{v}{\sqrt{2}}\,\frac{3 + 2\dfrac{m_1}{m_2} - 2e}{5 + 2\dfrac{m_1}{m_2}}}}\,, \qquad \underline{\underline{\bar{v}_{Cy} = \frac{v}{\sqrt{2}}}}\,, \qquad \underline{\underline{\bar{\omega}_1 = \frac{v}{l}\,\frac{6(1+e)}{5 + 2\dfrac{m_1}{m_2}}}}\,,$$

$$\underline{\underline{\bar{w}_{Cx} = \frac{v}{\sqrt{2}}\,\frac{2(1+e)\dfrac{m_1}{m_2}}{5 + 2\dfrac{m_1}{m_2}}}}\,, \qquad \underline{\underline{\bar{w}_{Cy} = 0}}\,, \qquad \underline{\underline{\bar{\omega}_2 = 0}}\,.$$

P6.17 **Problem 6.17** A homogeneous circular disk slides frictionless along a smooth plane and hits under angle $\alpha = 45°$ with speed v and angular velocity ω a *rough* rigid boundary (coefficient of restitution $e = 1/2$).

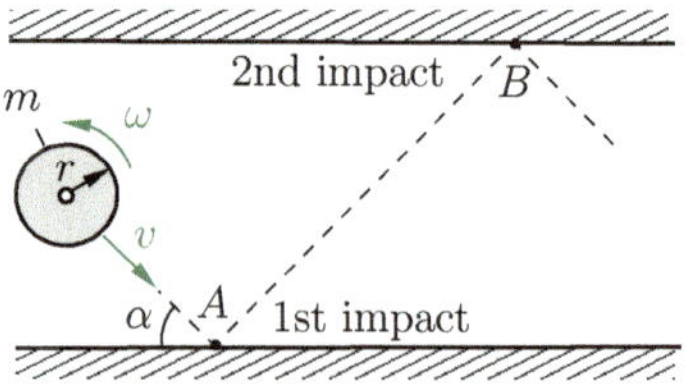

Determine the magnitude and direction of the velocity after the second impact at the opposite boundary. Assume that the disc does not slide during impact at the contact point.

Solution The first impact is described by the principles of impulse and momentum

$$\rightarrow: \quad m(\bar{v}_{Cx} - v_{Cx}) = -\widehat{H}_1 \ ,$$

$$\overset{\curvearrowleft}{C}: \quad \Theta_C(\bar{\omega} - \omega) = -r\widehat{H}_1 \ ,$$

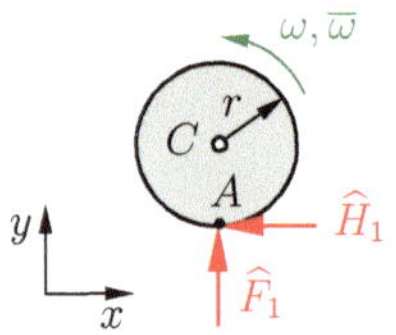

the impact hypothesis

$$e = -\frac{\bar{v}_{Ay}}{v_{Ay}} = -\frac{\bar{v}_{Cy}}{v_{Cy}}$$

and the condition of no sliding at point A

$$\bar{v}_{Ax} = \bar{v}_{Cx} + r\,\bar{\omega} = 0 \ .$$

From the first two equations follows

$$\bar{\omega} = \omega - \frac{mr}{\Theta_C}(v_{Cx} - \bar{v}_{Cx}) \ .$$

Hence, with $\Theta_C = mr^2/2$ and

$$v_{Cx} = v\cos\alpha = \frac{v}{\sqrt{2}} \ , \qquad v_{Cy} = -v\sin\alpha = -\frac{v}{\sqrt{2}}$$

the velocity components are obtained as

$$\bar{v}_{Cx} = -r\,\bar{\omega} = -\frac{r\,\omega}{3} + \frac{2}{3}v_{Cx} = \frac{1}{3}(\sqrt{2}v - r\,\omega) \ , \qquad \bar{v}_{Cy} = -ev_{Cy} = \frac{\sqrt{2}}{4}v \ .$$

Remarks:

- For $\omega > \sqrt{2}v/r$ follows $\bar{v}_{Cx} < 0$, i.e. after impact the disc moves to the left.
- The impulse $\widehat{F}_1$ can be determined from the principle of impulse and momentum in y-direction.

We denote the velocities *after* the *second impact* with two dashes. Then the principles of impulse and momentum read

$$\rightarrow: \quad m(\overline{\overline{v}}_{Cx} - \overline{v}_{Cx}) = -\widehat{H}_2 \ ,$$

$$\overset{\curvearrowleft}{C}: \quad \Theta_C(\overline{\overline{\omega}} - \overline{\omega}) = r\widehat{H}_2 \ .$$

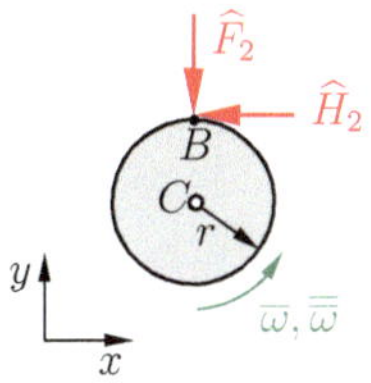

With the impact hypothesis

$$e = -\frac{\overline{\overline{v}}_{By}}{\overline{v}_{By}} = -\frac{\overline{\overline{v}}_{Cy}}{\overline{v}_{Cy}}$$

and the condition of no sliding at B (consider signs!)

$$\overline{\overline{v}}_{Bx} = \overline{\overline{v}}_{Cx} - r\,\overline{\overline{\omega}} = 0$$

follow

$$\overline{\overline{v}}_{Cx} = r\,\overline{\overline{\omega}} = \frac{r\,\overline{\omega}}{3} + \frac{2}{3}\,\overline{v}_{Cx} = \frac{1}{9}(\sqrt{2}\,v - r\omega) \ ,$$

$$\overline{\overline{v}}_{Sy} = -e\overline{v}_{Sy} = -\frac{\sqrt{2}}{8}\,v \ .$$

Hence, mangnitude and direction of the velocity are given by

$$\underline{\underline{\overline{\overline{v}}}} = \sqrt{\overline{\overline{v}}_{Cx}^2 + \overline{\overline{v}}_{Cy}^2}$$

$$= \underline{\underline{\sqrt{\left(\frac{2}{81} + \frac{2}{64}\right)v^2 - \frac{2\sqrt{2}}{9}\,vr\omega + \frac{r^2\omega^2}{81}}}} \ ,$$

$$\underline{\underline{\tan\beta}} = \frac{|\overline{\overline{v}}_{Cy}|}{\overline{\overline{v}}_{Cx}} = \underline{\underline{\frac{9\sqrt{2}}{8(\sqrt{2} - \frac{r\omega}{v})}}} \ .$$

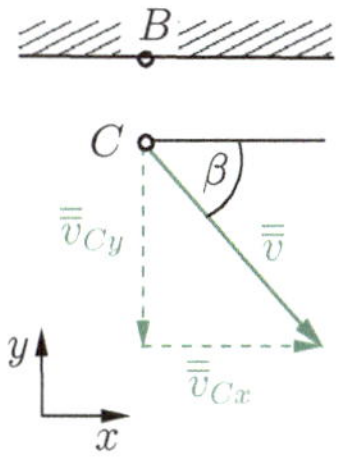

Remarks:

- For $\omega = 0$ we obtain $\overline{\overline{v}} = 0.24\ v$ and $\tan\beta = 9/8 \ \rightsquigarrow\ \beta = 48.4°$.
- For $\omega = \sqrt{2}\,v/r$ the disk rebounds orthogonal at A and therefore also at B.

Chapter 7

Vibrations

7

The following formulas and problems are restricted to vibrations of *linear* systems with *one* degree of freedom.

1. Free undamped vibration

The equation of motion (*vibration equation*)

$$\ddot{x} + \omega^2 x = 0$$

has the general solution

$$x(t) = A\cos\omega t + B\sin\omega t = C\cos(\omega t - \alpha)$$

where

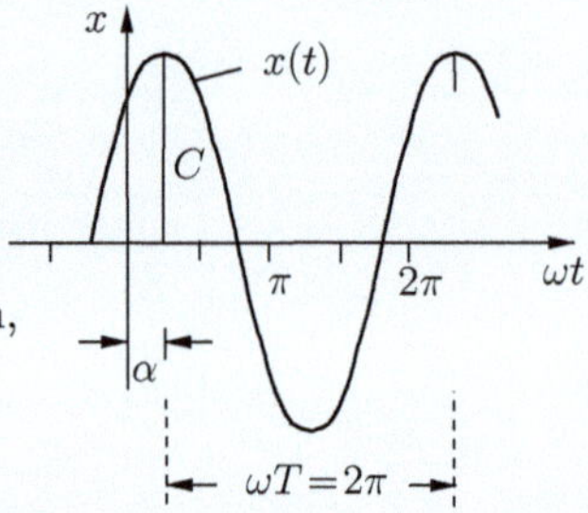

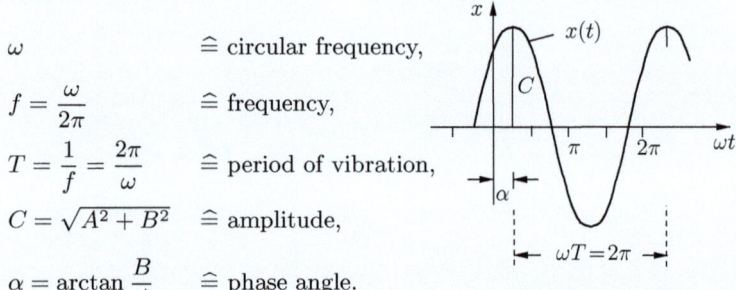

ω $\widehat{=}$ circular frequency,

$f = \dfrac{\omega}{2\pi}$ $\widehat{=}$ frequency,

$T = \dfrac{1}{f} = \dfrac{2\pi}{\omega}$ $\widehat{=}$ period of vibration,

$C = \sqrt{A^2 + B^2}$ $\widehat{=}$ amplitude,

$\alpha = \arctan\dfrac{B}{A}$ $\widehat{=}$ phase angle.

Remarks:

- The constants A, B, C and α follow from the initial conditions $x(0) = x_0$, $\dot{x}(0) = v_0$ as

$$A = x_0\,, \qquad B = \frac{v_0}{\omega}\,, \qquad C = \sqrt{x_0^2 + \left(\frac{v_0}{\omega}\right)^2}\,, \qquad \alpha = \arctan\frac{v_0}{x_0\omega}\,.$$

- A system which is described by the differentiial equation above is also called a *harmonic oscillator.*

- If the position coordinate is *not* counted from the equilibrium position, the equation of motion takes the form

  $$\ddot{x} + \omega^2 x = \omega^2 x_{st}$$

  and its solution reads

  $$x(t) = D\cos(\omega t - \psi) + x_{st}\,.$$

Examples of single mass vibration systems

All coordinates are counted from the respective static equilibrium position.

system	diff. equ.	eigenfrequency ω
spring-mass-system 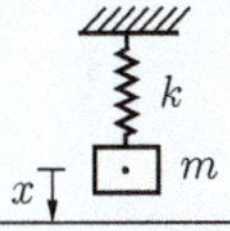	$m\ddot{x} + kx = 0$	$\sqrt{\frac{k}{m}}$
simple pendulum (small displacements) 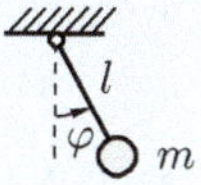	$l\ddot{\varphi} + g\varphi = 0$	$\sqrt{\frac{g}{l}}$
massless bar with end mass 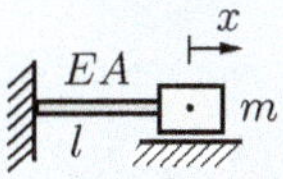	$m\ddot{x} + kx = 0$	$\sqrt{\frac{k}{m}}$ with $k = EA/l$
massless beam with end mass 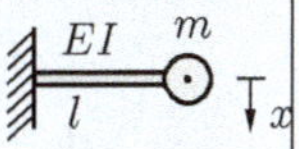	$m\ddot{x} + kx = 0$	$\sqrt{\frac{k}{m}}$ with $k = 3EI/l^3$
massless shaft with end disk 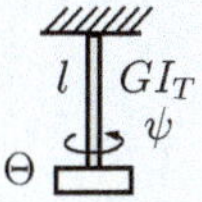	$\Theta\ddot{\psi} + k_T\psi = 0$	$\sqrt{\frac{k_T}{\Theta}}$ with $k_T = GI_T/l$

Spring constants

Springs in parallel:

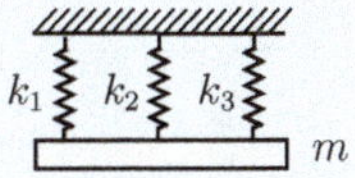

≙

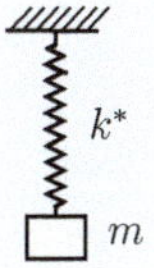

$$k^* = \sum k_i$$

Springs in series:

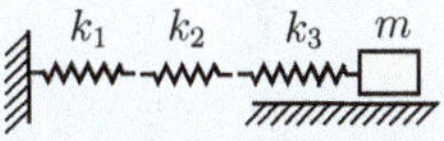

≙

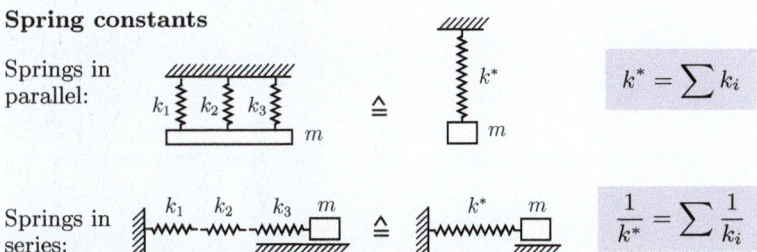

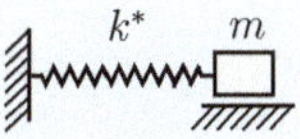

$$\frac{1}{k^*} = \sum \frac{1}{k_i}$$

Spring Compliance: The inverse spring stiffness is called spring compliance (flexibility): $c = 1/k$. The compliance of an elastic system can be determined by loading it by a virtual force „1“ at the position of the vibrating mass and calculating the dicplacement δ in direction of the force. Then $c = \delta$ and $k = 1/\delta$.

2. Free damped Vibration

a) Dry (Coulomb) Friction: The solution of the equation of motion

$$\ddot{x} + \omega^2 x = \mp\omega^2 r \quad \text{for} \quad \dot{x} \gtrless 0$$

is given by $x(t) = C\cos(\omega t - \varphi) \pm r$ für $\dot{x} \gtrless 0$.

Amplitude decrease: $x(\omega t) - x(\omega t + 2\pi) = 4r$.

b) Viscous Damping: The solution of the equation of motion

$$\ddot{x} + 2\xi\dot{x} + \omega^2 x = 0$$

reads for

$\underline{\zeta = \xi/\omega < 1}$ (underdamped system): $x(t) = C\,\mathrm{e}^{-\xi t}\cos(\omega_d t - \alpha)$

$$\begin{aligned} \xi &\;\widehat{=}\; \text{damping coefficient,} \\ \zeta = \xi/\omega &\;\widehat{=}\; \text{damping ratio,} \\ \omega_d = \omega\sqrt{1-\zeta^2} &\;\widehat{=}\; \text{circular frequency of damped vibration,} \\ T_d = 1/f_d = 2\pi/\omega_d &\;\widehat{=}\; \text{period.} \end{aligned}$$

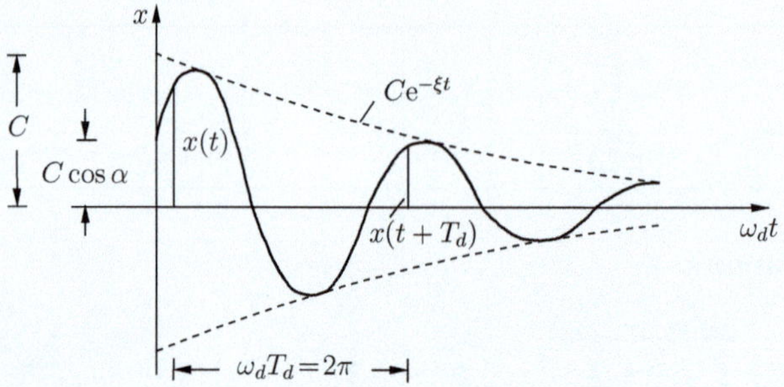

The amplitude decay is characterized by the *logarithmic decrement*

$$\delta = \ln\frac{x(t)}{x(t+T_d)} = \frac{2\pi\xi}{\omega_d} = \frac{2\pi\zeta}{\sqrt{1-\zeta^2}}\,.$$

For weak damping $\zeta \ll 1$ follows $\delta \approx 2\pi\zeta$.

$\underline{\zeta = \xi/\omega = 1}$ (critical damping): $x(t) = (A_1 + A_2 t)\,\mathrm{e}^{-\xi t}$.

$\underline{\zeta = \xi/\omega > 1}$ (overdamped system): $x(t) = \mathrm{e}^{-\xi t}(A_1\mathrm{e}^{\mu t} + A_2\mathrm{e}^{-\mu t})$

where $\mu = \omega\sqrt{\zeta^2 - 1}$.

3. Forced vibrations

The equation of motion for *harmonically excited* vibrations can be written as

$$\frac{1}{\omega^2}\,\ddot{x} + \frac{2\zeta}{\omega}\,\dot{x} + x = E x_0 \cos \Omega t$$

where

$$\begin{aligned}
\Omega &\quad \widehat{=} \quad \text{circular frequency of excitation,} \\
\eta = \frac{\Omega}{\omega} &\quad \widehat{=} \quad \text{frequency ratio,} \\
\zeta = \frac{\delta}{\omega} &\quad \widehat{=} \quad \text{damping ratio,} \\
E x_0 &\quad \widehat{=} \quad \text{excitation amplitude,}
\end{aligned}$$

$$E = \begin{cases} 1 & \text{excitation through a force or via a spring,} \\ 2D\eta & \text{excitation via a damper,} \\ \eta^2 & \text{excitation through unbalanced rotation.} \end{cases}$$

It has the *steady state* solution (particular solution)

$$x = x_0 V \cos(\Omega t - \varphi) \,.$$

Here are

$$V = \frac{E}{\sqrt{(1-\eta^2)^2 + 4\zeta^2\eta^2}} \quad \widehat{=} \text{ magnification, frequency response ,}$$

$$\tan\varphi = \frac{2\zeta\eta}{1-\eta^2} \quad \widehat{=} \text{ phase angle .}$$

Remarks:

- The general solution of the equation of motion is composed of the solution of the homogeneous differential equation (decaying motion, see page 172) and the particular solution.
- For undamped vibrations ($\zeta = 0$) the amplitude tends for $\eta \to 1$ ($\Omega \to \omega$) to infinity (*resonance*).
- For weakly damped systems ($\zeta \ll 1$) at resonance ($\eta \approx 1$) the maximum magnification is: $V_{\max} \approx E/2\zeta$.
- An excitation with $\eta < 1$ is called subcritical and with $\eta > 1$ supercritical.
- The phase angle φ represents the delay of the response x relative to the excitation.

P7.1 **Problem 7.1** Two drums support a homogeneous beam of weight $W = mg$. They rotate with different coefficients of kinetic friction μ_A, μ_B in opposite directions.

Show that the horizontal motion of the beam is a harmonic vibration and determine the natural frequency ω.

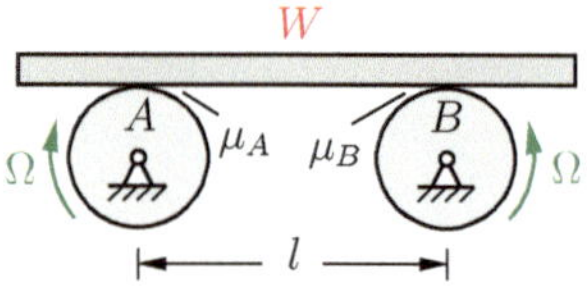

Solution We first determine the equilibrium distance a of point C of the beam and the support forces. From the equilibrium conditions and the friction law

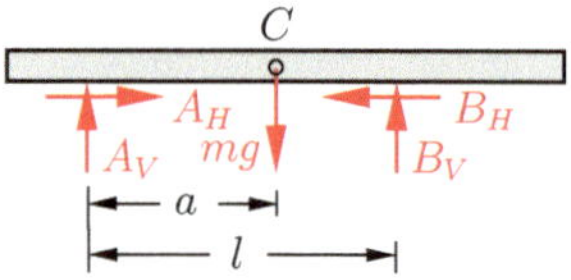

$$\rightarrow: \ A_H^{eq} = B_H^{eq}, \quad \uparrow: \ A_V^{eq} + B_V^{eq} = mg,$$

$$\overset{\curvearrowleft}{A}: \ B_V^{eq} l = mga\,,$$

$$A_H^{eq} = \mu_A A_V\,, \quad B_H^{eq} = \mu_B B_V$$

follow $a/l = 1/(1 + \mu_A/\mu_B)$ and

$$A_V^{eq} = \frac{l-a}{l}\, mg\,, \quad B_V^{eq} = \frac{a}{l}\, mg\,.$$

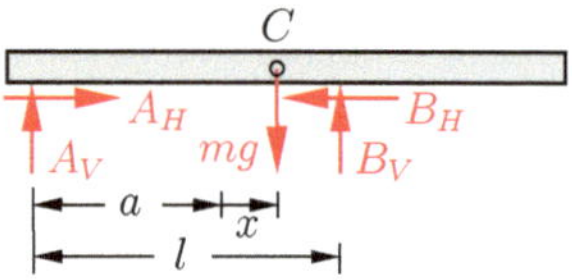

Now we consider an arbitrary displacement x from the equilibrium position. The support reactions then are

$$A_V = \frac{l-(a+x)}{l}\, mg\,, \quad B_V = \frac{a+x}{l}\, mg = B_V^{eq} + \frac{x}{l}\, mg\,,$$

$$A_H = \mu_A \frac{l-(a+x)}{l} = A_H^{eq} - \mu_A \frac{x}{l}\, mg\,, \quad B_H = B_H^{eq} + \mu_B \frac{x}{l}\, mg\,.$$

Thus, with $A_H^{eq} = B_H^{eq}$ the equation of motion is given by

$$\rightarrow: \quad m\ddot{x} = A_H - B_H = -(\mu_A + \mu_B)\frac{x}{l}\, mg\,,$$

which leads to the differential equation for harmonic vibrations

$$\underline{\underline{\ddot{x} + (\mu_A + \mu_B)\,\frac{g}{l}\,x = 0\,.}}$$

The natural frequency is given by

$$\underline{\underline{\omega = \sqrt{(\mu_A + \mu_B)\,\frac{g}{l}}}}\,.$$

Remark: In case of $\mu_A = \mu_B = \mu$, the result simplyfies to $\omega = \sqrt{2\mu g/l}$.

Problem 7.2 A small homogeneous disk (mass m, radius r) is attached to a big disk (mass M, radius R). In the positions a) and b), both systems are in the equilibrium position. P7.2

Determine the natural frequencies ω for both systems. Assume small rotational amplitudes.

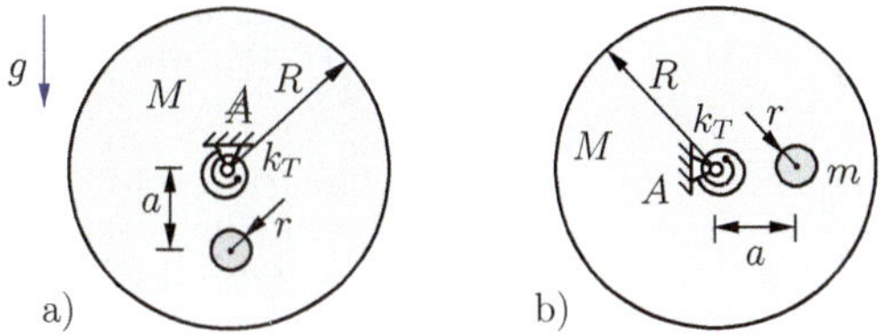

Solution To formulate the equation of motion it is advantageous in both cases to apply the principle of angular momentum.

case a) We obtain

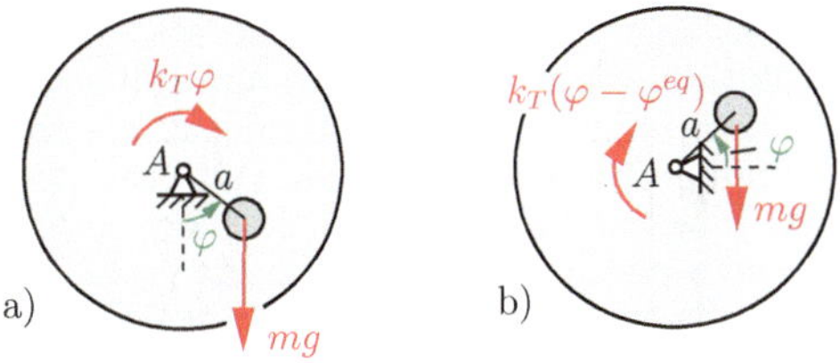

$$\overset{\curvearrowleft}{A}: \quad \Theta_A \ddot{\varphi} = -k_T \varphi - mga \sin\varphi \,.$$

For the assumed small amplitudes ($\varphi \ll 1$) this equation reduces with $\sin\varphi \approx \varphi$ to the differential equation for harmonic vibrations, which provides the natural frequency:

$$\ddot{\varphi} + \omega^2 \varphi = 0 \quad \text{with} \quad \omega_a^2 = \frac{k_T + mga}{\Theta_A} \quad \rightsquigarrow \quad \omega_a = \sqrt{\frac{k_T + mga}{\Theta_A}} \,.$$

case b) Since the torsion spring is in the equilibrium position already stretched by $\varphi^{eq} = mga/k_T$, its moment due to a displacement φ is given by $k_T(\varphi - \varphi^{eq})$. Thus, we obtain in this case

$$\overset{\curvearrowleft}{A}: \quad \Theta_A \ddot{\varphi} = -k_T(\varphi - \varphi^{eq}) - mga\cos\varphi = -k_T\varphi + mga(1 - \cos\varphi) \,,$$

which for small amplitudes, i.e. $\cos\varphi \approx 1$, reduces to

$$\ddot{\varphi} + \omega^2 \varphi = 0 \quad \text{with} \quad \omega_b^2 = \frac{k_T}{\Theta_A} \quad \rightsquigarrow \quad \omega_b = \sqrt{\frac{k_T}{\Theta_A}} \,.$$

Introducing the moment of inertia

$$\Theta_A = \frac{MR^2}{2} + \left[\frac{mr^2}{2} + ma^2\right]$$

the natural frequencies can be written as

$$\underline{\underline{\omega_a = \sqrt{\frac{2(k_T + mga)}{MR^2 + m(r^2 + 2a^2)}}}} \,, \qquad \underline{\underline{\omega_b = \sqrt{\frac{2k_T}{MR^2 + m(r^2 + 2a^2)}}}} \,.$$

P7.3 **Problem 7.3** The system shown consists of a homogeneous disk (mass m, radius r), a block (mass M) and a spring (stiffness k). The mass of the string and of the pulley B can be neglected.

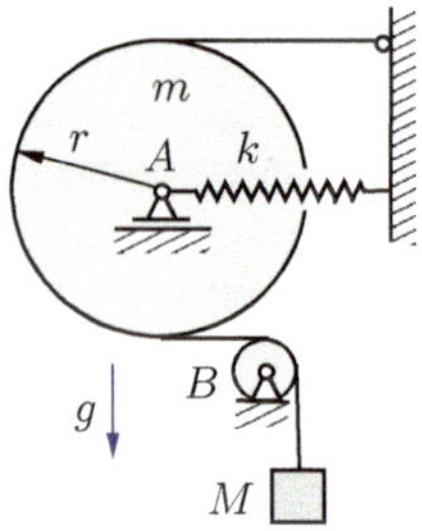

Determine the equation of motion and the period of the vibration. Specify the solution for the case that point A is horizontally displaced by a from equilibrium and then released with zero initial velocity.

Solution We separate the system and introduce coordinates with their origin at the equilibrium position of the system. With this choice, the weights of the disk and the block must not be considered in the free-body diagram. Thus, the equations of motion are given by

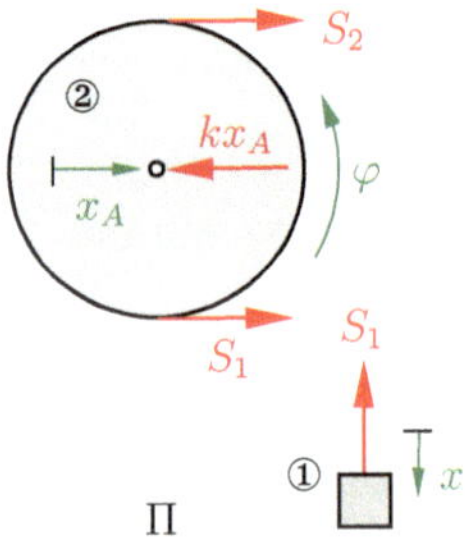

$$① \quad \downarrow: \quad M\ddot{x} = -S_1 ,$$

$$② \quad \rightarrow: \quad m\ddot{x}_A = S_1 + S_2 - kx_A ,$$

$$\overset{\curvearrowleft}{A}: \quad \Theta_A\ddot{\varphi} = rS_1 - rS_2 .$$

If we use the kinematic relations (Π $\widehat{=}$ instantaneous center of rotation)

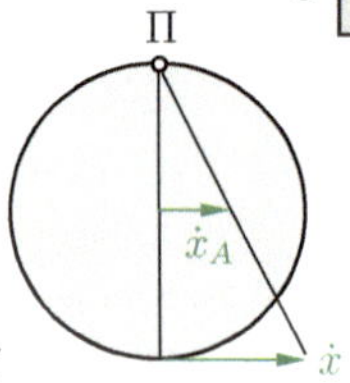

$$x_A = \frac{x}{2}, \quad 2r\varphi = x \quad \rightsquigarrow \quad \ddot{x}_A = \frac{\ddot{x}}{2}, \quad \ddot{\varphi} = \frac{\ddot{x}}{2r}$$

and introduce the moment of inertia $\Theta_A = mr^2/2$ we obtain, by solving for x, the equation of motion

$$\underline{\underline{\ddot{x} + \frac{k}{4M + \frac{3}{2}\,m}\,x = 0}} \qquad \text{where} \qquad \omega = \sqrt{\frac{k}{4M + \frac{3}{2}\,m}}$$

is the natural frequency. It is related with the period of the vibration by

$$T = \frac{2\pi}{\omega} \quad \rightsquigarrow \quad \underline{\underline{T = 2\pi\sqrt{\frac{k}{4M + \frac{3}{2}\,m}}}} .$$

From the general solution $x = A\cos\omega t + B\sin\omega t$ with $x_A = x/2$ and the initial conditions $x_A = a$, $\dot{x}_A = 0$ follows

$$\underline{\underline{x(t) = 2a\cos\omega t}} .$$

Problem 7.4 The system shown consists of two clamped beams (negligible masses, bending stiffnesses EI_i), a spring (spring constant k_3) and a block of mass m. P7.4

Determine the eigenfrequency ω.

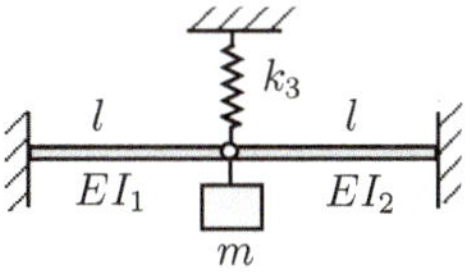

Solution **1. approach:** We separate the system and formulate the equation of motion for the block:

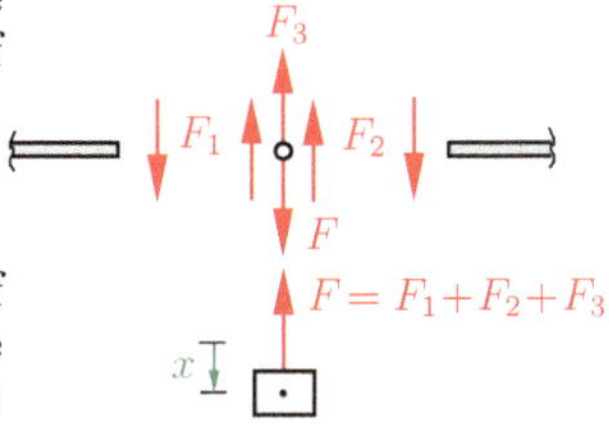

$$\downarrow: \quad m\ddot{x} = -F_1 - F_2 - F_3 \ .$$

The forces F_i and the displacements w_i of the beams and the spring, respectively, are related by (see volume 2, table of elastic lines)

$$w_1 = \frac{F_1 l^3}{3EI_1} \ , \qquad w_2 = \frac{F_2 l^3}{3EI_2} \ , \qquad w_3 = \frac{F_3}{k_3} \ .$$

Therefore, from the condition $w_1 = w_2 = w_3 = x$ follows

$$m\ddot{x} + \Big(\frac{3EI_1}{l^3} + \frac{3EI_2}{l^3} + k_3\Big)x = 0 \ .$$

Hence, we obtain for the eigenfrequency

$$\underline{\underline{\omega^2 = \frac{1}{m}\Big(\frac{3EI_1}{l^3} + \frac{3EI_2}{l^3} + k_3\Big)}} \ .$$

2. approach: Since all spring ends undergo the same displacement x, the springs are in parallel:

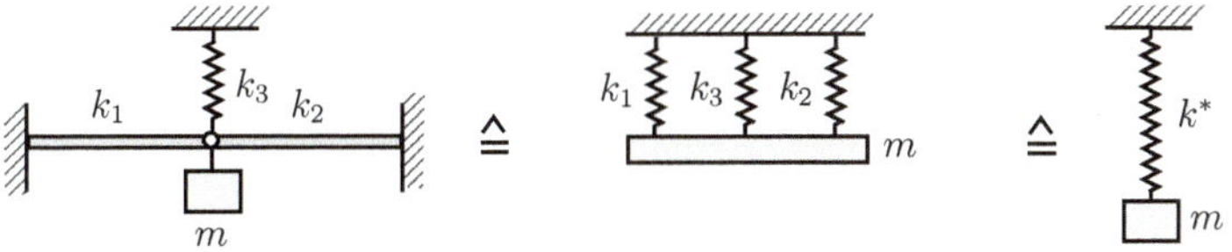

With the spring constants

$$k_1 = \frac{F_1}{x_1} = \frac{3EI_1}{l^3} \ , \qquad k_2 = \frac{F_2}{x_2} = \frac{3EI_2}{l^3}$$

we obtain according to page 171

$$k^* = k_1 + k_2 + k_3 = \frac{3EI_1}{l^3} + \frac{3EI_2}{l^3} + k_3$$

and therefore

$$\underline{\underline{\omega^2 = \frac{c^*}{m} = \frac{1}{m}\Big(\frac{3EI_1}{l^3} + \frac{3EI_2}{l^3} + k_3\Big)}} \ .$$

P7.5 **Problem 7.5** The system consists of a hinge supported beam, reinforced by three bars, and of a block of mass m.

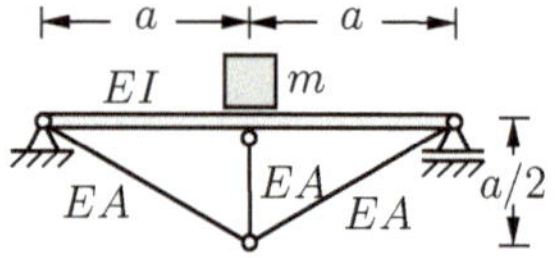

Determine the eigenfrequency for vertical vibrations. Assume that the mass of the beam and the bars is negligible.

Solution The elastic system of beam and bars can regarded as a system consisting of two springs in parallel, one representing the beam, one representing the bars. To determine the spring constant k_B of the beam we subject it by the unit force $F_B = 1$ at the location of the block. It produces the deflection (see volume 2, chapter 4)

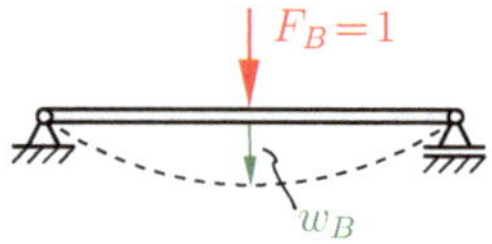

$$w_B = \frac{1 \cdot (2a)^3}{48EI} .$$

Hence, the spring constant is given by

$$k_B = \frac{1}{w_B} = \frac{48\,EI}{(2a)^3} = \frac{6\,EI}{a^3}.$$

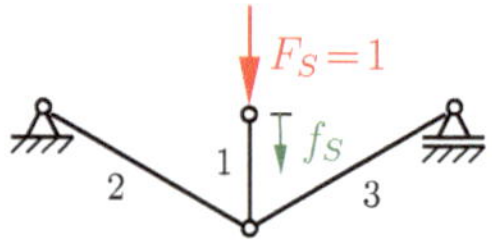

To find the spring constant k_S of the bar system, we apply a force $F_S = 1$ at the bar 1. It causes the displacement (see volume 2, chapter 6)

$$f_S = h_S = \sum \frac{\bar{S}_i^2 l_i}{EA} \quad (\bar{S}_i = \text{forces in bars, } l_i = \text{lengths of bars}) .$$

With

$$\bar{S}_1 = -1 , \qquad \bar{S}_2 = \bar{S}_3 = \frac{\sqrt{5}}{2} , \qquad l_1 = \frac{a}{2} , \qquad l_2 = l_3 = \frac{\sqrt{5}}{2}\, a$$

we obtain

$$f_S = \frac{1}{EA}\Big[(-1)^2 \cdot \frac{a}{2} + 2 \cdot \Big(\frac{\sqrt{5}}{2}\Big)^2 \frac{\sqrt{5}}{2}\, a\Big] = \Big(1 + \frac{5\sqrt{5}}{2}\Big)\frac{a}{2EA}$$

$$\rightsquigarrow \quad k_S = \frac{4EA}{(2+5\sqrt{5})a} .$$

From the spring constants of the two springs in parallel follows the constant k^* of the equivalent single spring:

$$k^* = k_B + k_S = \frac{6\,EI}{l^3} + \frac{EA}{(1+\sqrt{2})l} .$$

Thus, the eigenfrequency is given by

$$\underline{\underline{\omega = \sqrt{\frac{k^*}{m}} = \frac{1}{l}\sqrt{\frac{1}{ml}\Big(6\,EI + \frac{EAl^2}{1+\sqrt{2}}\Big)}}} .$$

Problem 7.6 For the two systems ① and ② the equivalent spring constants for vertical vibrations of the body of mass m shall be determined. The mass of the beams can be neglected. **P7.6**

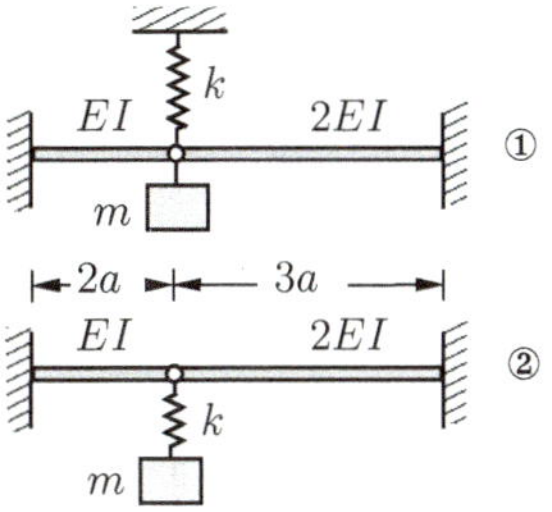

Solution In system ① all three springs are directly connected with the vibrating body of mass m and experience the same displacements. Thus, the springs are in parallel:

$$k^* = \sum k_i \, .$$

The spring constants k_L and k_R of the left and the right beam follow from the end displacement of a cantilever beam (stiffness EI, length l) subjected to a unit force „1“

$$w = \frac{1 \cdot l^3}{3EI} = k \quad \text{and} \quad c = \frac{1}{k}$$

as

$$k_L = \frac{1}{c_L} = \frac{3EI}{(2a)^3} \, , \qquad k_R = \frac{1}{c_R} = \frac{3(2EI)}{(3a)^3} \, .$$

Hence, the stiffness of the equivalent spring is

$$\underline{\underline{k^*}} = k_L + k_R + k = \frac{43}{72}\,\frac{EI}{a^3} + k = \underline{\underline{\frac{43EI + 72ka^3}{72a^3}}} \, .$$

In system ② both beams are in parallel. Their equivalent spring with constant $\bar{k}$ is in series with the spring with constant k. Thus, the total equivalent constant k^* is calculated as follows:

$$\bar{k} = k_L + k_R = \frac{43}{72}\,\frac{EI}{a^3} \, ,$$

$$\frac{1}{k^*} = \frac{1}{\bar{k}} + \frac{1}{k} = \frac{72a^3}{43EI} + \frac{1}{k}$$

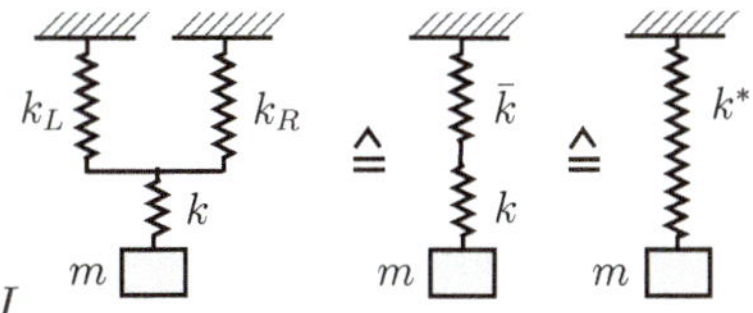

$$\leadsto \quad \underline{\underline{k^*}} = \frac{43EIk}{43EI + 72ka^3} = \underline{\underline{\frac{43EI}{72a^3 + 43\,\dfrac{EI}{k}}}} \, .$$

Remark: The second system has a smaller stiffness and therefore vibrates with a lower frequency.

P7.7

Problem 7.7 A wheel (mass m, moment of inertia Θ_C, radius r) rolls without slipping on a flat surface. On its top, the wheel carries a horizontally guided beam (mass M) and moves it without slipping.

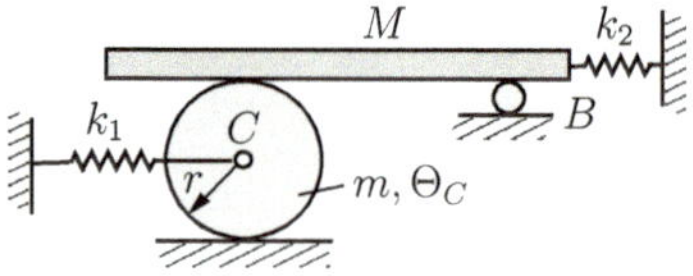

Determine the eigenfrequency of the system. Neglect the mass of the guiding roll B.

Solution We separate the system and introduce the coordinates x_1, x_2 and φ, measured from the equilibrium position. Then the equations of motion are

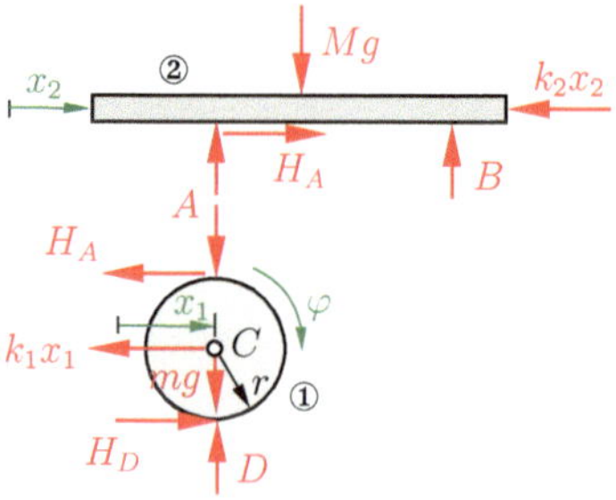

$$① \rightarrow : \quad m\ddot{x}_1 = -k_1 x_1 - H_A + H_D\,,$$

$$\overset{\curvearrowright}{C} : \quad \Theta_C\,\ddot{\varphi} = -r\,H_A - rH_D\,,$$

$$② \rightarrow : \quad M\,\ddot{x}_2 = -k_2 x_2 + H_A\,.$$

With the kinematic relations

$$x_1 = r\varphi\,, \quad x_2 = 2x_1 \rightsquigarrow \quad \ddot{x}_1 = r\ddot{\varphi}\,, \quad \ddot{x}_2 = 2\ddot{x}_1$$

we now have five equations for the five unknowns x_1, x_2, φ, H_A and H_D. Solving for $x_1(t)$ yields the equation of motion

$$\ddot{x}_1\left(m + 4M + \frac{\Theta_C}{r^2}\right) + (k_1 + 4k_2)\,x_1 = 0$$

or in standard form

$$\ddot{x}_1 + \omega^2 x_1 = 0$$

with the eigenfrequency

$$\underline{\underline{\omega = \sqrt{\frac{k_1 + 4k_2}{m + 4M + \dfrac{\Theta_C}{r^2}}}}}\,.$$

Problem 7.8 For a connecting rod of weight mg, the periods T_a and T_b are measured for the hangings a) and b).

Determine the moment of inertia Θ_C and the distance a of the center of mass C.

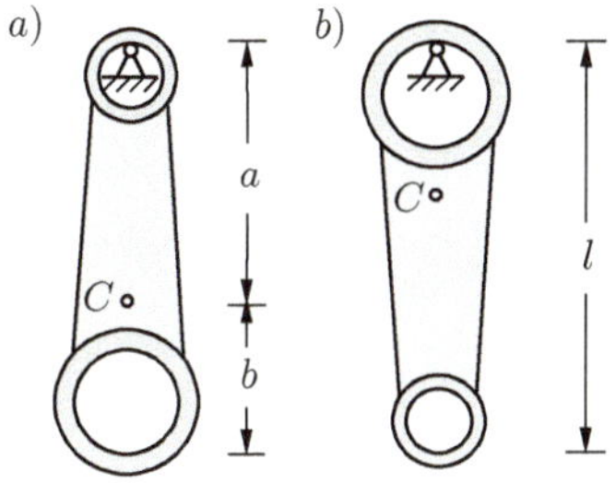

P7.8

Solution The equation of motion for small amplidudes for the hanging a) is given by

$$\Theta_A \ddot{\varphi} + mga\,\varphi = 0$$

and the respective period follows as

$$T_a^2 = \frac{(2\pi)^2}{\omega^2} = \frac{4\pi^2\Theta_A}{mga} .$$

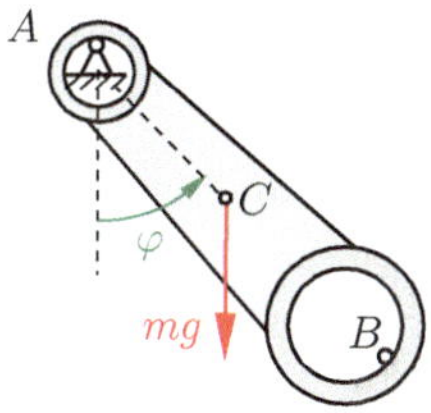

Analogous for hanging b) results

$$T_b^2 = \frac{4\pi^2\Theta_B}{mgb} .$$

With the parallel axis theorem

$$\Theta_A = \Theta_C + ma^2 , \qquad \Theta_B = \Theta_C + mb^2$$

and with $a + b = l$ we obtain

$$\underline{\underline{\Theta_C = \frac{T_a^2 mga}{4\pi^2} - ma^2}}$$

and

$$\underline{\underline{a = \frac{T_b^2 g - 4\pi^2 l}{(T_a^2 + T_b^2)g - 8\pi^2 l}\, l}} .$$

Remarks:

- A swinging rigid body (here a connecting rod) which cannot be regarded as point mass at a massless cord, is called a *compound pendulum* or *physical pendulum*.
- Analogeous to a simple pendulum, the equation of motion of a compound pendulum is often written as

  $$\ddot{\varphi} + \frac{g}{l_{\text{red}}}\,\varphi = 0 \qquad \text{with} \qquad l_{\text{red}} = \frac{\Theta_A}{ma} ,$$

  where a = distance between pivot A and center C of mass.

P7.9 **Problem 7.9** A body (mass m) is pulled by a spring (stiffness k) to a smooth horizontal path. In vertical position the spring is stretched by the distance a and its length in the unstretched state is l.

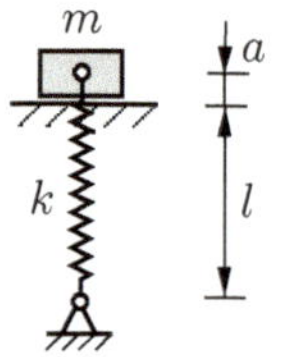

a) Which condition must be fulfilled by the horizontal displacement, that the vibration about the equilibrium position is harmonic?

b) Determine the period for this case.

Solution **a)** The equation of motion is given by

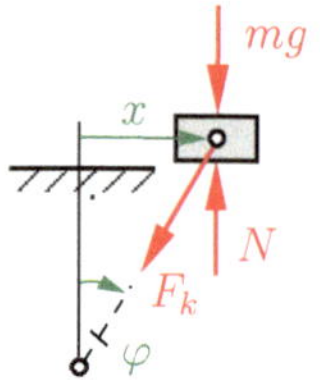

$$\rightarrow: \quad m\ddot{x} = -F_k \sin\varphi$$

with

$$F_k = k\,\Delta_k\,, \quad \sin\varphi = \frac{x}{\sqrt{(l+a)^2 + x^2}}$$

and the spring elongation

$$\Delta_k = a + \left[\sqrt{(l+a)^2 + x^2} - (l+a)\right].$$

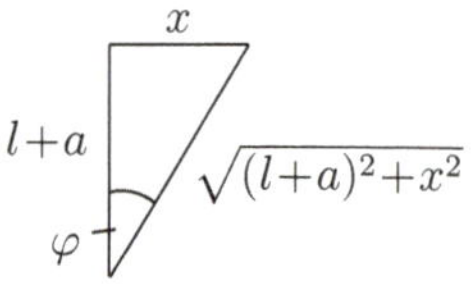

The vibration is harmonic if the equation of motion is of the type $\ddot{x} = -\omega^2 x$. Consequently, $F_k \sin\varphi$ must be linear in x! This is only possible if the term x^2 in the square roots can be neglected in comparison with $(l+a)^2$, i.e. if the condition

$$x^2 \ll (l+a)^2 \quad \rightsquigarrow \quad \underline{\underline{|x| \ll l + a}}$$

is fulfilled. Therefore, the displacement x must always be sufficiently small. In this case, in a first approximation, $\Delta_k = a = \text{const}$ and $\sin\varphi = x/(l+a)$ hold and the equation of motion is given by

$$m\ddot{x} + \frac{k\,a}{l+a}\,x = 0\,.$$

b) We write the equation of motion in its standard form

$$\ddot{x} + \omega^2 x = 0 \quad \text{with} \quad \omega^2 = \frac{k}{m}\,\frac{a}{l+a}$$

and obtain with $\omega = 2\pi/T$ for the period

$$\underline{\underline{T = 2\pi\sqrt{\frac{m(l+a)}{k\,a}}}}\,.$$

Problem 7.10 A simple pendulum is connected with a spring (stiffness k) and a dashpot (damping coefficient d). **P7.10**

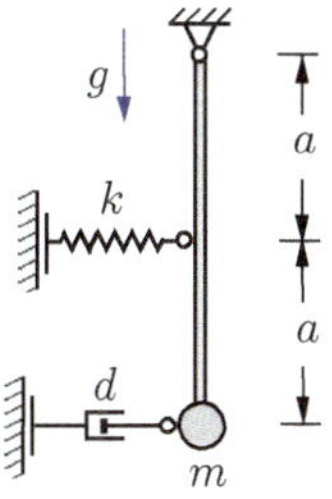

a) Determine the damping coefficient d such that vibrations are possible. Assume small amplitudes.

b) What damping ratio ζ must be chosen such that the amplitude is reduced to 1/100 of its initial value after 20 full cycles. Calculate the corresponding period T_d.

Solution **a)** Assuming small amplitudes, i.e. $\cos\varphi \approx 1$, $\sin\varphi \approx \varphi$, the equation of motion follows from the principle of angular momentum:

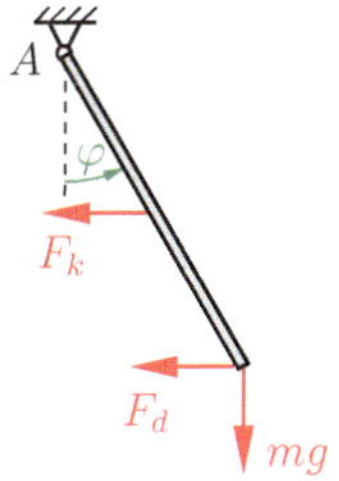

$$\overset{\curvearrowleft}{A}: \quad \Theta_A \ddot{\varphi} = -F_k a - F_d 2a - mg2a\varphi \,.$$

Introducing

$$\Theta_A = m\,(2a)^2\,, \qquad F_k = k\,a\,\varphi\,, \qquad F_d = d\,2a\,\dot{\varphi}$$

we obtain

$$\ddot{\varphi} + \frac{d}{m}\,\dot{\varphi} + \left(\frac{k}{4m} + \frac{g}{2a}\right)\varphi = 0 \quad \rightsquigarrow \quad \ddot{\varphi} + 2\xi\dot{\varphi} + \omega^2\varphi = 0\,,$$

where

$$\xi = \frac{d}{2m}\,, \qquad \omega^2 = \frac{k}{4m} + \frac{g}{2a}\,.$$

To ensure vibrations, the system must be underdamped, i.e. $\xi < \omega$:

$$\frac{d}{2m} < \sqrt{\frac{k}{4m} + \frac{g}{2a}} \qquad \rightsquigarrow \qquad \underline{\underline{d < \sqrt{km + 2\frac{gm^2}{a}}}}\,.$$

b) The necessary damping ζ ratio follows with $x_{n+20} = x_n/100$ from the logarithmic decrement (see page 172):

$$20\frac{2\pi\zeta}{\sqrt{1-\zeta^2}} = \ln\frac{x_n}{x_{n+20}} = \ln 100 \quad \rightsquigarrow \quad \underline{\underline{\zeta}} = \sqrt{\frac{1}{\left(\frac{40\pi}{\ln 100}\right)^2 + 1}} = \underline{\underline{0.037}}\,.$$

This leads to the period

$$\underline{\underline{T_d}} = \frac{2\pi}{\omega\sqrt{1-\zeta^2}} \approx \frac{2\pi}{\omega} = \underline{\underline{2\pi\sqrt{\frac{4am}{ak + 2gm}}}}\,.$$

P7.11 **Problem 7.11** Determine the eigenfrequency for the displayed system with viscous damping. The mass of the cantilever beam and the bar can be neglected.

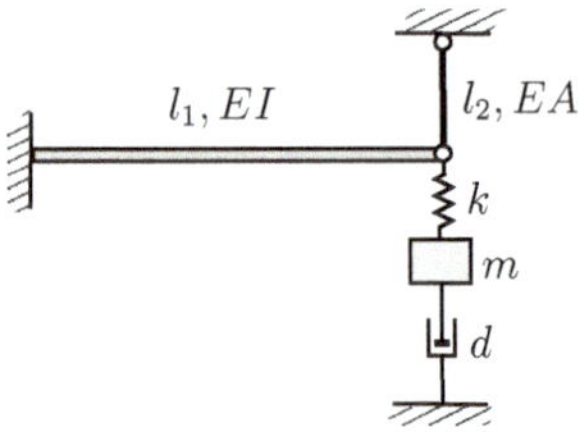

Solution We replace the stiffnesses of the beam, the bar and the spring by an equivalent stiffness k^*. When the body (mass m) is displaced by x from its equilibrium position, it is loaded by a spring force $F_k = k^* x$ and a damping force $F_d = \mathrm{d}\dot{x}$. Thus, the equation of motion is given by

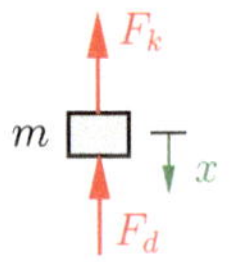

$$\downarrow: \quad m\ddot{x} = -F_k - F_d \, .$$

The stiffness of the equivalent spring follows from the spring stiffnesses in parallel of the beam and the bar

$$k_{12} = k_1 + k_2$$

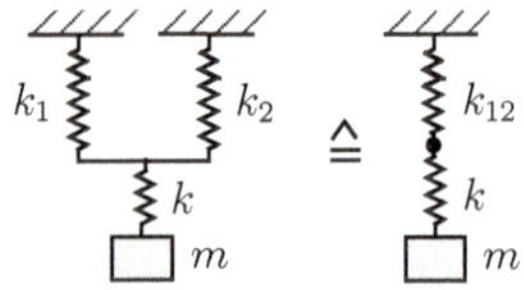

where

$$k_1 = \frac{3EI}{l_1^3} \, , \qquad k_2 = \frac{EA}{l_2}$$

and the spring stiffnesses in series of the spring k_{12} and the spring k as

$$\frac{1}{k^*} = \frac{1}{k_{12}} + \frac{1}{k} \quad \leadsto$$

$$k^* = \frac{k\left(\dfrac{3EI}{l_1^3} + \dfrac{EA}{l_2}\right)}{k + \dfrac{3EI}{l_1^3} + \dfrac{EA}{l_2}} \, .$$

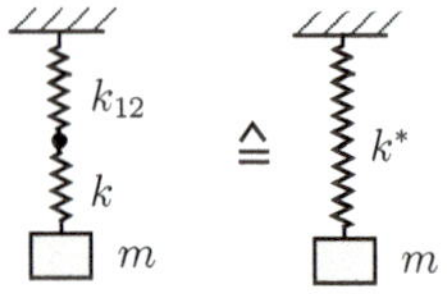

Hence, we obtain

$$m\ddot{x} + \mathrm{d}\dot{x} + k^* x = 0 \qquad \leadsto \qquad \ddot{x} + 2\xi\dot{x} + \omega^2 x = 0$$

where

$$\xi = \frac{\mathrm{d}}{2m} \, , \qquad \omega^2 = \frac{k^*}{m} \, .$$

Therefore, the eigenfrequency is given by

$$\underline{\underline{\omega_d}} = \sqrt{\omega^2 - \delta^2} = \underline{\underline{\sqrt{\frac{k^*}{m} - \left(\frac{\mathrm{d}}{2m}\right)^2}}} \, .$$

Problem 7.12 A bar (weight mg, length l) which is connected with a spring (spring constant k) vibrates in a viscous fluid about point A. The viscous drag force F_d is proportional to the local velocity (proportionality factor α). **P7.12**

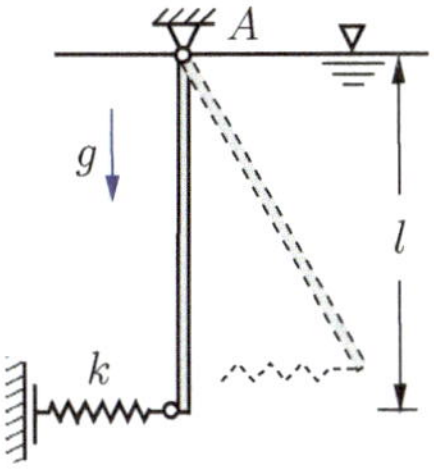

a) Derive the equation of motion under the assumption of small amplitudes.

b) Calculate the critical value α^*, separating vibrations from a creep motion.

Solution **a)** An element of length $\mathrm{d}x$ of the bar is subjected to the drag force

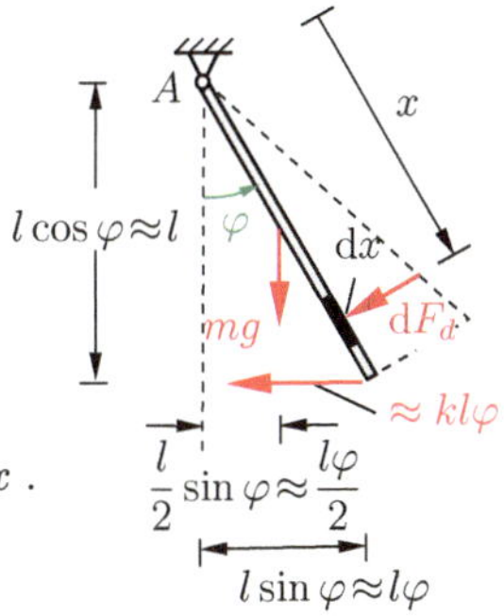

$$\mathrm{d}F_d = \alpha\, v(x)\,\mathrm{d}x = \alpha x\dot{\varphi}\,\mathrm{d}x\,.$$

Thus, considering small amplitudes $\varphi \ll 1$ ($\sin\varphi \approx \varphi$, $\cos\varphi \approx 1$), the principle of angular momentum yields

$$\overset{\curvearrowleft}{A}: \quad \Theta_A\ddot{\varphi} = -mg\frac{l}{2}\,\varphi - kl^2\varphi - \int_0^l \alpha x^2\dot{\varphi}\,\mathrm{d}x\,.$$

Evaluating the integral and introducing $\Theta_A = ml^2/3$ leads to the equation of motion

$$\underline{\underline{\ddot{\varphi} + \frac{\alpha l}{m}\,\dot{\varphi} + \frac{3k}{m}\left(1 + \frac{mg}{2kl}\right)\varphi = 0}} \quad \rightsquigarrow \quad \ddot{\varphi} + 2\xi\dot{\varphi} + \omega^2\varphi = 0$$

where

$$\xi = \frac{\alpha l}{2m}\,, \qquad \omega^2 = \frac{3k}{m}\left(1 + \frac{mg}{2kl}\right).$$

b) Damped vibrations are separated from an aperiodic motion by the critical damping

$$\xi = \omega \qquad \text{or} \qquad \zeta = 1\,.$$

From this condition follows

$$\frac{\alpha^* l}{2m} = \sqrt{\frac{3k}{m}\left(1 + \frac{mg}{2kl}\right)} \quad \rightsquigarrow \quad \underline{\underline{\alpha^* = \sqrt{\frac{12km}{l^2}}\sqrt{1 + \frac{mg}{2kl}}}}\,.$$

P7.13

Problem 7.13 Determine for the displayed underdamped system

a) the equation of motion for vibrations around the equilibrium position,

b) the circular frequency for the case $r_2 = R_2/4$, $k_2 = 2k_1$, $m_2 = 4m_1$ and

c) the solution $x_1(t)$ for the initial conditions $x_1(0) = 0$, $\dot{x}_1(0) = v_0$.

The rope and rolls can be considered as massless.

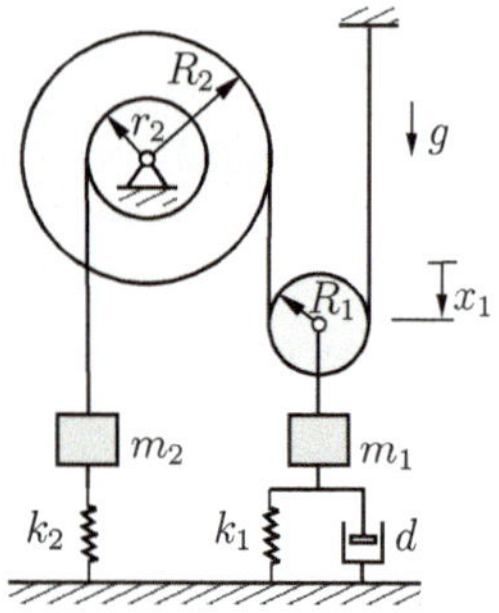

Solution **a)** We introduce all coordinates with their origin at the equilibrium position. Then the weights of the bodies must not be considered. From the kinematic relations at the rolls ① and ②

$$x_3 = 2x_1\,, \qquad \frac{x_3}{x_2} = \frac{R_2}{r_2}$$

it first follows

$$x_2 = 2\frac{r_2}{R_2}\,x_1\,, \qquad \ddot{x}_2 = 2\frac{r_2}{R_2}\,\ddot{x}_1\,.$$

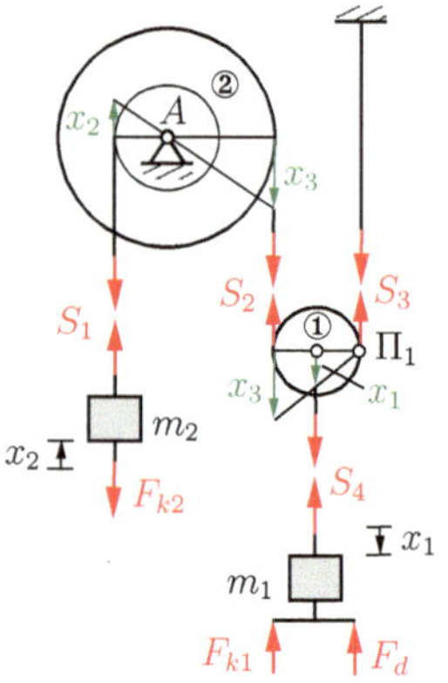

Now we separate the system and formulate the equations of motion for the bodies (mass m_1, m_2) and the rolls. Since the rolls are massless, the latter are reduced to the equilibrium conditions:

$$m_1 \;\downarrow:\quad m_1\ddot{x}_1 = -S_4 - F_{k1} - F_d\,,$$

$$m_2 \;\uparrow:\quad m_2\ddot{x}_2 = S_1 - F_{k2}\,,$$

$$① \;\overset{\curvearrowleft}{\Pi}_1:\quad R_1 S_4 = 2R_1 S_2\,,$$

$$② \;\overset{\curvearrowleft}{A}:\quad r_1 S_1 = R_2 S_2\,.$$

In conjunction with the spring and damper laws

$$F_{k1} = k_1 x_1\,, \quad F_d = d\,\dot{x}_1\,, \quad F_{k2} = k_2 x_2$$

we now have 9 equations for the 9 unknowns x_1, x_2, S_1, S_2, S_3, S_4, F_{k1}, F_d and F_{k2}. Solving for x_1 leads to the differential equation

$$\left[m_1 + m_2\left(2\frac{r_2}{R_2}\right)^2\right]\ddot{x}_1 + d\,\dot{x}_1 + \left[c_1 + c_2\left(2\frac{r_2}{R_2}\right)^2\right]x_1 = 0$$

or in standard form $\ddot{x}_1 + 2\xi\,\dot{x}_1 + \omega^2\,x_1 = 0$ where

$$2\xi = \frac{d}{m_1 + m_2\left(2\frac{r_2}{R_2}\right)^2}, \qquad \omega^2 = \frac{k_1 + k_2\left(2\frac{r_2}{R_2}\right)^2}{m_1 + m_2\left(2\frac{r_2}{R_2}\right)^2}.$$

b) The circular frequency of the underdamped vibration is calculated from $\omega_d = \sqrt{\omega^2 - \xi^2}$. For the parameters $r_2 = R_2/4$, $k_2 = 2k_1$, $m_2 = 4m_1$ we obtain

$$2\xi = \frac{d}{2\,m_1} \quad \rightsquigarrow \quad \xi^2 = \frac{d^2}{16\,m_1^2}, \qquad \omega^2 = \frac{3\,k_1}{4\,m_1},$$

and it follows

$$\omega_d = \sqrt{\frac{3\,k_1}{4\,m_1} - \frac{d^2}{16\,m_1^2}}\,.$$

c) The general solution for the vibration of an underdamped system reads

$$x_1(t) = \mathrm{e}^{-\xi t}\,(A\cos\omega_d t + B\sin\omega_d t)\,,$$

from which it follows by differentiation

$$\dot{x}_1(t) = \mathrm{e}^{-\xi t}\,[(-A\,\xi + B\,\omega_d)\cos\omega_d t - (A\,\omega_d + B\,\xi)\sin\omega_d t]\,.$$

The initial conditions lead to

$$x_1(0) = 0 \quad \rightsquigarrow \quad A = 0\,,$$

$$\dot{x}_1(0) = v_0 \quad \rightsquigarrow \quad -A\,\delta + B\,\omega_d = v_0 \quad \rightsquigarrow \quad B = \frac{v_0}{\omega_d}\,,$$

and we finally obtain

$$x_1(t) = \frac{v_0}{\omega_d}\,\mathrm{e}^{-\xi t}\sin\omega_d t\,.$$

P7.14 **Problem 7.14** A car (mass m), simplified modelled by a spring-mass-system, drives with constant horizontal velocity v_0 through sine-shaped periodic bumps (amplitude U_0, wave length L).

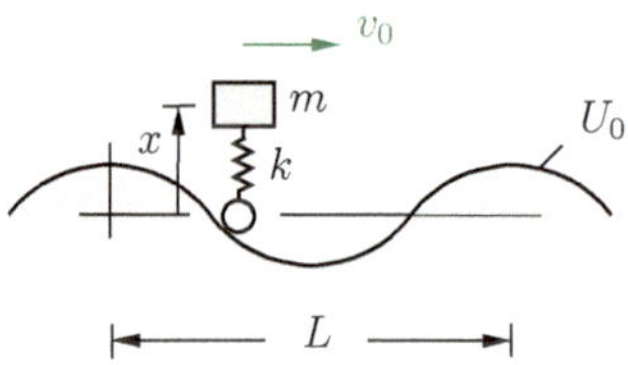

a) Derive the equation of motion of the car in vertical direction and determine the exciting frequency Ω.

b) Determine the vertical amplitude x_0 of the car in dependence of the velocity v_0.

c) Calculate the critical velocity v_c (resonance!).

Solution a) We denote the vertical displacement of the car by x and describe the shape of the bumps by u. Then from Newton's law follows

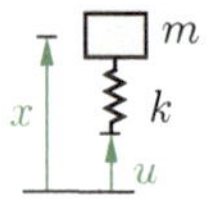

$$\uparrow: \quad m\ddot{x} = -k\,(x-u)\ .$$

With the position of the car $s = v_0 t$ in horizontal direction, the function u is represented as

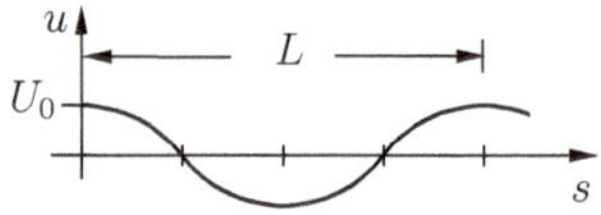

$$u = U_0 \cos\frac{2\pi s}{L} = U_0 \cos\frac{2\pi v_0 t}{L} = U_0 \cos\Omega t\ ,$$

and we obtain the equation of motion and the exciting frequency

$$\underline{\underline{m\,\ddot{x} + k\,x = k\,U_0\cos\Omega t}} \qquad \text{where} \qquad \underline{\underline{\Omega = \frac{2\pi v_0}{L}}}\ .$$

b) The steady state solution of the equation of motion is of the type of its right-hand side. Thus, the ansatz $x = x_0 \cos\Omega t$ leads with $\omega^2 = k/m$ to the amplitude of the steady state vibrations

$$\underline{\underline{x_0 = \frac{U_0}{1-\dfrac{\Omega^2}{\omega^2}} = \frac{U_0}{1-\dfrac{4\pi^2 v_0^2}{L^2}\dfrac{m}{k}}}}\ .$$

c) The amplitude x_0 tends to infinty when Ω approaches ω (resonance):

$$\Omega^2 = \omega^2 \quad \rightsquigarrow \quad \frac{4\pi^2 v_k^2}{L^2}\,\frac{m}{k} = 1 \quad \rightsquigarrow \quad \underline{\underline{v_k = \frac{L}{2\pi}\sqrt{\frac{k}{m}}}}\ .$$

Problem 7.15 A pressure gauge consists of a piston ① (mass m_1, cross section A), a bar ② (mass m_2), a thin needle ③ (mass m_3) and a spring (stiffness k). **P7.15**

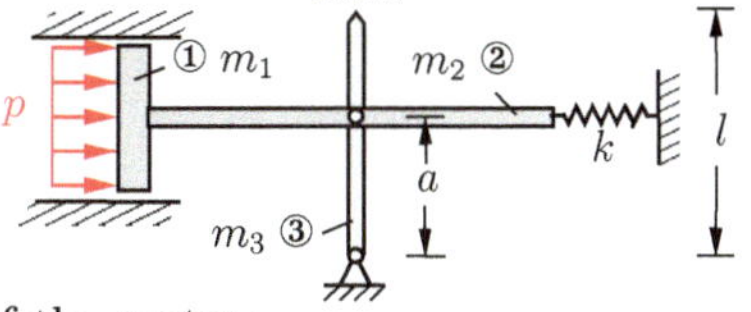

a) Determine the eigenfrequency of the system.

b) Calculate the amplitude Q_0 (small displacements!) of the needle tip in the steady state case if the pressure is given by $p = p_0 \cos \Omega t$.

Solution **a)** We separate the system and make all acting forces visible. Then, we have for the parts ① + ② and ③

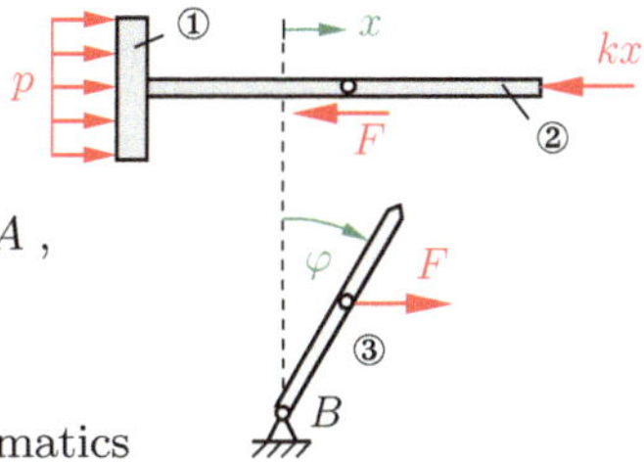

$$\rightarrow: \quad (m_1 + m_2)\ddot{x} = -F - kx + p(t)A \, ,$$

$$\overset{\curvearrowright}{B}: \quad \Theta_B \ddot{\varphi} = aF \, .$$

Hence, with $\Theta_B = m_3 l^2/3$ and the kinematics

$$x = a\varphi \quad \leadsto \quad \ddot{x} = a\ddot{\varphi}$$

follows the equation of motion

$$\left(m_1 + m_2 + \frac{m_3 l^2}{3a^2}\right)\ddot{x} + kx = p_0 A \cos \Omega t \, .$$

For the eigenfrequency, we directly obtain from this equation

$$\underline{\underline{\omega = \sqrt{\frac{k}{m_1 + m_2 + \dfrac{m_3 l^2}{3a^2}}}}} \, .$$

b) The steady state solution is described by an ansatz (of the right hand side type) $x = x_0 \cos \Omega t$. Substituting into the differential equation yields

$$x_0 = \frac{p_0 A}{k\left(1 - \dfrac{\Omega^2}{\omega^2}\right)} \, .$$

With the leverage we obtain for the amplitude of the needle tip

$$\underline{\underline{Q_0}} = x_0 \frac{l}{a} = \underline{\underline{\frac{1}{1 - \dfrac{\Omega^2}{\omega^2}} \frac{p_0 A}{k} \frac{l}{a}}} \, .$$

P7.16

Problem 7.16 A homogeneous wheel of mass m is attached to a spring (spring constant k). The wheel rolls without slipping on a rough surface which moves horizontally according to $u = u_0 \cos \Omega t$.

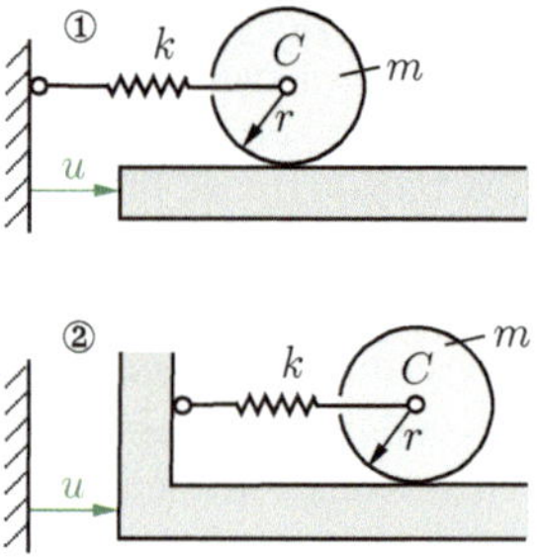

a) Derive the differential equations for the vibrations in case ① and ②.

b) Determine the amplitudes of the steady state vibrations.

Solution a) The equations of motion read in case ①

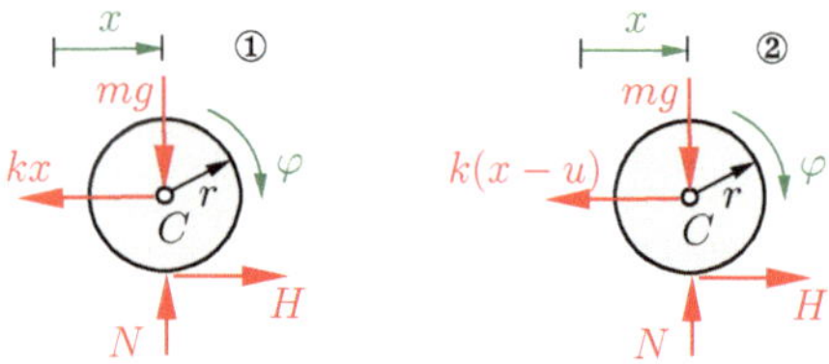

$$\rightarrow: \quad m\ddot{x} = -kx + H\,,$$

$$\overset{\curvearrowright}{C}: \quad \Theta_C\,\ddot{\varphi} = -r\,H\,.$$

With the kinematic relation

$$x = u + r\varphi \qquad \leadsto \qquad \ddot{x} = \ddot{u} + r\ddot{\varphi} = -u_0\,\Omega^2 \cos \Omega t + r\ddot{\varphi}$$

and $\Theta_S = \frac{1}{2} m r^2$ we obtain the differential equation for forced vibrations

$$\ddot{x} + \frac{2}{3}\,\frac{k}{m}\,x = -\frac{1}{3}\,u_0\,\Omega^2 \cos \Omega t\,.$$

In case ② the equations of motion are given by

$$\rightarrow: \quad m\ddot{x} = -k(x-u)x + H\,, \qquad \overset{\curvearrowright}{C}: \quad \Theta_C\,\ddot{\varphi} = -r\,H$$

and the kinematic relation again reads

$$x = u + r\varphi \qquad \leadsto \qquad \ddot{x} = \ddot{u} + r\ddot{\varphi} = -u_0\,\Omega^2 \cos \Omega t + r\ddot{\varphi}$$

which leads to the differential equation

$$\ddot{x} + \frac{2}{3}\,\frac{k}{m}\,x = \frac{1}{3}\left(2\,\frac{k}{m} - \Omega^2\right) u_0 \cos \Omega t\,.$$

b) The steady state solution is of the type of the right-hand side. Thus, the ansatz $x = x_0 \cos \Omega t$ leads to the amplitudes

$$\text{case ①} \quad \underline{\underline{|x_0| = u_0\,\frac{\Omega^2}{|2k/m - 3\Omega^2|}}}\,, \qquad \text{case ②} \quad \underline{\underline{|x_0| = u_0\,\frac{|2k/m - \Omega^2|}{|2k/m - 3\Omega^2|}}}\,.$$

Note that resonance in both cases occurs for $\Omega^2 = 2k/3m$, but that the amplitudes for $\Omega^2 \ll 2k/3m$ and for $\Omega^2 \gg 2k/3m$ are very different.

Problem 7.17 In a soil compactor (housing weight mg), the drive (mass M) is resiliently mounted. In the drive two unbalances (each mass $m_2/2$) counter-rotate with constant number n of revolutions per minute. **P7.17**

How must the spring and damper be designed so that the device runs in resonance and the base plate does not lift-off from the ground (small damping ratio!)?

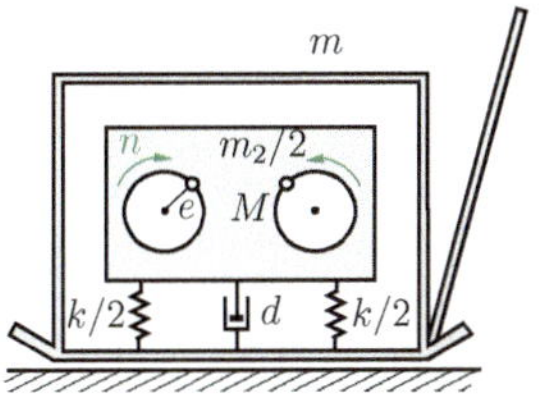

Solution We replace the drive by the displayed model. With the displacement x of the drive from the equilibrium position we obtain for the unbalanced mass

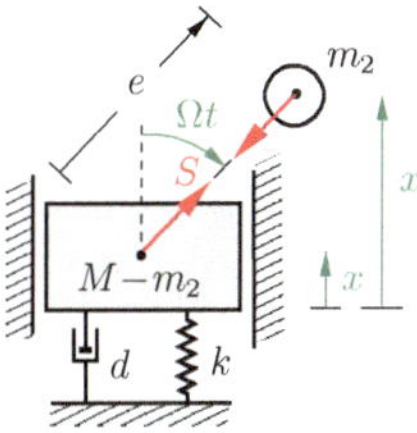

$$x_2 = x + e\cos\Omega t \quad \rightsquigarrow \quad \ddot{x}_2 = \ddot{x} - e\Omega^2\cos\Omega t\,.$$

The equations of motion in vertical direction for both masses are given by:

$$(M-m_2)\ddot{x} = -d\dot{x} - kx + S\cos\Omega t\,, \qquad m_2\ddot{x}_2 = -S\cos\Omega t\,.$$

Introduction of $\ddot{x}_2$ and elimination of S yields

$$\frac{\ddot{x}}{\omega^2} + \frac{2\xi\dot{x}}{\omega^2} + x = x_0\eta^2\cos\Omega t\,,$$

where $\omega^2 = k/M$, $\xi = \mathrm{d}/2M$, $x_0 = em_2/M$, $\eta = \Omega/\omega$. The particular solution (steady state) reads

$$x = x_0 V\cos(\Omega t - \varphi)\,, \qquad V = \frac{\eta^2}{\sqrt{(1-\eta^2)^2 + 4\zeta^2\eta^2}}\,, \qquad \zeta = \frac{\xi}{\omega}\,.$$

Resonance occurs for *small damping ratio* ζ at $\eta \approx 1$:

$$\Omega = \omega \quad \rightsquigarrow \quad \frac{\pi n}{30} = \sqrt{\frac{k}{M}} \quad \rightsquigarrow \quad \underline{\underline{k = \left(\frac{\pi n}{30}\right)^2 M}}\,.$$

Furthermore, in resonance the magnification is $V \approx 1/2\zeta$.

No lift-off occurs, if the following condition at maximum spring displacement $x_{\max}$ (then the damping force is zero!) is fulfilled:

$$N = (M+m)g - kx_{\max} = (M+m)g - \frac{kx_0}{2\zeta} \geq 0\,.$$

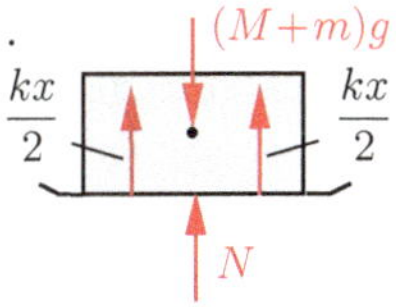

Hence, it follows

$$(M+m)g \geq \frac{kx_0}{2\zeta} \quad \rightsquigarrow \quad \underline{\underline{\zeta \geq \frac{kx_0}{2(M+m)g}}}\,.$$

P7.18 **Problem 7.18** A single story frame is modelled by a rigid beam of mass m which is supportet by clamped massless beams (stiffness EI) and damped by a dashpot. Due to an earthquake, the ground moves with a horizontal acceleration $\ddot{u}_E = b_0 \cos \Omega t$ which is known from measurements.

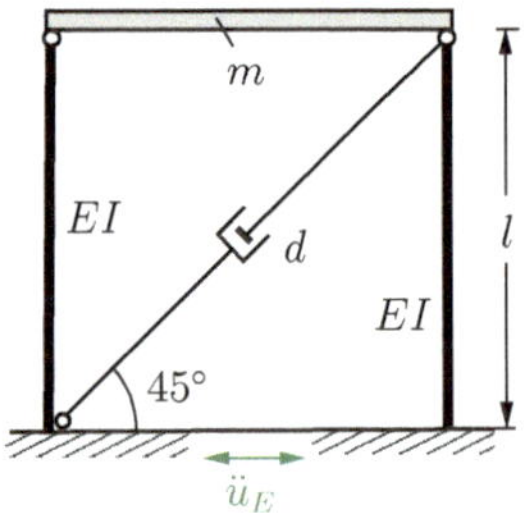

Determine the maximum amplitude of the steady state vibrations by assuming small amplitudes and a weakly damped system.

Solution We separate the system and replace the vertical beams by equivalent springs with the spring constant $k = 3EI/l^3$. When the beams are deflected by $x - u_E$, the elongation of the diagonal, assuming small amplitudes, is

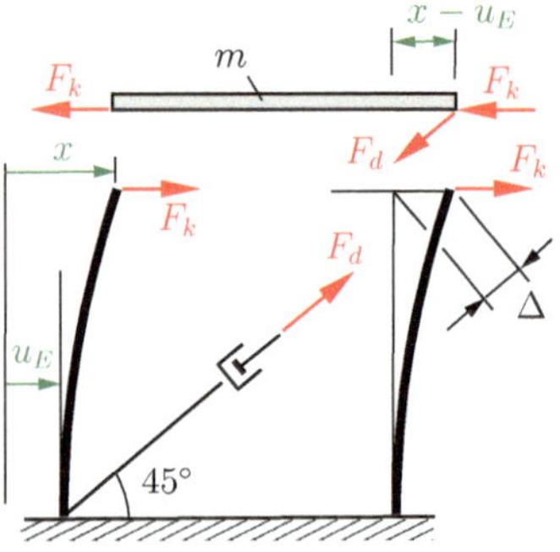

$$\Delta = (x - u_E)/\sqrt{2}\,.$$

Hence, the spring forces and the damping force can be written as

$$F_k = k(x - u_E)\,, \qquad F_d = d\,\dot{\Delta} = d\,(\dot{x} - \dot{u}_E)/\sqrt{2}\,.$$

Then, the equation of motion of the horizontal beam is given by

$$\rightarrow: \; m\ddot{x} = -\frac{\sqrt{2}}{2} F_d - 2F_k \quad \leadsto \quad m\ddot{x} + \frac{d}{2}(\dot{x} - \dot{u}_E) + 2k(x - u_E) = 0\,.$$

Thus, the relative displacement $y = x - u_E$ is described by

$$m\ddot{y} + \frac{d}{2}\,\dot{y} + 2ky = m\,b_0 \cos \Omega t \quad \leadsto \quad \frac{1}{\omega^2}\,\ddot{y} + \frac{2\zeta}{\omega}\,\dot{y} + y = y_0 \cos \Omega t\,,$$

where

$$\omega^2 = \frac{2k}{m}\,, \qquad \zeta = \frac{d}{2}\sqrt{\frac{1}{8k\,m}}\,, \qquad y_0 = \frac{m\,b_0}{2k}\,.$$

The maximum amplitude A occurs for resonance, i.e. for $\eta = \omega/\Omega \approx 1$. In case of weak damping ($\zeta \ll 1$) it is given by

$$\underline{\underline{A}} = y_0 V_{max} \approx \frac{y_0}{2\zeta} = \underline{\underline{2\sqrt{2}\,\frac{b_0}{d}\sqrt{\frac{m^3}{4k}}}}\,.$$

Problem 7.19 A wheel is connected with a rigid frame via a spring and a damper. When the frame experiences a displacement $u = v_0 t$, the initially ($t = 0$) resting wheel starts rolling without slip. **P7.19**

a) How must the damper be designed so that for free vibrations critical damping occurs? Which form has the equation of motion in this case?

b) Determine the solution for the given initial conditions.

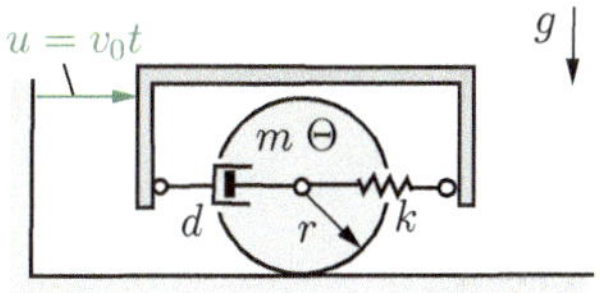

Solution a) We count x and φ from the equilibrium position of the wheel at $t = 0$. Then the equations of motion

$$\rightarrow: \quad m\ddot{x} = -d(\dot{x} - \dot{u}) - k(x - u) - H\,, \qquad \overset{\curvearrowright}{C}: \quad \Theta\,\ddot{\varphi} = r\,H$$

lead with $\dot{u} = v_0$ and $x = r\,\varphi$ by eliminating H and $\ddot{\varphi}$ to the differential equation

$$\left(m + \frac{\Theta}{r^2}\right)\ddot{x} + d\,\dot{x} + k\,x = d\,\dot{u} + k\,u\,.$$

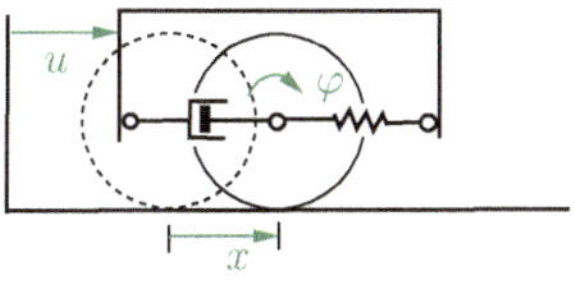

It can be written in the standard form

$$\ddot{x} + 2\xi\,\dot{x} + \omega^2 x = 2\,\xi\,v_0 + \omega^2 v_0\,t\,,$$

where

$$2\,\xi = \frac{d}{m + \Theta/r^2}\,, \qquad \omega^2 = \frac{k}{m + \Theta/r^2}\,.$$

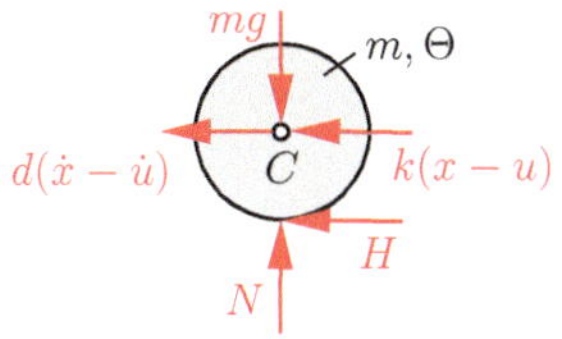

The condition for critical damping of free vibrations is given by $\xi/\omega = 1$, or $\xi^2 = \omega^2$. This leads for the damping coefficient to

$$\frac{d^2}{4\,(m + \Theta/r^2)^2} = \frac{k}{m + \Theta/r^2} \quad \rightsquigarrow \quad \underline{\underline{d = 2\sqrt{k\,(m + \Theta/r^2)}}}\,.$$

In this case, the equation of motion takes the following form:

$$\underline{\underline{\ddot{x} + 2\omega\,\dot{x} + \omega^2 x = 2\,\omega\,v_0 + \omega^2 v_0\,t}}\,.$$

b) The solution of the differential equation is composed of the solution of the homogeneous differential equation (= free vibration at critical damping)

$$x_h = (A_1 + A_2 t)\,\mathrm{e}^{-\omega t}$$

and the particular solution x_p. To find the latter one, we choose an ansatz of the type of the right hand side:

$$x_p = a + b\,t\,.$$

Introducing it into the differential equation and comparing the coefficients yields $a = 0$ and $b = v_0$. Thus, the general solution is given by

$$x(t) = x_h + x_p = (A_1 + A_2 t)\,\mathrm{e}^{-\omega t} + v_0 t\,.$$

The constants A_1 and A_2 are determined from the initial conditions:

$$x(0) = 0 \quad \leadsto \quad A_1 = 0\,, \qquad \dot{x}(0) = 0 \quad \leadsto \quad A_2 = -v_0\,.$$

Hence, we obtain the specific solution

$$\underline{\underline{x(t)}} = -v_0 t\,\mathrm{e}^{-\omega t} + v_0 t = \underline{\underline{v_0 t\,(1 - \mathrm{e}^{-\omega t})}}\,.$$

It can be seen that the motion of the wheel exponentially approaches the motion of the frame. For $\omega t \gg 1$ the motion of the frame and the wheel are the same.

Chapter 8

Non-Inertial Reference Frames

8

Fixed and moving Reference Frame

It is often advantageous to describe the motion of a point P not in reference to a fixed coordinate system (x, y, z) but in reference to a moving system (ξ, η, ζ).

Kinematics of a point for a translating and rotating reference system

$$\boldsymbol{v} = \boldsymbol{v}_f + \boldsymbol{v}_r \ ,$$

$$\boldsymbol{a} = \boldsymbol{a}_f + \boldsymbol{a}_c + \boldsymbol{a}_r \ ,$$

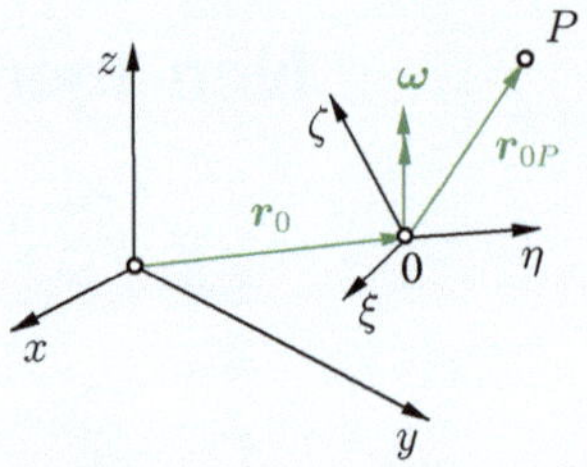

where

absolute velocity	$\boldsymbol{v}$,
absolute acceleration	$\boldsymbol{a}$,
fictitious velocity	$\boldsymbol{v}_f = \dot{\boldsymbol{r}}_0 + \boldsymbol{\omega} \times \boldsymbol{r}_{0P}$,
relative velocity	$\boldsymbol{v}_r = \overset{*}{\boldsymbol{r}}_{0P}$,
fictitious acceleration	$\boldsymbol{a}_f = \ddot{\boldsymbol{r}}_0 + \dot{\boldsymbol{\omega}} \times \boldsymbol{r}_{0P} + \boldsymbol{\omega} \times (\boldsymbol{\omega} \times \boldsymbol{r}_{0P})$,
relative acceleration	$\boldsymbol{a}_r = \overset{**}{\boldsymbol{r}}_{0P}$,
Coriolis acceleration	$\boldsymbol{a}_c = 2\boldsymbol{\omega} \times \boldsymbol{v}_r$.

and

$()^{\cdot} \mathrel{\widehat{=}}$ time derivative with respect to the fixed system (x, y, z),

$()^{*} \mathrel{\widehat{=}}$ time derivative with respect to the moving system (ξ, η, ζ).

Remarks:

- The equations simplify for a pure translation of the reference system ($\boldsymbol{\omega} = 0$).
- The *Coriolis acceleration* $\boldsymbol{a}_c$ is orthogonal to $\boldsymbol{\omega}$ and $\boldsymbol{v}_r$.

Equation of motion in a moving reference system

In addition to the real forces $\boldsymbol{F}$ acting on the point mass, the *fictitious force* $\boldsymbol{F}_f$ and the *Coriolis force* $\boldsymbol{F}_c$ appear in the equation of motion:

$$m\boldsymbol{a}_r = \boldsymbol{F} + \boldsymbol{F}_f + \boldsymbol{F}_c$$

where

$$\boldsymbol{F}_f = -m\boldsymbol{a}_f = -m[\ddot{\boldsymbol{r}}_0 + \dot{\boldsymbol{\omega}} \times \boldsymbol{r}_{0P} + \boldsymbol{\omega} \times (\boldsymbol{\omega} \times \boldsymbol{r}_{0P})] \ ,$$

$$\boldsymbol{F}_c = -m\boldsymbol{a}_c = -2m\boldsymbol{\omega} \times \boldsymbol{v}_r \ .$$

Problem 8.1 Point A of a simple pendulum (mass m, length l) moves with a constant acceleration a_0 obliquely upwards. P8.1

Derive the equation of motion and determine the force in the wire.

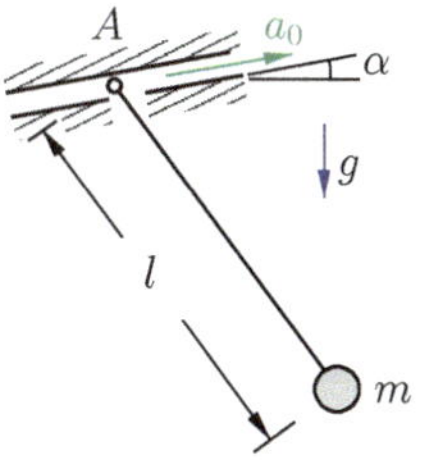

Solution We introduce the ξ, η coordinate system that moves translatoric with point A. Then the equations of motion in the moving system read

$$m\xi^{**} = F_\xi + F_{f\xi}\,, \qquad m\eta^{**} = F_\eta + F_{f\eta}$$

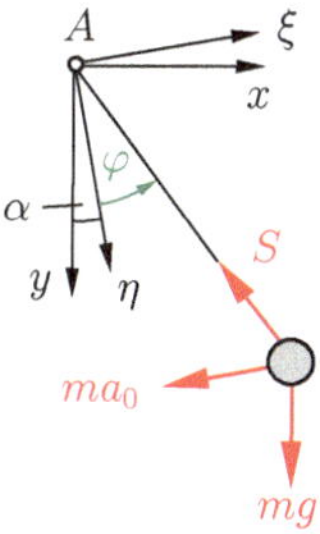

where, with $a_{f\xi} = a_0$, and $a_{f\eta} = 0$, the (real) forces and fictitious forces are given by

$$F_\xi = -S\sin\varphi - mg\sin\alpha\,, \qquad F_\eta = -S\cos\varphi + mg\cos\alpha\,,$$
$$F_{f\xi} = -ma_{f\xi} = -ma_0\,, \qquad F_{f\eta} = -ma_{f\eta} = 0\,.$$

The relative accelerations $a_{r\xi}$, $a_{r\eta}$ follow from the coordinates of the point mass in the moving system through differentiation (note that the time derivatives of φ in the moving and the fixed system are the same: $\varphi^* = \dot\varphi$)

$$\xi = l\sin\varphi\,, \qquad \eta = l\cos\varphi\,,$$
$$v_{r\xi} = \xi^* = l\dot\varphi\cos\varphi\,, \qquad v_{r\eta} = \eta^* = -l\dot\varphi\sin\varphi\,,$$
$$a_{r\xi} = \xi^{**} = l\ddot\varphi\cos\varphi - l\dot\varphi^2\sin\varphi\,, \quad a_{r\eta} = \eta^{**} = -l\ddot\varphi\sin\varphi - l\dot\varphi^2\cos\varphi\,.$$

Introducing them into the equations of motion yields

$$l\ddot\varphi\cos\varphi - l\dot\varphi^2\sin\varphi = -S\sin\varphi - mg\sin\alpha - ma_0\,,$$
$$-l\ddot\varphi\sin\varphi - l\dot\varphi^2\cos\varphi = -S\cos\varphi + mg\cos\alpha\,.$$

These are two equations for the unknowns φ and S. Solving for φ and subsequently for S leads to the equation of motion and the force in the wire:

$$\underline{\underline{l\ddot\varphi + g\sin(\alpha+\varphi) + a_0\cos\varphi = 0}}\,,$$

$$\underline{\underline{S = m[l\dot\varphi^2 + g\cos(\alpha+\varphi) - a_0\sin\varphi]}}\,.$$

Remark: For $\ddot\varphi = 0$ the equation of motion reduces to $\tan\varphi_0 = a_0/g\cos\alpha$ $-tan\alpha$ characterizing the equilibrium position of the pendulum.

P8.2 **Problem 8.2** A point P moves on a plate with constant relative velocity v_r and initial condition $\varphi(0) = 0$ along a circular path. The plate moves rectilinear with the velocity $v = a_0 t$.

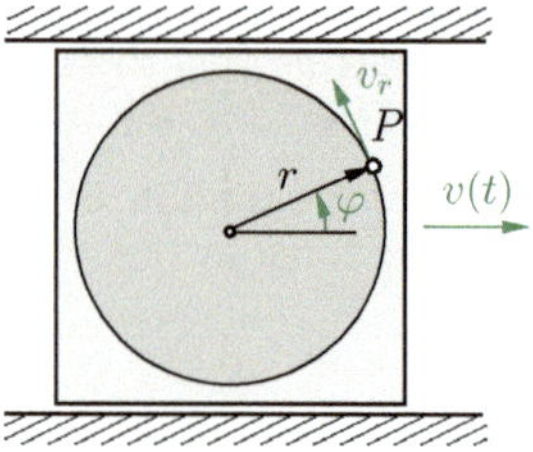

Determine the magnitude of absolute velocity and acceleration of P as functions of angle φ.

Solution We introduce the fixed x, y-coordinates and the translatoric moving reference frame ξ, η. In the moving system, introducing $\varphi^* = \dot{\varphi} = v_r/r$, the components of the relative velocity and acceleration are

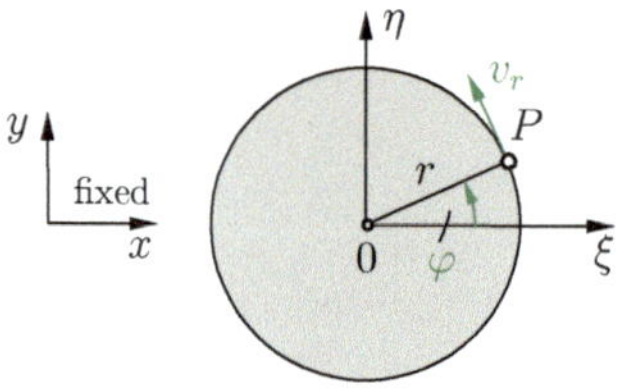

$$v_{r\xi} = \xi^* = -v_r \sin\varphi \,,$$

$$v_{r\eta} = \eta^* = v_r \cos\varphi \,,$$

$$a_{r\xi} = \xi^{**} = -v_r \varphi^* \cos\varphi = -\frac{v_r^2}{r}\cos\varphi \,,$$

$$a_{r\eta} = \eta^{**} = -v_r \varphi^* \sin\varphi = -\frac{v_r^2}{r}\sin\varphi \,.$$

The reference frame undergoes a translation in x-direction with velocity $v = a_0 t$ and acceleration a, where time t on account of $\varphi = \dot{\varphi} t = v_r t/r$ can be replaced by φ. Accordingly, the absolute velocity and acceleration are given by

$$v_x = a_0 t + v_{r\xi} = a_0 t - v_r \sin\varphi = \frac{a_0 r}{v_r}\varphi - v_r \sin\varphi \,,$$

$$v_y = v_{r\eta} = v_r \cos\varphi \,,$$

$$a_x = a_0 + a_{r\xi} = a_0 - \frac{v_r^2}{r}\cos\varphi \,,$$

$$a_y = a_{r\eta} = -\frac{v_r^2}{r}\sin\varphi \,.$$

Thus, the magnitudes of velocity and acceleration follow as

$$\underline{\underline{v(\varphi)}} = \sqrt{v_x^2 + v_y^2} = \underline{\underline{\sqrt{\frac{a_0^2 r^2}{v_r^2}\varphi^2 - 2a_0 r\,\varphi + v_r^2}}} \,,$$

$$\underline{\underline{a(\varphi)}} = \sqrt{a_x^2 + a_y^2} = \underline{\underline{\sqrt{a_0^2 + \frac{v_r^4}{r^2} - 2a_0\,\frac{v_r^2}{r}\cos\varphi}}} \,.$$

Problem 8.3 A point P moves on a disk with constant relative velocity v_r along a circular path. The disk rotates with constant angular velocity ω about A. P8.3

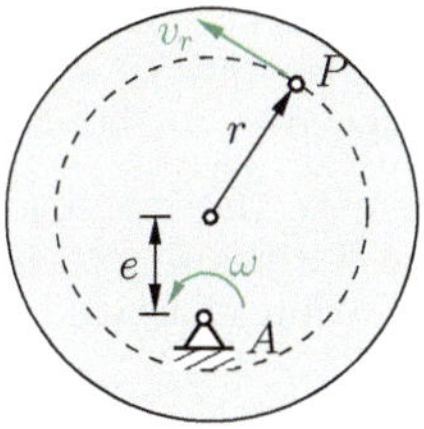

Determine the absolute velocity and the absolute acceleration of P.

Solution We introduce the moving coordinate system ξ, η, ζ with its origin in the center 0 of the disk. Relative to this system, point P undergoes a circular motion. With the relative velocity v_r and the magnitude $a_r = v_r^2/r$ of the relative acceleration and its direction (from P to 0) we can write

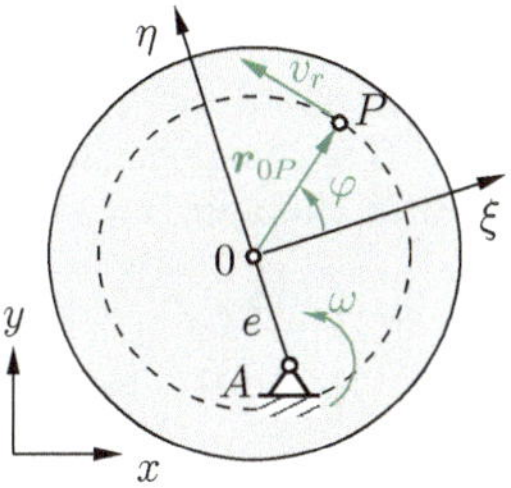

$$\boldsymbol{v}_r = v_r(-\boldsymbol{e}_\xi \sin\varphi + \boldsymbol{e}_\eta \cos\varphi)\,,$$

$$\boldsymbol{a}_r = -\frac{v_r^2}{r}\,(\boldsymbol{e}_\xi \cos\varphi + \boldsymbol{e}_\eta \sin\varphi)\,.$$

With

$$\boldsymbol{\omega} = \omega \boldsymbol{e}_\zeta\,, \qquad \dot{\boldsymbol{\omega}} = 0\,, \qquad \boldsymbol{r}_{0P} = \boldsymbol{e}_\xi r\cos\varphi + \boldsymbol{e}_\eta r \sin\varphi\,,$$

$$\boldsymbol{r}_0 = e\,\boldsymbol{e}_\eta\,, \qquad \dot{\boldsymbol{r}}_0 = -e\,\omega \boldsymbol{e}_\xi \qquad \ddot{\boldsymbol{r}}_0 = \boldsymbol{a}_0 = -e\,\omega^2 \boldsymbol{e}_\eta$$

we obtain

$$\begin{aligned}
\boldsymbol{v}_f &= \dot{\boldsymbol{r}}_0 + \boldsymbol{\omega}\times\boldsymbol{r}_{0P} = -(e + r\omega\sin\varphi)\boldsymbol{e}_\xi + r\omega\cos\varphi\boldsymbol{e}_\eta\,,\\
\boldsymbol{a}_f &= \boldsymbol{a}_0 + \boldsymbol{\omega}\times(\boldsymbol{\omega}\times\boldsymbol{r}_{0P})\\
&= -e\,\omega^2\boldsymbol{e}_\xi + r\omega^2[\boldsymbol{e}_\zeta\times(\boldsymbol{e}_\zeta\times\boldsymbol{e}_\xi\cos\varphi) + \boldsymbol{e}_\zeta\times(\boldsymbol{e}_\zeta\times\boldsymbol{e}_\eta\sin\varphi)]\\
&= -(e + r\cos\varphi)\omega^2\boldsymbol{e}_\xi - r\omega^2\sin\varphi\boldsymbol{e}_\eta\,,\\
\boldsymbol{a}_c &= 2\boldsymbol{\omega}\times\boldsymbol{v}_r = 2\omega v_r[\boldsymbol{e}_\zeta\times(-\boldsymbol{e}_\xi\sin\varphi) + \boldsymbol{e}_\zeta\times\boldsymbol{e}_\eta\cos\varphi]\\
&= -2\omega v_r(\boldsymbol{e}_\xi\cos\varphi + \boldsymbol{e}_\eta\sin\varphi)\,.
\end{aligned}$$

Thus, the absolute velocity and acceleration are found as

$$\underline{\underline{\boldsymbol{v}}} = \boldsymbol{v}_f + \boldsymbol{v}_r = \underline{\underline{-[e + (v_r + r\omega)\sin\varphi]\boldsymbol{e}_\xi + (v_r + r\omega)\cos\varphi\boldsymbol{e}_\eta}}\,,$$

$$\begin{aligned}
\underline{\underline{\boldsymbol{a}}} &= \boldsymbol{a}_f + \boldsymbol{a}_r + \boldsymbol{a}_c\\
&= -[e\omega^2 + (r\omega^2 + \frac{v_r^2}{r} + 2\omega v_r)\cos\varphi]\boldsymbol{e}_\xi - [r\omega^2 + \frac{v_r^2}{r} + 2\omega v_r]\sin\varphi\,\boldsymbol{e}_\eta\\
&= \underline{\underline{-[e\omega^2 + r(\omega + \frac{v_r}{r})^2\cos\varphi]\boldsymbol{e}_\xi - r(\omega + \frac{v_r}{r})^2\sin\varphi\,\boldsymbol{e}_\eta}}\,.
\end{aligned}$$

P8.4 **Problem 8.4** On the rotating earth (radius $R = 6370\ km$) a point P moves with speed $v_r = 150\ km/h$ northwards.

Determine the magnitudes and directions of the fictitious acceleration and Coriolis acceleration at latitude $\varphi = 30°$.

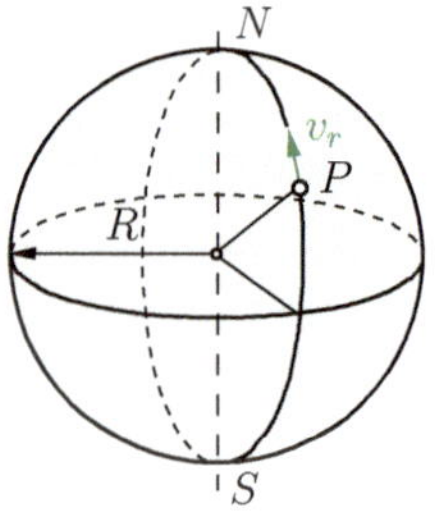

Solution The earth-fixed system ξ, η, ζ rotates with the angular velocity $\omega = 2\pi/(24\cdot 3600) \approx 73\cdot 10^{-6}\ s^{-1}$ around the ζ-axis. If we neglect the motion of the earth around the sun ($\ddot{\boldsymbol{r}}_0 = 0$), we obtain with

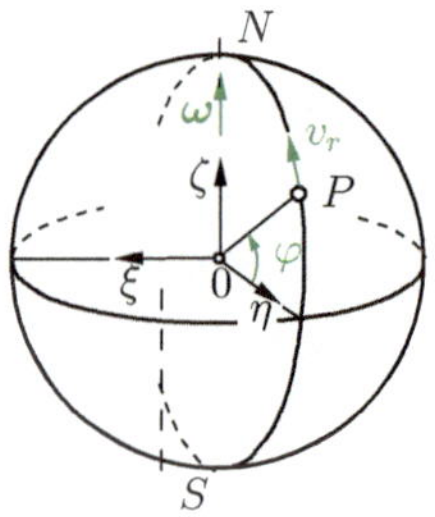

$$\boldsymbol{\omega} = \omega\, \boldsymbol{e}_\zeta\ ,$$
$$\boldsymbol{r}_{0P} = R\cos\varphi\ \boldsymbol{e}_\eta + R\sin\varphi\ \boldsymbol{e}_\zeta\ ,$$
$$\dot{\boldsymbol{\omega}} = 0\ ,$$
$$\boldsymbol{v}_r = -v_r\sin\varphi\ \boldsymbol{e}_\eta + v_r\cos\varphi\ \boldsymbol{e}_\zeta$$

for $\boldsymbol{a}_f$ and $\boldsymbol{a}_c$

$$\boldsymbol{a}_f = \boldsymbol{\omega}\times(\boldsymbol{\omega}\times\boldsymbol{r}_{0P}) = \omega^2 R[\boldsymbol{e}_\zeta\times(\boldsymbol{e}_\zeta\times\cos\varphi\ \boldsymbol{e}_\eta) + \boldsymbol{e}_\zeta\times(\boldsymbol{e}_\zeta\times\sin\varphi\ \boldsymbol{e}_\zeta)]$$
$$= -\omega^2 R\cos\varphi\ \boldsymbol{e}_\eta\ ,$$

$$\boldsymbol{a}_c = 2\boldsymbol{\omega}\times\boldsymbol{v}_r = 2\omega v_r[\boldsymbol{e}_\zeta\times(-\sin\varphi\ \boldsymbol{e}_\eta) + \boldsymbol{e}_\zeta\times(\cos\varphi\ \boldsymbol{e}_\zeta)]$$
$$= 2\omega v_r\sin\varphi\ \boldsymbol{e}_\xi\ .$$

Thus, the magnitudes of the accelerations for $\varphi = 30°$ are

$$\underline{\underline{a_f}} = \omega^2 R\cos\varphi = (73)^2\cdot 10^{-12}\cdot 6370\cdot 10^3\cos 30° = \underline{\underline{0,029\ \mathrm{m/s^2}}}\ ,$$

$$\underline{\underline{a_c}} = 2\omega v_r\sin\varphi = 2\cdot 73\cdot 10^{-6}\cdot 150\cdot\frac{1}{3,6}\cdot\sin 30° = \underline{\underline{0,003\ \mathrm{m/s^2}}}\ .$$

The fictitious acceleration is perpendicular to the axis of rotation of the earth and the Coriolis acceleration points tangential to the latitude to the west.

Remarks:

- With reference to the moving system, the motion of point P is a pure circular motion.
- The Coriolis acceleration has its maximum at the north pole.

Problem 8.5 Along the upper part of an angled rod which rotates with angular velocity ω, a knuckle (weight $W = mg$) may fritionless slide. **P8.5**

After which time t_1 the knuckle touches the rod end B if it is released at A with zero relative velocity?

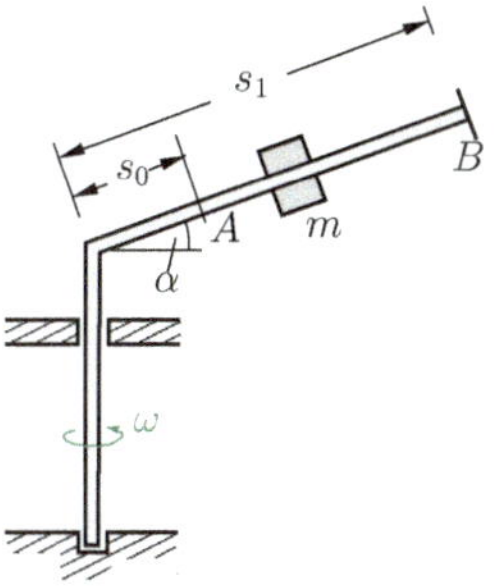

Solution We introduce the rod-fixed rotating reference system ξ, η, ζ and the coordinate s. Then we obtain with

$$\boldsymbol{\omega} = \omega \boldsymbol{e}_\zeta \, ,$$

$$\boldsymbol{r}_{0P} = s \cos\alpha \, \boldsymbol{e}_\xi + s \sin\alpha \, \boldsymbol{e}_\zeta \, ,$$

$$\boldsymbol{v}_r = v_r \cos\alpha \, \boldsymbol{e}_\xi + v_r \sin\alpha \, \boldsymbol{e}_\zeta$$

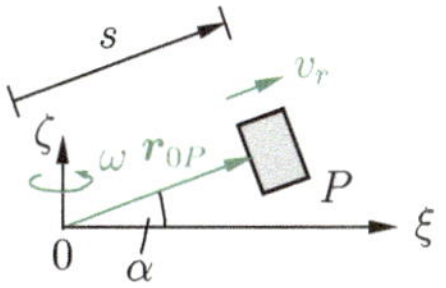

and $\dot{\boldsymbol{\omega}} = 0$, $\ddot{\boldsymbol{r}}_0 = 0$ for the fictitious force and Coriolis force:

$$\boldsymbol{F}_f = -m\boldsymbol{\omega} \times (\boldsymbol{\omega} \times \boldsymbol{r}_{0P}) = m\omega^2 s \cos\alpha \, \boldsymbol{e}_\xi \, ,$$

$$\boldsymbol{F}_c = -2m\boldsymbol{\omega} \times \boldsymbol{v}_r = -2m\omega v_r \cos\alpha \, \boldsymbol{e}_\eta \, .$$

Thus, the equation of motion in $s-$direction reads

$$\nearrow: \quad m\ddot{s} = -mg\sin\alpha + m\omega^2 s\cos^2\alpha$$

$$\rightsquigarrow \quad \ddot{s} - \kappa^2 s = -g\sin\alpha \quad \text{with} \quad \kappa = \omega\cos\alpha .$$

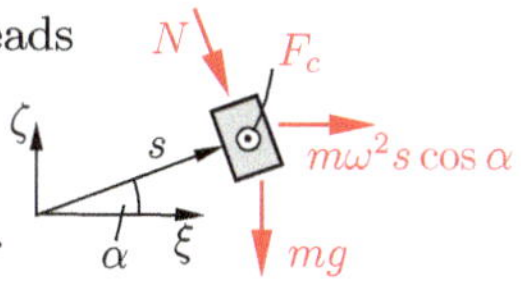

From the general solution of this differential equation

$$s(t) = A\cosh\kappa t + B\sinh\kappa t + \frac{g\sin\alpha}{\kappa^2}$$

in conjunction with the initial conditions

$$s(t=0) = s_0 \quad \rightsquigarrow \quad A = s_0 - \frac{g\sin\alpha}{\kappa^2} \, ,$$

$$v_r(t=0) = \dot{s}(t=0) = 0 \quad \rightsquigarrow \quad B = 0 \, ,$$

$$s(t_1) = s_1 \quad \rightsquigarrow \quad s_1 = A\cosh\kappa t_1 + \frac{g\sin\alpha}{\kappa^2}$$

follows

$$\underline{\underline{t_1 = \frac{1}{\kappa}\operatorname{arcosh}\left(\frac{s_1 - \dfrac{g\sin\alpha}{\kappa^2}}{s_0 - \dfrac{g\sin\alpha}{\kappa^2}}\right)}} .$$

P8.6

Problem 8.6 The cantilever of a whirlgig moves about the x-axis with the time-dependent angle $\vartheta(t)$. At its end a circular disk is fixed which rotates with constant angular velocity ω_0 about an axis perpendicular to the cantilever.

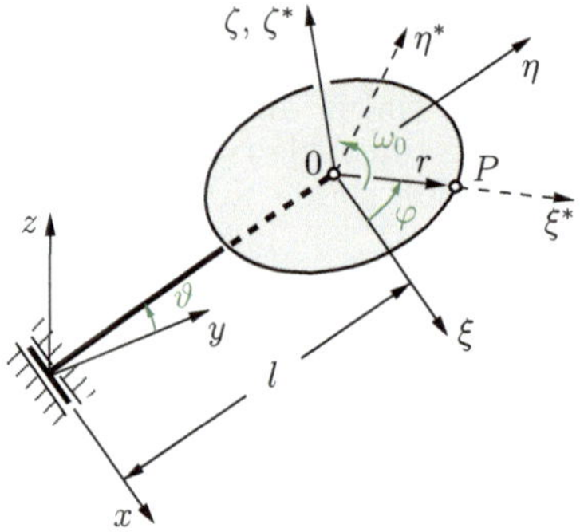

Determine for point P of the disk the absolute velocity and absolute acceleration by using

a) the cantilever-fixed systems ξ, η, ζ and

b) the disk-fixed systems ξ^*, η^*, ζ^*.

Solution The absolute velocity and acceleration are determined from the general relations

$$\boldsymbol{v} = \dot{\boldsymbol{r}}_0 + \boldsymbol{\omega} \times \boldsymbol{r}_{0P} + \boldsymbol{v}_r \,,$$
$$\boldsymbol{a} = \ddot{\boldsymbol{r}}_0 + \dot{\boldsymbol{\omega}} \times \boldsymbol{r}_{0P} + \boldsymbol{\omega} \times (\boldsymbol{\omega} \times \boldsymbol{r}_{0P}) + 2\,\boldsymbol{\omega} \times \boldsymbol{v}_r + \boldsymbol{a}_r \,.$$

a) For the coordinate system ξ, η, ζ with the unit vectors $\boldsymbol{e}_\xi$, $\boldsymbol{e}_\eta$, $\boldsymbol{e}_\zeta$ we first have

$$\boldsymbol{r}_0 = l\,\boldsymbol{e}_\eta \,, \qquad \boldsymbol{\omega} = \dot{\vartheta}\,\boldsymbol{e}_\xi \,, \qquad \boldsymbol{r}_{0P} = r\cos\varphi\,\boldsymbol{e}_\xi + r\sin\varphi\,\boldsymbol{e}_\eta \,.$$

From these relations, by introducing ($\dot{\boldsymbol{e}}_i = \boldsymbol{\omega} \times \boldsymbol{e}_i$)

$$\dot{\boldsymbol{e}}_\xi = 0 \,, \qquad \dot{\boldsymbol{e}}_\eta = \dot{\vartheta}\,\boldsymbol{e}_\zeta \,, \qquad \dot{\boldsymbol{e}}_\zeta = -\dot{\vartheta}\,\boldsymbol{e}_\eta \,, \qquad \dot{\varphi} = \omega_0 \,,$$

we obtain

$$\dot{\boldsymbol{r}}_0 = l\,\dot{\vartheta}\,\boldsymbol{e}_\zeta \,, \qquad \ddot{\boldsymbol{r}}_0 = l\,\ddot{\vartheta}\,\boldsymbol{e}_\zeta - l\,\dot{\vartheta}^2\boldsymbol{e}_\eta \,, \qquad \dot{\boldsymbol{\omega}} = \ddot{\vartheta}\,\boldsymbol{e}_\xi \,.$$

With reference to this system, the movement of P is a circular motion with constant angular velocity ω_0, i.e. we have

$$\boldsymbol{v}_r = r\,\omega_0(-\sin\varphi\,\boldsymbol{e}_\xi + \cos\varphi\,\boldsymbol{e}_\eta) \,, \qquad \boldsymbol{a}_r = -r\,\omega_0^2(\cos\varphi\,\boldsymbol{e}_\xi + \sin\varphi\,\boldsymbol{e}_\eta) \,.$$

Introducing these relations with

$$\boldsymbol{\omega} \times \boldsymbol{r}_{0P} = r\,\dot{\vartheta}\sin\varphi\,\boldsymbol{e}_\zeta \,, \qquad \dot{\boldsymbol{\omega}} \times \boldsymbol{r}_{0P} = r\,\ddot{\vartheta}\sin\varphi\,\boldsymbol{e}_\zeta \,,$$

$$\boldsymbol{\omega} \times (\boldsymbol{\omega} \times \boldsymbol{r}_{0P}) = -r\,\dot{\vartheta}^2\sin\varphi\,\boldsymbol{e}_\eta \,, \qquad \boldsymbol{\omega} \times \boldsymbol{v}_r = r\,\omega_0\,\dot{\vartheta}\cos\varphi\,\boldsymbol{e}_\zeta$$

finally leads to the results

$$\underline{\underline{v = -r\,\omega_0 \sin\varphi\, e_\xi + r\,\omega_0 \cos\varphi\, e_\eta + (l + r\,\sin\varphi)\dot\vartheta\, e_\zeta}}\ ,$$

$$\underline{\underline{a = -r\,\omega_0^2 \cos\varphi\, e_\xi - \Big[(l + r\,\sin\varphi)\,\dot\vartheta^2 + r\,\omega_0^2 \sin\varphi\Big]\, e_\eta}}$$

$$\underline{\underline{+\Big[(l + r\,\sin\varphi)\,\ddot\vartheta + 2\,r\,\omega_0\,\dot\vartheta\cos\varphi\Big]\, e_\zeta}}\ .$$

b) For the coordinate system ξ^*, η^*, ζ^* from the representations

$$r_0 = l\,(\sin\varphi\, e_\xi^* + \cos\varphi\, e_\eta^*)\ , \qquad r_{0P} = r\, e_\xi^*\ ,$$

$$\omega = \dot\vartheta\,(\cos\varphi\, e_\xi^* - \sin\varphi\, e_\eta^*) + \omega_0\, e_\zeta^*\ ,$$

with

$$\dot e_\xi^* = \omega_0\, e_\eta^* + \dot\vartheta\,\sin\varphi\, e_\zeta^*\ , \qquad \dot e_\eta^* = -\omega_0\, e_\xi^* + \dot\vartheta\,\cos\varphi\, e_\zeta^*\ ,$$

$$\dot e_\zeta^* = -\dot\vartheta\,(\sin\varphi\, e_\xi^* + \cos\varphi\, e_\eta^*)\ , \qquad \dot\varphi = \omega_0$$

follow the relations

$$\dot r_0 = l\,\dot\vartheta\, e_\zeta^*\ , \qquad \ddot r_0 = -l\,\dot\vartheta^2(\sin\varphi\, e_\xi^* + \cos\varphi\, e_\eta^*) + l\,\ddot\vartheta\, e_\zeta^*\ ,$$

$$\dot\omega = (\ddot\vartheta\,\cos\varphi - \dot\vartheta\,\omega_0\sin\varphi) e_\xi^* - (\ddot\vartheta\,\sin\varphi + \dot\vartheta\,\omega_0\cos\varphi) e_\eta^*\ .$$

Since P with reference to the system ξ^*, η^*, ζ^* is at rest, we now have $v_r = a_r = \mathbf{0}$. With

$$\omega \times r_{0P} = r\,\omega_0\, e_\eta^* + r\,\dot\vartheta\sin\varphi\, e_\zeta^*\ ,$$

$$\dot\omega \times r_{0P} = r(\ddot\vartheta\,\sin\varphi + \dot\vartheta\,\omega_0\cos\varphi) e_\zeta^*\ ,$$

$$\omega \times (\omega \times r_{0P}) = -r(\dot\omega_0^2 + \dot\vartheta^2\,\sin^2\varphi) e_\xi^* - r\,\dot\vartheta^2\,\sin\varphi\cos\varphi\, e_\eta^* + r\,\omega_0\,\dot\vartheta\cos\varphi\, e_\zeta^*$$

we then obtain

$$\underline{\underline{v = r\,\omega_0\, e_\eta^* + \dot\vartheta\,(l + r\,\sin\varphi)\, e_\zeta^*}}\ ,$$

$$\underline{\underline{a = \Big[r\,\omega_0^2 - \dot\vartheta^2(l + r\,\sin\varphi)\,\sin\varphi\Big]\, e_\xi^* - \dot\vartheta^2(l + r\,\sin\varphi)\,\cos\varphi\, e_\eta^*}}$$

$$\underline{\underline{+\Big[\ddot\vartheta(l + r\,\sin\varphi) + 2\,r\,\omega_0\,\dot\vartheta\cos\varphi\Big]\, e_\zeta^*}}\,.$$

Remark: The representations in a) and b) can be transformed into each other using $e_\xi = e_\xi^*\,\cos\varphi - e_\eta^*\,\sin\varphi$, $e_\eta = e_\xi^*\,\sin\varphi + e_\eta^*\,\cos\varphi$.

P8.7

Problem 8.7 At the lever ①, rotating with constant angular velocity ω about the fixed axis $\overline{AA}$, a cylinder ② is mounted. The cylinder rotates about the axis $\overline{BB}$ parallel to $\overline{AA}$ with constant angular velocity ω^* with respect to the lever. Fixed at the cylinder is a tube where a point P moves with speed $v(t)$ with respect to the cylinder.

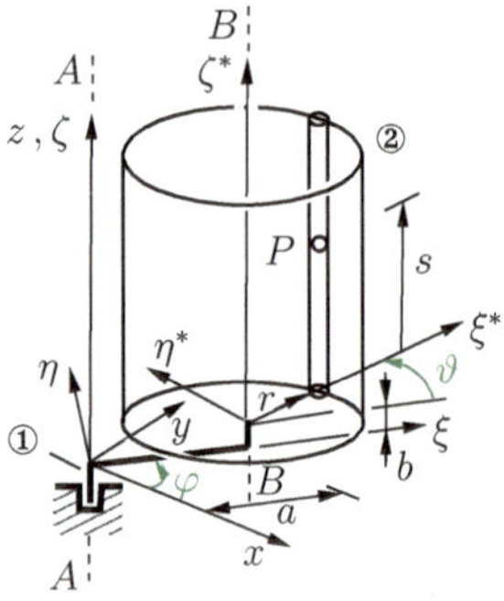

Determine the absolute velocity and absolute acceleration of P by using

a) the space fixed system x, y, z,

b) the lever-fixed system ξ, η, ζ,

c) the cylinder-fixed system ξ^*, η^*, ζ^*.

Solution **a)** With the angular velocity $\omega_2 = \omega + \omega^*$ of the cylinder with respect to the space-fixed system and with

$$\varphi = \omega t\,, \qquad \vartheta = \omega^* t\,, \qquad \varphi + \vartheta = \omega_2 t\,,$$

the position vector of P in the space-fixed system x, y, z is given by

$$\begin{aligned}\boldsymbol{r}_P &= (a\cos\varphi + r\cos(\varphi+\vartheta))\boldsymbol{e}_x + (a\sin\varphi + r\sin(\varphi+\vartheta))\boldsymbol{e}_y + (b+s)\boldsymbol{e}_z \\ &= (a\cos\omega t + r\cos\omega_2 t)\boldsymbol{e}_x + (a\sin\omega t + r\sin\omega_2 t)\boldsymbol{e}_y + (b+s)\boldsymbol{e}_z\,.\end{aligned}$$

From its derivatives, by considering $\dot{s} = v$, follow

$$\underline{\underline{\boldsymbol{v}_P = -(a\omega\sin\omega t + r\omega_2\sin\omega_2 t)\boldsymbol{e}_x + (a\omega\cos\omega t + r\omega_2\cos\omega_2 t)\boldsymbol{e}_y + v\boldsymbol{e}_z}}\,,$$

$$\underline{\underline{\boldsymbol{a}_P = -(a\omega^2\cos\omega t + r\omega_2^2\cos\omega_2 t)\boldsymbol{e}_x - (a\omega^2\sin\omega t + r\omega_2^2\sin\omega_2 t)\boldsymbol{e}_y + \dot{v}\boldsymbol{e}_z}}\,.$$

b) The general relations for $\boldsymbol{v}_P$ and $\boldsymbol{a}_P$ are given by

$$\begin{aligned}\boldsymbol{v}_P &= \dot{\boldsymbol{r}}_0 + \boldsymbol{\omega}\times\boldsymbol{r}_{0P} + \boldsymbol{v}_r\,,\\ \boldsymbol{a}_P &= \ddot{\boldsymbol{r}}_0 + \dot{\boldsymbol{\omega}}\times\boldsymbol{r}_{0P} + \boldsymbol{\omega}\times(\boldsymbol{\omega}\times\boldsymbol{r}_{0P}) + 2\,\boldsymbol{\omega}\times\boldsymbol{v}_r + \boldsymbol{a}_r\,.\end{aligned}$$

For the ξ, η, ζ system we have

$$\boldsymbol{\omega} = \omega\boldsymbol{e}_\zeta\,, \quad \dot{\boldsymbol{\omega}} = \boldsymbol{0}\,, \quad \boldsymbol{r}_0 = \dot{\boldsymbol{r}}_0 = \ddot{\boldsymbol{r}}_0 = \boldsymbol{0}\,, \quad \vartheta = \omega^* t$$

$$\boldsymbol{r}_{0P} = (a + r\cos\omega^* t)\boldsymbol{e}_\xi + r\sin\omega^* t\,\boldsymbol{e}_\eta + (b+s)\,\boldsymbol{e}_\zeta\,,$$

$$\boldsymbol{\omega} \times \boldsymbol{r}_{0P} = \omega(a + r\cos\omega^* t)\boldsymbol{e}_\eta - r\omega\sin\omega^* t\,\boldsymbol{e}_\xi\,,$$

$$\boldsymbol{v}_r = -r\omega^*\sin\omega^* t\,\boldsymbol{e}_\xi + r\omega^*\cos\omega^* t\,\boldsymbol{e}_\eta + v\,\boldsymbol{e}_\zeta\,,$$

$$\boldsymbol{\omega} \times (\boldsymbol{\omega} \times \boldsymbol{r}_{0P}) = -\omega^2(a + r\cos\omega^* t)\boldsymbol{e}_\xi - \omega^2 r\sin\omega^* t\,\boldsymbol{e}_\eta\,,$$

$$2\boldsymbol{\omega} \times \boldsymbol{v}_r = -2r\omega\omega^*\sin\omega^* t\,\boldsymbol{e}_\eta - 2r\omega\omega^*\cos\omega^* t\,\boldsymbol{e}_\xi\,,$$

$$\boldsymbol{a}_r = -r\omega^{*2}\cos\omega^* t\,\boldsymbol{e}_\xi - r\omega^{*2}\sin\omega^* t\,\boldsymbol{e}_\eta + \dot{v}\,\boldsymbol{e}_\zeta\,,$$

and by considering $\omega + \omega^* = \omega_2$ we obtain

$$\underline{\underline{\boldsymbol{v}_P = -r\omega_2\sin\omega^* t\,\boldsymbol{e}_\xi + (a\omega + r\omega_2\cos\omega^* t)\boldsymbol{e}_\eta + v\boldsymbol{e}_\zeta,}}$$

$$\underline{\underline{\boldsymbol{a}_P = -(a\omega^2 + r\omega_2^2\cos\omega^* t)\boldsymbol{e}_\xi - r\omega_2^2\sin\omega^* t\,\boldsymbol{e}_\eta + \dot{v}\boldsymbol{e}_\zeta.}}$$

c) For the ξ^*, η^*, ζ^* system we have

$$\boldsymbol{\omega} = \omega_2\boldsymbol{e}_\zeta^*\,, \quad \dot{\boldsymbol{\omega}} = \boldsymbol{0}\,,$$

$$\boldsymbol{r}_0 = a\cos\omega^* t\,\boldsymbol{e}_\xi^* - a\sin\omega^* t\,\boldsymbol{e}_\eta^* + b\,\boldsymbol{e}_\zeta^*\,,$$

$$\dot{\boldsymbol{r}}_0 = -a\omega^*\sin\omega^* t\,\boldsymbol{e}_\xi^* - a\omega^*\cos\omega^* t\,\boldsymbol{e}_\eta^*\,,$$

$$\ddot{\boldsymbol{r}}_0 = -a\omega^{*2}\cos\omega^* t\,\boldsymbol{e}_\xi^* + a\omega^{*2}\sin\omega^* t\,\boldsymbol{e}_\eta^*\,,$$

$$\boldsymbol{r}_{0P} = r\,\boldsymbol{e}_\xi^* + s\,\boldsymbol{e}_\zeta^*\,,$$

$$\boldsymbol{v}_r = v\,\boldsymbol{e}_\zeta^*\,,$$

$$\boldsymbol{\omega} \times \boldsymbol{r}_{0P} = r\omega_2\boldsymbol{e}_\eta^*$$

$$\boldsymbol{\omega} \times (\boldsymbol{\omega} \times \boldsymbol{r}_{0P}) = -r\omega_2^2\boldsymbol{e}_\xi^*\,,$$

$$2\boldsymbol{\omega} \times \boldsymbol{v}_r = \boldsymbol{0}\,.$$

Thus, it follows

$$\underline{\underline{\boldsymbol{v}_P = -a\omega^*\sin\omega^* t\,\boldsymbol{e}_\xi^* - (a\omega^*\cos\omega^* t + r\omega_2)\boldsymbol{e}_\eta^* + v\boldsymbol{e}_\zeta^*,}}$$

$$\underline{\underline{\boldsymbol{a}_P = -(a\omega^{*2}\cos\omega^* t + r\omega_2^2)\,\boldsymbol{e}_\xi^* + a\omega^{*2}\sin\omega^* t\,\boldsymbol{e}_\eta^* + \dot{v}\boldsymbol{e}_\zeta^*.}}$$

P8.8 **Problem 8.8** In the frictionless channel of a disk, rotating with the angular velocity ω, a slider of mass m is fixed at springs.

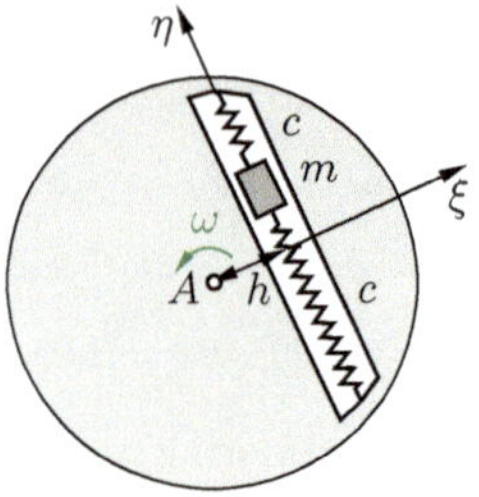

Formulate the equation of motion with respect to the moving system ξ, η.

Determine the force exerted from the channel on the slider.

Solution The coordinate system carries out a circular motion about A. With ($\zeta \perp$ to ξ, η)

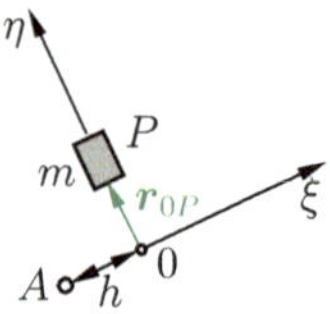

$$\boldsymbol{\omega} = \omega \boldsymbol{e}_\zeta , \qquad \dot{\boldsymbol{\omega}} = 0 , \qquad \boldsymbol{r}_{0P} = \eta \boldsymbol{e}_\eta ,$$

$$\ddot{\boldsymbol{r}}_0 = \boldsymbol{a}_0 = -h\omega^2 \boldsymbol{e}_\xi , \qquad \boldsymbol{v}_r = \eta' \boldsymbol{e}_\eta$$

the fictitious acceleration and Coriolis acceleration are given by

$$\boldsymbol{a}_f = \boldsymbol{a}_0 + \dot{\boldsymbol{\omega}} \times \boldsymbol{r}_{0P} + \boldsymbol{\omega} \times (\boldsymbol{\omega} \times \boldsymbol{r}_{0P}) = -h\omega^2 \boldsymbol{e}_\xi + \eta\omega^2 [\boldsymbol{e}_\zeta \times (\boldsymbol{e}_\xi \times \boldsymbol{e}_\eta)]$$

$$= -h\omega^2 \boldsymbol{e}_\xi + \eta\omega^2 [\boldsymbol{e}_\zeta \times (-\boldsymbol{e}_\xi)] = -h\omega^2 \boldsymbol{e}_\xi - \eta\omega^2 \boldsymbol{e}_\eta ,$$

$$\boldsymbol{a}_c = 2\boldsymbol{\omega} \times \boldsymbol{v}_r = 2\omega\eta' (\boldsymbol{e}_\zeta \times \boldsymbol{e}_\eta) = -2\omega\eta' \boldsymbol{e}_\xi .$$

In the equations of motion must be considered in addition to the external forces (spring force $2c\eta$, channel force N), the fictitious force $\boldsymbol{F}_f = -m\boldsymbol{a}_f$ and Coriolis force $\boldsymbol{F}_c = -m\boldsymbol{a}_c$. Thus, with $\xi'' = 0$ we obtain for the equation of motion and the channel force

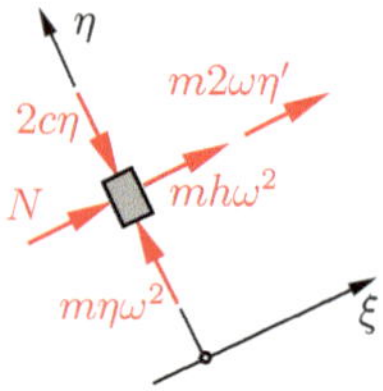

$$\nwarrow: \quad m\eta'' = -2c\eta + m\eta\omega^2 \quad \leadsto \quad \underline{\underline{\eta'' + \left(\frac{2c}{m} - \omega^2\right)\eta = 0}} ,$$

$$\nearrow: \quad 0 = N + mh\omega^2 + m2\omega\eta' \quad \leadsto \quad \underline{\underline{N = -m\omega(h\omega + 2\eta')}} .$$

Remark: The equation of motion has the solution (see page 170) $\eta(t) = A\cos\Omega t + B\sin\Omega t$ with the angular frequency $\Omega = \sqrt{2c/m - \omega^2}$. For $\omega^2 = 2c/m$ the slider rotates with the disk without vibrating.

Problem 8.9 A little sphere (point mass m) oscillates frictionless in a circular channel of a horizontal disk which rotates about an axis through A with constant angular velocity Ω. **P8.9**

Determine the circular frequency ω of the sphere if small amplitudes φ are assumed.

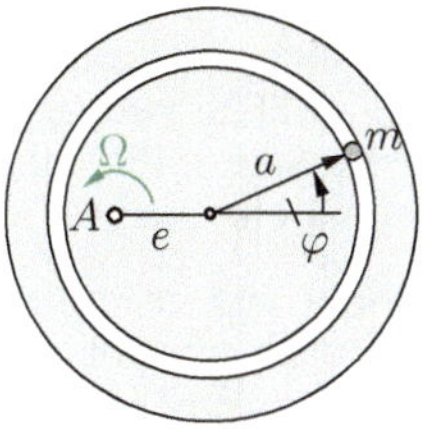

Solution We introduce the rotating ξ, η, ζ-coordinate system with its origin 0 at the center of the disk. Then we can write

$$\boldsymbol{\Omega} = \Omega\,\boldsymbol{e}_\zeta\,, \qquad \boldsymbol{a}_0 = \ddot{\boldsymbol{r}}_0 = -e\Omega^2\boldsymbol{e}_\xi\,,$$
$$\dot{\boldsymbol{\Omega}} = 0\,, \qquad \boldsymbol{r}_{0P} = a\cos\varphi\,\boldsymbol{e}_\xi + a\sin\varphi\,\boldsymbol{e}_\eta\,.$$

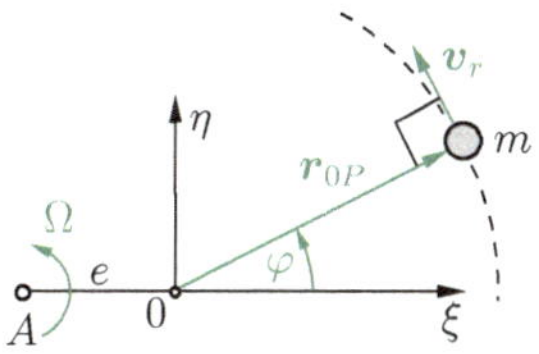

The relative velocity can be expressed by the relative angular velocity φ^* where $(\cdot)^*$ denotes the time derivative relative to the moving system):

$$v_r = a\varphi^* \qquad \rightsquigarrow \qquad \boldsymbol{v}_r = -a\varphi^*\sin\varphi\,\boldsymbol{e}_\xi + a\varphi^*\cos\varphi\,\boldsymbol{e}_\eta\,.$$

Thus, the fictitious forces $\boldsymbol{F}_f$ and $\boldsymbol{F}_c$ are

$$\boldsymbol{F}_f = -m\boldsymbol{a}_0 - m\boldsymbol{\Omega}\times(\boldsymbol{\Omega}\times\boldsymbol{r}_{0P}) = m(e\Omega^2 + a\Omega^2\cos\varphi)\boldsymbol{e}_\xi$$
$$+m\Omega^2 a\sin\varphi\,\boldsymbol{e}_\eta\,,$$
$$\boldsymbol{F}_c = -2m\boldsymbol{\Omega}\times\boldsymbol{v}_r = 2m\Omega a\varphi^*(\boldsymbol{e}_\xi\cos\varphi + \boldsymbol{e}_\eta\sin\varphi)\,.$$

With the tangential relative acceleration $a_{rt} = a\varphi^{**}$, the equation of motion in tangential direction is obtained as

$$\nwarrow: \quad ma\varphi^{**} = m(a\Omega^2 + 2a\varphi^*\Omega)\sin\varphi\,\cos\varphi$$
$$-m[e\Omega^2 + (a\Omega^2$$
$$+2a\varphi^*\Omega)\cos\varphi]\sin\varphi$$
$$= -me\Omega^2\sin\varphi\,.$$

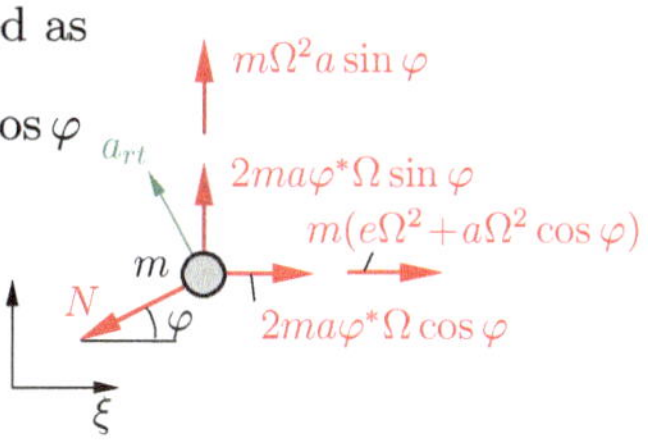

Assuming small amplitudes ($\sin\varphi \approx \varphi$), this leads to the equation for harmonic vibrations

$$\varphi^{**} + \frac{e\Omega^2}{a}\,\varphi = 0\,.$$

Hence, the circular frequency of the oscillations is

$$\underline{\underline{\omega = \sqrt{e/a}\;\Omega}}\,.$$

P8.10 **Problem 8.10** The displayed system rotates about the vertical axis with constant angular velocity ω. The two bodies are connected by an inextensible cable and they can frictionless slide along the parts of the elbow.

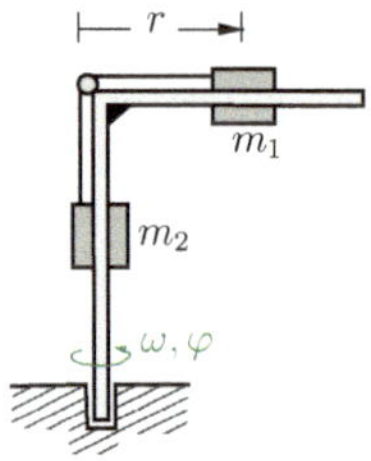

Determine the reaction force acting on m_1 and the path $r(\varphi)$ of m_1, if at time $t = 0$ the initial conditions are given by $r(0) = r_0$ and $r^*(0) = 0$.

Solution We introduce the rotating elbow-fixed coordinate system ξ, η, ζ and draw the free-body diagrams for both bodies with the external forces, reaction forces and fictitious force and Coriolis force. There is no relative motion of m_1 in ζ- and in ξ-direction. Thus, the reaction forces follow directly from the respective 'equilibrium conditions' as

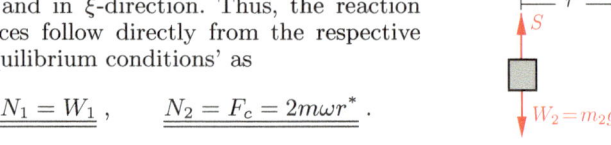

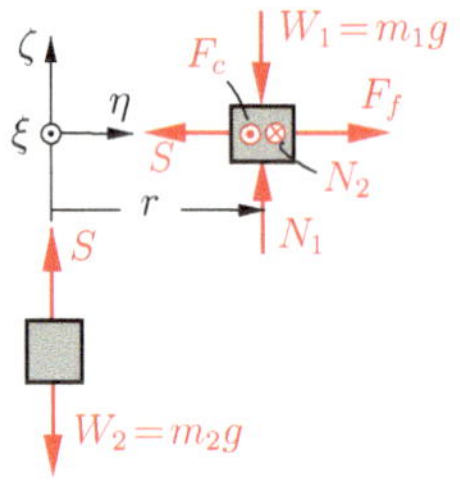

$$\underline{\underline{N_1 = W_1}}\,, \qquad \underline{\underline{N_2 = F_c = 2m\omega r^*}}\,.$$

The equations of motion

$$m_2\zeta_2^{**} = S - W_2\,, \qquad m_1 r^{**} = F_f - S = m_1 r\omega^2 - S$$

lead with $\zeta_2^{**} = r^{**}$ (inextensible cable!) to

$$(m_1 + m_2)r^{**} - m_1\omega^2 r = -W_2\,.$$

This inhomogeneous differential equation has the solution

$$r(t) = r_h + r_p = Ae^{\lambda t} + Be^{-\lambda t} + \frac{m_2 g}{m_1\omega^2} \quad \text{where} \quad \lambda = \omega\sqrt{\frac{m_1}{m_1 + m_2}}.$$

The initial conditions

$$r^*(0) = 0 \;\rightsquigarrow\; A - B = 0\,, \qquad r(0) = r_0 \;\rightsquigarrow\; r_0 = A + B + \frac{m_2 g}{m_1\omega^2}$$

lead to $A = B = \frac{1}{2}[r_0 - m_2 g/(m_1\omega^2)]$ and herewith, considering $\omega t = \varphi$, we finally obtain the path equation

$$\underline{\underline{r(\varphi) = \left(r_0 - \frac{m_2 g}{m_1\omega^2}\right)\cosh\left(\varphi\sqrt{\frac{m_1}{m_1 + m_2}}\right) + \frac{m_2 g}{m_1\omega^2}}}\,.$$

Remark: For $r_0 m_1\omega^2 = S = m_2 g$ we have 'equilibrium' ($r^{**} = 0$) in this position, which, however, is unstable: for any small diplacement (disturbance) the system starts to move!

Problem 8.11 In the radial channel of a wobbling disk, which rotates with constant angular velocity ω, slides frictionless a knuckle of mass m. **P8.11**

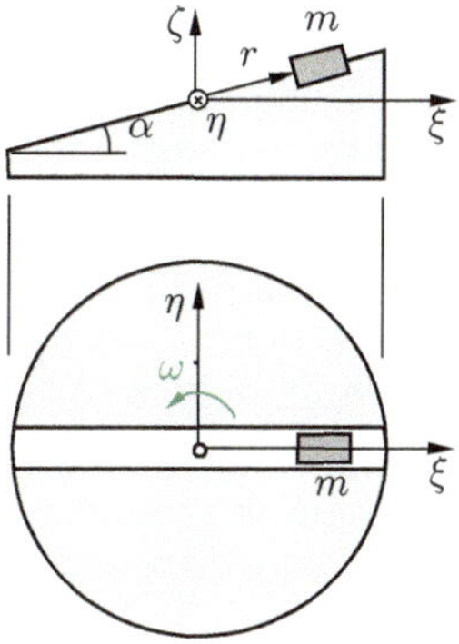

a) Determine the force $K(r)$ that must exerted to the knuckle in channel direction, that it moves according to the law $r(t) = r_0 \sin \omega t$ (the weight shall be neglected).

b) Determine the lateral contact force $N_\eta(r)$ between the knuckle and the channel.

Solution **a)** We use the rotating reference system ξ, η, ζ and draw the free-body diagram by considering the fictitious force and Coriolis force

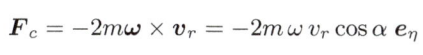

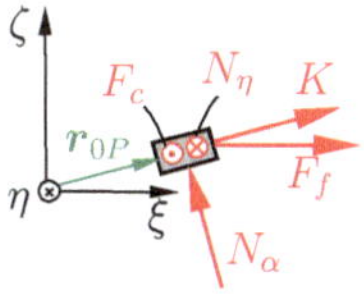

$$\boldsymbol{F}_f = -m\,\boldsymbol{\omega} \times (\boldsymbol{\omega} \times \boldsymbol{r}_{0P}) = mr\omega^2 \cos\alpha\, \boldsymbol{e}_\xi \ ,$$

$$\boldsymbol{F}_c = -2m\boldsymbol{\omega} \times \boldsymbol{v}_r = -2m\,\omega\, v_r \cos\alpha\, \boldsymbol{e}_\eta \ ,$$

where $v_r = r^* = r_0\omega \cos \omega t$.

With $a_r = -r_0\omega^2 \sin \omega t = -\omega^2 r$ the equation of motion in channel direction reads

$$ma_r = K + mr\omega^2 \cos^2 \alpha \, .$$

It leads to the force K:

$$\underline{\underline{K(r)}} = ma_r - mr\omega^2 \cos^2 \alpha = \underline{\underline{-m\omega^2 r\,(1 + \cos^2 \alpha)}} \, .$$

b) The lateral contact force N_η is calculated from the 'equilibrium condition' (no relative acceleration in η-direction)

$$N_\eta = 2m\omega v_r \cos\alpha = 2m\omega^2 r_0 \cos\alpha \cos\omega t$$

or with $\cos\omega t = \sqrt{1 - \sin^2 \omega t} = \sqrt{1 - (r/r_0)^2}$ as

$$\underline{\underline{N_\eta(r) = 2m\omega^2 \cos\alpha \sqrt{r_0^2 - r^2}}} \, .$$

P8.12

Problem 8.12 The arm of an automatic assembly machine starts moving from rest with constant acceleration a_0 along a straight track. At the same time, the arm begins to rotate with constant angular velocity ω_0 and the slider P begins to move towards point A with constant relative acceleration a_r. The initial positions of the arm and the slider are given by $\varphi_0 = 0$ and s_0.

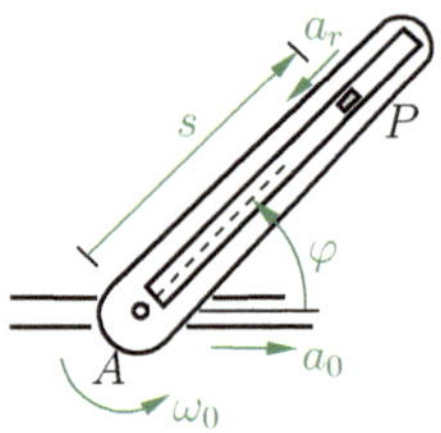

Determine the absolute velocity and acceleration of the slider P in dependence of time t.

Solution We use the fixed coordinate system x, y, z, where the x-axis coincides with the track. In addition, we introduce the moving coordinate system ξ, η, ζ, where the ξ-axis rotates with the arm. Then, the general equations for the absolute velocity and acceleration are

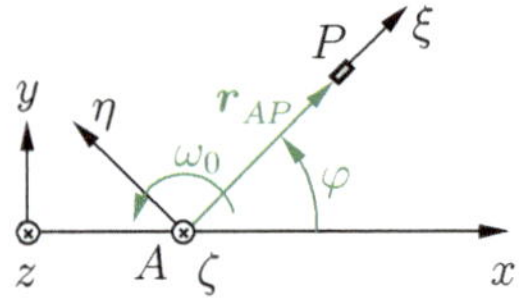

$$\boldsymbol{v} = \boldsymbol{v}_A + \boldsymbol{\omega} \times \boldsymbol{r}_{AP} + \boldsymbol{v}_r \,,$$
$$\boldsymbol{a} = \boldsymbol{a}_A + \dot{\boldsymbol{\omega}} \times \boldsymbol{r}_{AP} + \boldsymbol{\omega} \times (\boldsymbol{\omega} \times \boldsymbol{r}_{AP}) + 2\,\boldsymbol{\omega} \times \boldsymbol{v}_r + \boldsymbol{a}_r \,.$$

We start counting time t from the beginning of the motion. then, by using the given accelerations, angular velocity and initial conditions, we obtain

$$\boldsymbol{a}_A = a_0 \boldsymbol{e}_x \quad \rightsquigarrow \quad \boldsymbol{v}_0 = a_0 t\, \boldsymbol{e}_x \,,$$
$$\boldsymbol{\omega} = \omega_0 \boldsymbol{e}_\zeta \,, \quad \dot{\boldsymbol{\omega}} = 0 \,,$$
$$\boldsymbol{a}_r = -a_r \boldsymbol{e}_\xi \quad \rightsquigarrow \quad \boldsymbol{v}_r = -a_r t\, \boldsymbol{e}_\xi \quad \rightsquigarrow \quad \boldsymbol{r}_{AP} = (-\tfrac{1}{2} a_r t^2 + s_0)\boldsymbol{e}_\xi \,,$$

and

$$\boldsymbol{\omega} \times \boldsymbol{r}_{AP} = \omega(-\tfrac{1}{2} a_r t^2 + s_0)\boldsymbol{e}_\eta \,, \qquad 2\,\boldsymbol{\omega} \times \boldsymbol{v}_r = -2\,\omega_0\, a_r t\, \boldsymbol{e}_\eta \,,$$
$$\boldsymbol{\omega} \times (\boldsymbol{\omega} \times \boldsymbol{r}_{AP}) = -\omega_0^2(-\tfrac{1}{2} a_r t^2 + s_0)\boldsymbol{e}_\xi \,.$$

Taking into account the relations $\boldsymbol{e}_x = \boldsymbol{e}_\xi \cos\varphi - \boldsymbol{e}_\eta \sin\varphi$, where $\varphi = \omega_0\, t$, we finally obtain

$$\underline{\underline{\boldsymbol{v} = [a_0 t \cos\omega_0 t - a_r t]\boldsymbol{e}_\xi + [-a_0 t \sin\omega_0 t + \omega_0(-\tfrac{1}{2} a_r t^2 + s_0)]\boldsymbol{e}_\eta}} \,,$$

$$\underline{\underline{\boldsymbol{a} = [a_0 \cos\omega_0 t - \omega_0^2(-\tfrac{1}{2} a_r t^2 + s_0) - a_r]\boldsymbol{e}_\xi - [a_0 \sin\omega_0 t + 2\,\omega_0\, a_r t]\boldsymbol{e}_\eta}} \,.$$

Chapter 9

Principles of Mechanics

9

It is often advantageous to determine the equations of motion not by using NEWTON'S axioms (principles of linear and angular momentum) but by using equivalent laws which are called *Principles of Mechanics.*

Formal Reduction of Kinetics to Statics

Rewriting NEWTON'S law of the motion for a point mass (or the center of mass of a rigid body) in the form

$$m\boldsymbol{a} = \sum \boldsymbol{F} \qquad \rightsquigarrow \qquad \sum \boldsymbol{F} - m\boldsymbol{a} = 0$$

and introducing D'ALEMBERT'S *inertial force* (*pseudo force, fictitious force*)

$$\boldsymbol{F}_I = -m\boldsymbol{a}$$

leads to the 'dynamic equilibrium condition of forces'

$$\sum \boldsymbol{F} + \boldsymbol{F}_I = 0$$

Accordingly, a point mass or the center of mass of a rigid body moves such that the sum of *external forces* $\boldsymbol{F}$ and the *inertial force* $\boldsymbol{F}_I$ is equal to zero. In case of a *plane motion* of a rigid body, the pseudo moment $M_{IA} = -\Theta_A\dot{\omega}$ must be taken into account in the 'dynamic equilibrium condition of the moments'. Instead of the equations of motion according to page 102 we then obtain the 'equilibrium conditions'

$$\sum F_x - ma_x = 0\,, \quad \sum F_y - ma_y = 0\,, \quad \sum M_A - \Theta_A\dot{\omega} = 0$$

where $A \widehat{=}$ fixed point or center of mass.

Remarks:

- The inertial force (pseudo force) and the pseudo moment are directed opposite to the positive acceleration and angular acceleration, respectively.
- When solving problems, the pseudo forces and pseudo moments must be drawn into the free-body diagram with the respective sign.

d'ALEMBERT's Principle

A system moves such that for a virtual displacement the sum of the works δU of the *external forces* (moments) and δU_I of the *pseudo forces* (pseudo moments) vanishes at all times:

$$\delta U + \delta U_I = 0$$

Remarks:

- Virtual displacements are infinitesimally small, fictitious and kinematically admissible.
- The work done by constraint forces is zero for rigid constraints.
- D'ALEMBERT's principle may preferably be applied for systems with several constraints, if the constraint forces shall not be determined.
- In case of statics, the principle reduces to $\delta U = 0$ (see volume 1, chapter 7).

LAGRANGE Equations of the 2nd kind

The motion of a system with n degrees of freedom is described by

$$\frac{\mathrm{d}}{\mathrm{d}t}\left(\frac{\partial T}{\partial \dot{q}_j}\right) - \frac{\partial T}{\partial q_j} = Q_j \qquad (j = 1, \ldots, n)$$

where
$T \;\hat{=}\;$ kinetic energy,
$q_j \;\hat{=}\;$ generalized coordinates,
$\dot{q}_j \;\hat{=}\;$ generalized velocities,
$Q_j \;\hat{=}\;$ generalized forces.

For *conservative forces* (having a potential) the equations of motion simplify to

$$\frac{\mathrm{d}}{\mathrm{d}t}\left(\frac{\partial L}{\partial \dot{q}_j}\right) - \frac{\partial L}{\partial q_j} = 0 \qquad (j = 1, \ldots, n)$$

where
$L = T - V \;\hat{=}\;$ Lagrangian,
$V \;\hat{=}\;$ potential energy.

Remarks:

- The numbers of generalized coordinates (i.e. equations) and of degrees of freedom are equal.
- Generalized coordinates are linearly independent and may be lengths or angles.
- Generalized forces act in the direction of generalized coordinates and may be e.g. forces or moments.

P9.1

Problem 9.1 Two blocks of weights $W_1 = m_1 g$ and $W_2 = m_2 g$ are suspended by a rope drum (moment of inertia Θ_A). The block of mass m_1 slides frictionless on an inclined plane.

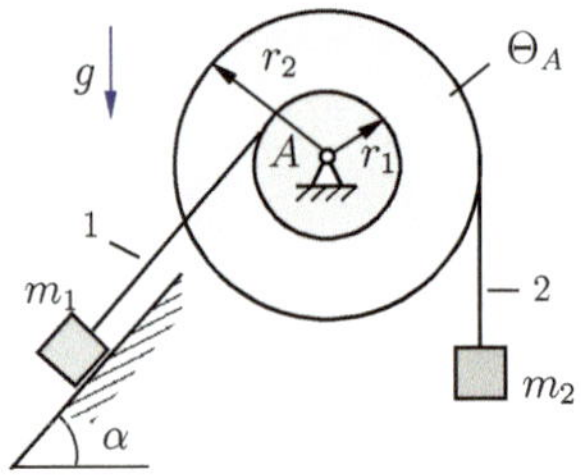

Determine the angular acceleration of the drum and the force in rope 2 by using dynamic equilibrium conditions.

Solution We separate the system, introduce coordinates x_1,x_2, φ, describing the motion of the system, and draw the free-body diagram. Since we solve the problem by applying dynamic equilibrium conditions, the inertia forces $m_i\ddot{x}_i$ and the pseudo moment must be considered; they point in negative coordinate directions. Then, the equilibrium conditions yield

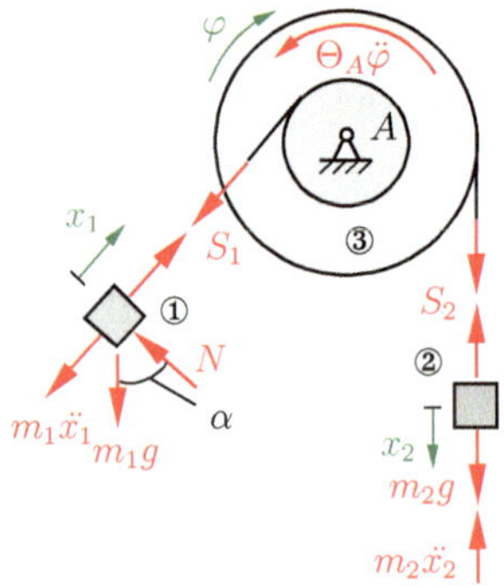

$$① \nearrow: \quad S_1 - m_1\ddot{x}_1 - m_1 g\sin\alpha = 0\,,$$

$$② \downarrow: \quad m_2 g - m_2\ddot{x}_2 - S_2 = 0\,,$$

$$③ \overset{\curvearrowright}{A}: \quad -r_1 S_1 + r_2 S_2 - \Theta_A\ddot{\varphi} = 0\,.$$

Using the kinematic relations

$$x_1 = r_1\varphi \quad \rightsquigarrow \quad \ddot{x}_1 = r_1\ddot{\varphi}\,,$$

$$x_2 = r_2\varphi \quad \rightsquigarrow \quad \ddot{x}_2 = r_2\ddot{\varphi}\,,$$

we obtain the angular acceleration

$$\underline{\underline{\ddot{\varphi} = \frac{r_2 m_2 - r_1 m_1\sin\alpha}{r_1^2 m_1 + r_2^2 m_2 + \Theta_A}\,g}}$$

and the force in the rope

$$\underline{\underline{S_2}} = m_2(g + r_2\ddot{\varphi}) = \underline{\underline{m_2 g\,\frac{r_1(r_1 + r_2\sin\alpha)m_1 + \Theta_A}{r_1^2 m_1 + r_2^2 m_2 + \Theta_A}}}\,.$$

Remark: For $r_2 m_2 > r_1 m_1\sin\alpha$ the drum rotates clockwise, for $r_2 m_2 < r_1 m_1\sin\alpha$ it rotates counterclockwise. In the special case $r_2 m_2 = r_1 m_1\sin\alpha$, the system is in static equilibrium: $\ddot{\varphi} = 0$.

P9.2

Problem 9.2 The displayed pendulum consists of a homogeneous cylinder of mass $m_1 = 2m$ and a point mass of weight $G = m_2 g = mg$ which is rigidly fixed by a massless rod. The cylinder rolls without slip on the rough plane.

Formulate the equation of motion.

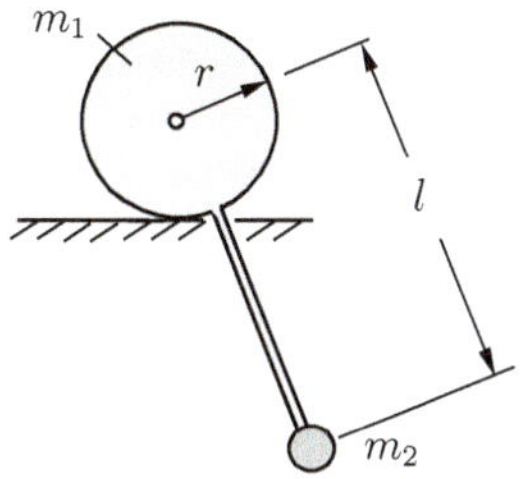

Solution We choose the reference system x, y in the undisplaced position. The center of mass C lies in the distance

$$e = \frac{l m_1}{m_1 + m_2} = \frac{2}{3}l$$

from the point mass m_2. In the displaced position we draw all external forces, inertial forces and the pseudo moment into the free-body diagram. The 'equilibrium conditions' then read

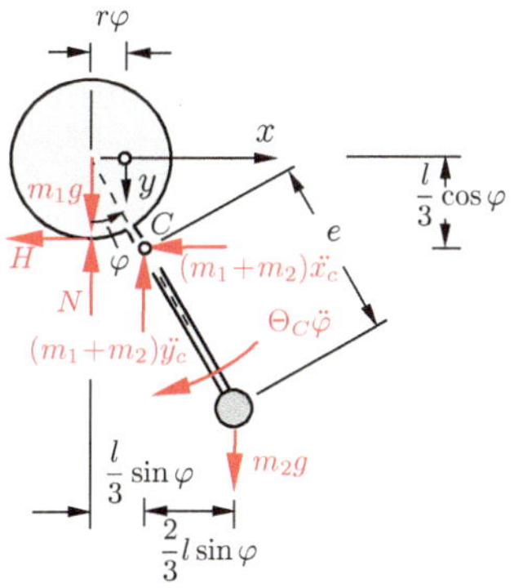

$$\rightarrow: \quad -H - (m_1 + m_2)\ddot{x}_C = 0\,,$$

$$\downarrow: \quad (m_1 + m_2)g - N - (m_1 + m_2)\ddot{y}_C = 0\,,$$

$$\overset{\curvearrowleft}{C}: \quad m_1 g \frac{l}{3}\sin\varphi + H\left(\frac{l}{3}\cos\varphi - r\right) - N\frac{l}{3}\sin\varphi - m_2 g\frac{2l}{3}\sin\varphi - \Theta_C\ddot{\varphi} = 0$$

where

$$\Theta_C = \left[\frac{m_1 r^2}{2} + m_1\left(\frac{l}{3}\right)^2\right] + m_2\left(\frac{2l}{3}\right)^2 = m\left(r^2 + \frac{2}{3}l^2\right).$$

For an angle change φ, the center of the cylinder is displaced by $r\varphi$ to the left. Thus, we find for the center of mass

$$x_C = -r\varphi + \frac{l}{3}\sin\varphi \quad \rightsquigarrow \quad \ddot{x}_C = -r\ddot{\varphi} + \frac{l}{3}\ddot{\varphi}\cos\varphi - \frac{l}{3}\dot{\varphi}^2\sin\varphi\,,$$

$$y_C = \frac{l}{3}\cos\varphi \quad \rightsquigarrow \quad \ddot{y}_C = -\frac{l}{3}\ddot{\varphi}\sin\varphi - \frac{l}{3}\dot{\varphi}^2\cos\varphi\,.$$

Solving the equations yields

$$\underline{\underline{\ddot{\varphi}(l^2 + 4r^2 - 2lr\cos\varphi) + lr\dot{\varphi}^2\sin\varphi + gl\sin\varphi = 0}}\,.$$

P9.3 **Problem 9.3** A wheel (weight $m_1 g$, moment of inertia Θ_A) and a block (weight $m_2 g$), both on inclined planes, are connected by a rope. The wheel rolls without slip while the block slips frictionless.

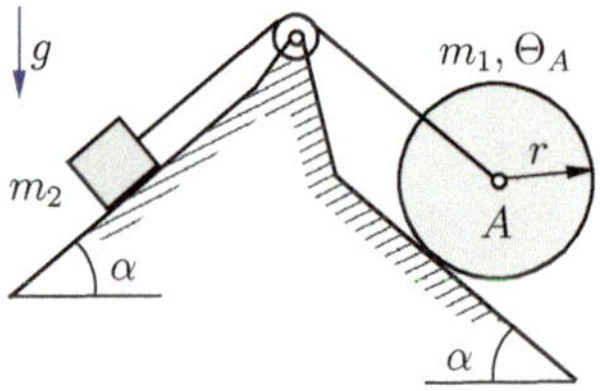

Determine the acceleration of the block applying d'Alembert's principle. Neglect the masses of the rope and the pulley.

Solution Since the constraint forces (force in the rope, contact forces) need not to bee determined, it is advantageous to apply d'Alembert's principle. To describe the motion we choose the coordinates x_i, φ. In addition to the real forces, the inertial forces $m_i\ddot{x}_i$ and the pseudo moment $\Theta_A\ddot{\varphi}$ (acting opposite to the chosen coordinates) are drawn into the sketch of the system. Then, d'Alembert's principle requires that the virtual work of all forces vanishes:

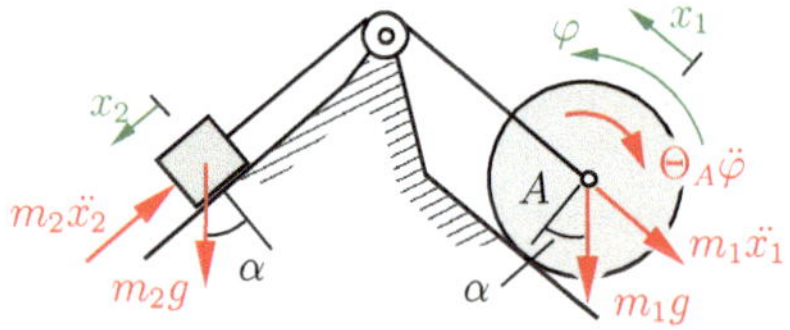

$$\delta U + \delta U_I = 0$$

This leads to

$$-m_1\ddot{x}_1\delta x_1 - m_1 g\sin\alpha\delta x_1 - \Theta_A\ddot{\varphi}\delta\varphi + m_2 g\sin\alpha\delta x_2 - m_2\ddot{x}_2\delta x_2 = 0\,.$$

With the kinematic relations

$$x_1 = x_2 = r\varphi = x \qquad \leadsto \qquad \begin{cases} \delta x_1 = \delta x_2 = r\delta\varphi = \delta x \\ \ddot{x}_1 = \ddot{x}_2 = r\ddot{\varphi} = \ddot{x} \end{cases}$$

we obtain

$$\left[-m_1\ddot{x} - m_1 g\sin\alpha - \frac{\Theta_A}{r^2}\,\ddot{x} + m_2 g\sin\alpha - m_2\ddot{x}\right]\delta x = 0\,.$$

Since $\delta x \neq 0$, the expression in the bracket must vanish. Thus,

$$\underline{\underline{\ddot{x} = \ddot{x}_2 = g\,\frac{(m_2 - m_1)\sin\alpha}{m_1 + m_2 + \dfrac{\Theta_A}{r^2}}}}\,.$$

Problem 9.4 Two drums are connected by a rope and carry blocks of weights $m_1 g$ and $m_2 g$. Drum ② is driven by the moment M_0. P9.4

Determine the acceleration of block ① using d'Alembert's principle. Neglect the mass of the ropes.

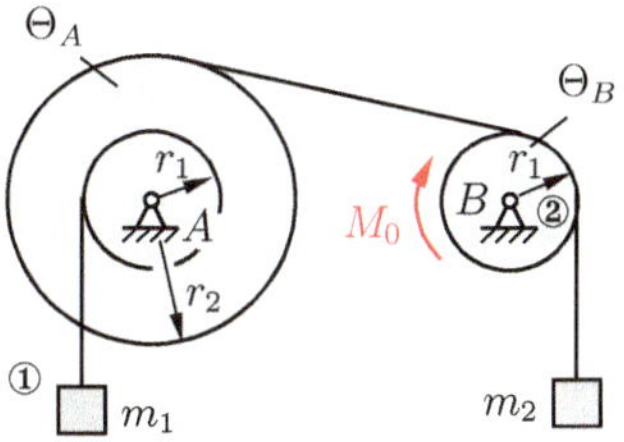

Solution We introduce the inertial forces $m_i\ddot{x}_i$ and the pseudo moments $\Theta_A\ddot{\varphi}_1$, $\Theta_B\ddot{\varphi}_2$. They act in opposite directions to the chosen positive coordinate directions. Then, d'Alembert's principle requires

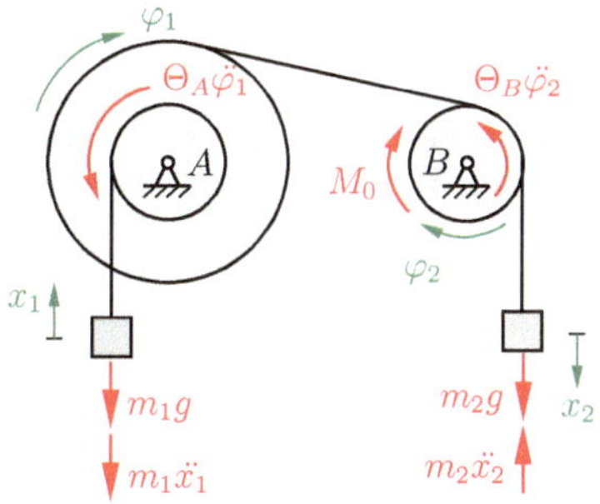

$$\delta U + \delta U_I = 0\,,$$

which leads to

$$-m_1(g+\ddot{x}_1)\delta x_1 + m_2(g-\ddot{x}_2)\delta x_2$$

$$+M_0\delta\varphi_2 - \Theta_A\ddot{\varphi}_1\delta\varphi_1 - \Theta_B\ddot{\varphi}_2\delta\varphi_2 = 0.$$

With the kinematic relations

$$\left.\begin{aligned} x_1 &= r_1\varphi_1 \\ x_2 &= r_1\varphi_2 \\ r_2\varphi_1 &= r_1\varphi_2 \end{aligned}\right\} \quad \leadsto \quad \begin{aligned} \ddot{\varphi}_1 &= \frac{\ddot{x}_1}{r_1}, \quad & \ddot{\varphi}_2 &= \frac{r_2}{r_1^2}\ddot{x}_1, \quad & \ddot{x}_2 &= \frac{r_2}{r_1}\ddot{x}_1, \\ \delta\varphi_1 &= \frac{\delta x_1}{r_1}, \quad & \delta\varphi_2 &= \frac{r_2}{r_1^2}\delta x_1, \quad & \delta x_2 &= \frac{r_2}{r_1}\delta x_1 \end{aligned}$$

we obtain

$$\left\{-m_1(g+\ddot{x}_1) + m_2\left(g - \frac{r_2}{r_1}\ddot{x}_1\right)\frac{r_2}{r_1} + \frac{r_2 M_0}{r_1^2} - \frac{\Theta_A}{r_1^2}\ddot{x}_1 - \frac{r_2^2\Theta_B}{r_1^4}\ddot{x}_1\right\}\delta x_1 = 0.$$

Since $\delta x_2 \neq 0$, the expression in the curly bracket must vanish. Thus, the acceleration of block ① is

$$\ddot{x}_1 = g\,\frac{1 - \dfrac{m_2 r_2}{m_1 r_1} - \dfrac{r_2 M_0}{r_{12} m_1 g}}{1 + \dfrac{m_2}{m_1}\left(\dfrac{r_2}{r_1}\right)^2 + \dfrac{\Theta_A}{m_1 r_1^2} + \dfrac{r_2^2\Theta_B}{m_1 r_1^4}}\,.$$

P9.5 **Problem 9.5** The displayed system shows a lever (weight $m_1 g$, moment of inertia Θ_D), pivoted at D. Attached at the lever ends are a turnable wheel ②, rolling along the rigid half-circular cylinder ①, and a counterweight $W_3 = m_3 g$.

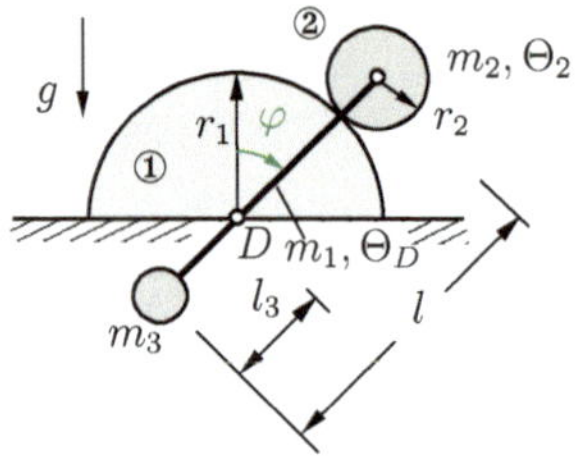

Determine the eigenfrequency for small displacements φ.

Solution We will derive the equation of motion by applying two different approaches. First we use d'ALEMBERT's principle. For this purpose, in addition to the external forces, all inertial forces and pseudo moments are drawn into the free-body diagram in the displaced position. The virtual work of all forces and moments must vanish (notice, $\sin\varphi \approx \varphi$):

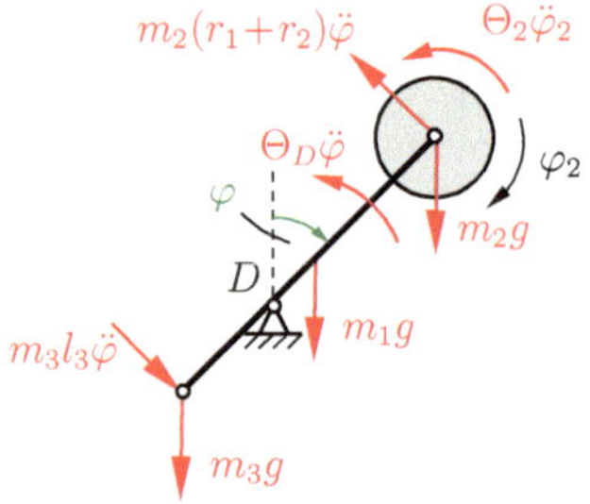

$$-\Theta_D\ddot{\varphi}\delta\varphi - \Theta_2\ddot{\varphi}_2\delta\varphi_2 - m_2(r_1+r_2)^2\ddot{\varphi}\delta\varphi - m_3 l_3^2\ddot{\varphi}\delta\varphi$$

$$-m_3 g l_3\varphi\delta\varphi + m_1 g\Big(\frac{l}{2} - l_3\Big)\varphi\delta\varphi + m_2 g(r_1+r_2)\varphi\delta\varphi = 0\,.$$

Because the wheel rolls, the kinematic relation reads

$$r_2\varphi_2 = (r_1+r_2)\,\varphi \quad \rightsquigarrow \quad \begin{cases} \ddot{\varphi}_2 = (1 + r_1/r_2)\,\ddot{\varphi}\,, \\ \delta\varphi_2 = (1 + r_1/r_2)\,\delta\varphi\,. \end{cases}$$

Introducing these quantities and considering $\delta\varphi \neq 0$, we obtain the equation of motion

$$\Big[\Theta_D + \Theta_2\Big(1 + \frac{r_1}{r_2}\Big)^2 + m_2(r_1+r_2)^2 + m_3 l_3^2\Big]\ddot{\varphi}$$

$$+\Big[m_3 l_3 - m_1\Big(\frac{l}{2} - l_3\Big) - m_2(r_1+r_2)\Big]g\varphi = 0\,,$$

from which the eigenfrequency is directly found as

$$\omega = \sqrt{\frac{m_3 l_3 - m_1\Big(\dfrac{l}{2} - l_3\Big) - m_2(r_1+r_2)}{\Theta_D + \Theta_2\Big(1 + \dfrac{r_1}{r_2}\Big)^2 + m_2(r_1+r_2)^2 + m_3 l_3^2}\,g}\,.$$

In the second approach we apply the LAGRANGE equations of the 2nd kind. The system is conservative and, by choosing the zero potential at the level of D, we obtain

$$V = m_1 g\Big(\frac{l}{2} - l_3\Big)\cos\varphi + m_2 g(r_1 + r_2)\cos\varphi - m_3 g\, l_3 \cos\varphi\ ,$$

$$T = \frac{1}{2}\Theta_D\dot{\varphi}^2 + \Big[\frac{1}{2}m_2(r_1+r_2)^2\dot{\varphi}^2 + \frac{1}{2}\Theta_2\dot{\varphi}_2^2\Big] + \frac{1}{2}l_3^2\, m_3\dot{\varphi}^2\ ,$$

$$L = T - V\ .$$

With the kinematic relations

$$r_2\varphi_2 = (r_1 + r_2)\varphi \qquad \rightsquigarrow \qquad r_2\dot{\varphi}_2 = (r_1+r_2)\dot{\varphi}$$

follow

$$\frac{\partial L}{\partial \dot{\varphi}} = \Big[\Theta_D + m_2(r_1+r_2)^2 + \Theta_2\Big(1+\frac{r_1}{r_2}\Big)^2 + m_3 l_3^2\Big]\dot{\varphi}\ ,$$

$$\frac{\mathrm{d}}{\mathrm{d}t}\Big(\frac{\partial L}{\partial \dot{\varphi}}\Big) = \Big[\Theta_D + m_2(r_1+r_2)^2 + \Theta_2\Big(1+\frac{r_1}{r_2}\Big)^2 + m_3 l_3^2\Big]\ddot{\varphi}\ ,$$

$$\frac{\partial L}{\partial \varphi} = \Big[m_1\Big(\frac{l}{2} - l_3\Big) + m_2(r_1+r_2) - m_3 l_3\Big] g\sin\varphi\ .$$

Thus, from

$$\frac{\mathrm{d}}{\mathrm{d}t}\Big(\frac{\partial L}{\partial \dot{\varphi}}\Big) - \frac{\partial L}{\partial \varphi} = 0$$

with $\sin\varphi \approx \varphi$ (small displacements) we obtain the already known result for the equation of motion

$$\Big[\Theta_D + \Theta_2\Big(1+\frac{r_1}{r_2}\Big)^2 + m_2(r_1+r_2)^2 + m_3 l_3^2\Big]\ddot{\varphi}$$

$$+\Big[m_3 l_3 - m_1\Big(\frac{l}{2} - l_3\Big) - m_2(r_1+r_2)\Big] g\varphi = 0\ ,$$

and accordingly for the eigenfrequency. It can be seen that vibrations are only possible if the nominator in the square root is positive, i.e. if

$$m_3 l_3 > m_1\Big(\frac{l}{2} - l_3\Big) + m_2(r_1+r_2)\ .$$

Remark: The system has only one degree of freedom; its position can uniquely be described by the generalized coordinate φ.

P9.6 **Problem 9.6** A pivoted homogeneous bar (mass m) is held by springs and carries an attached point mass $m/4$.

Determine the equation of motion for small displacements from the equilibrium position by using:
a) the energy conservation law,
b) d'ALEMBERT's principle,
c) the angular momentum theorem.

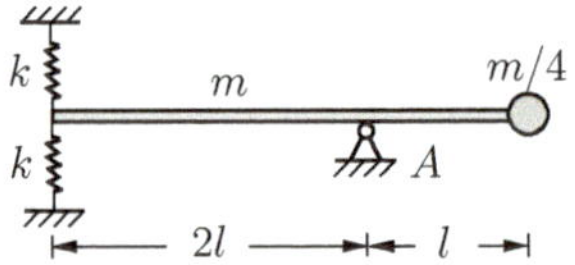

Solution The motion of the system is a rotation about point A, which appropriately is described by the rotation angle φ. Thereby, assuming small displacements, each spring experiences a length change $2l\varphi$.

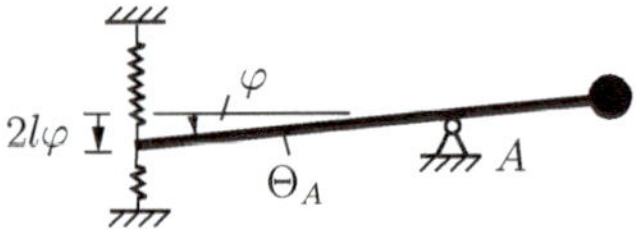

a) The total energy

$$E = V + T = 2\,\frac{1}{2}k(2l\varphi)^2 + \frac{1}{2}\Theta_A\dot{\varphi}^2$$

of the conservative system must be constant at all times. This leads to

$$\frac{\mathrm{d}E}{\mathrm{d}t} = 0 \quad \rightsquigarrow \quad 2k(2l\varphi)2l\dot{\varphi} + \Theta_A\dot{\varphi}\ddot{\varphi} = 0 \quad \rightsquigarrow \quad \underline{\underline{\Theta_A\ddot{\varphi} + 8kl^2\varphi = 0}}\,.$$

b) When applying d'ALEMBERT's principle $\delta U + \delta U_I = 0$, the work of the pseudo moment $\Theta_A\ddot{\varphi}$ (opposite to the positive direction of motion) must be taken into account:

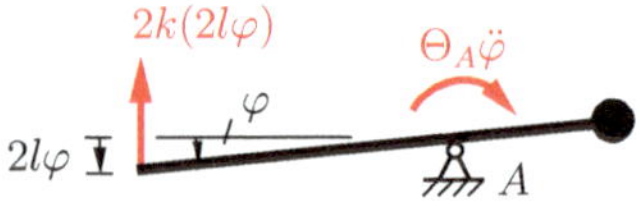

$$-\Theta_A\ddot{\varphi}\delta\varphi - 2l\,2\,k(2l\varphi)\delta\varphi = 0 \quad \rightsquigarrow \quad \underline{\underline{\Theta_A\ddot{\varphi} + 8kl^2\varphi = 0}}\,.$$

c) Application of the angular momentum theorem with respect to the fixed point A directly yields

$$\overset{\curvearrowleft}{A}: \quad \Theta_A\ddot{\varphi} = -2l\,2\,k(2l\varphi)$$

$$\rightsquigarrow \quad \underline{\underline{\Theta_A\ddot{\varphi} + 8kl^2\varphi = 0}}\,.$$

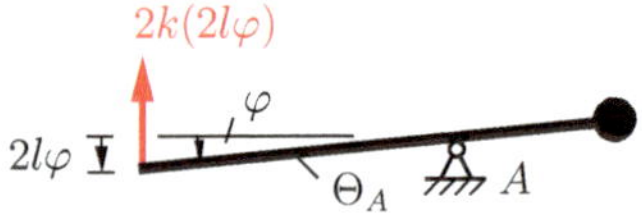

The results are (as expected) in all cases the same. With

$$\Theta_A = \frac{1}{3}\Big(\frac{2}{3}m\Big)(2l)^2 + \frac{1}{3}\Big(\frac{1}{3}m\Big)l^2 + \frac{1}{4}\,ml^2 = \frac{5}{4}\,ml^2$$

the equation of motion can also be written as

$$\ddot{\varphi} + \frac{32\,k}{5\,m}\,\varphi = 0\,.$$

Problem 9.7 A symmetric disk with a half-circular boundary rolls without slip on the flat surface. P9.7

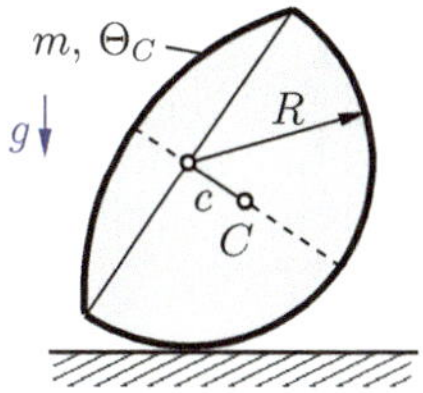

Derive the equation of motion using the Lagrange formalism.

Given: R, $c = \kappa R$, m, $\Theta_C = \alpha\, mR^2$

Solution The system is conservative and has one degree of freedom. To describe the motion we choose as generalized coordinate the angle φ. Then, with the kinematic relations

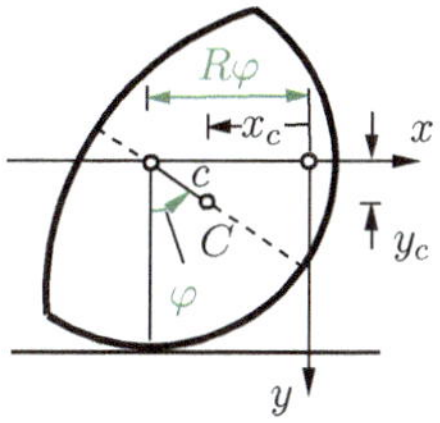

$$x_C = -R\varphi + c\sin\varphi = -R(\varphi - \kappa\sin\varphi)$$
$$\dot{x}_C = -R\dot{\varphi} + c\dot{\varphi}\cos\varphi = -R\dot{\varphi}(1 - \kappa\cos\varphi)\ ,$$
$$y_C = c\cos\varphi = \kappa R\cos\varphi\ ,$$
$$\dot{y}_C = -c\dot{\varphi}\sin\varphi = -\kappa R\dot{\varphi}\sin\varphi\ .$$

we can formulate the energies, the Lagrangean and the required derivatives:

$$V = -mgy_c = -mg\kappa R\cos\varphi\ ,$$
$$T = \frac{1}{2}m(\dot{x}_C^2 + \dot{y}_C^2) + \frac{1}{2}\Theta_C\dot{\varphi}^2 = \frac{1}{2}mR^2\dot{\varphi}^2\Big[(1-\kappa\cos\varphi)^2 + (\kappa\sin\varphi)^2 + \alpha\Big] = \frac{1}{2}mR^2\dot{\varphi}^2\Big(1 - 2\kappa\cos\varphi + \kappa^2 + \alpha\Big)\ ,$$
$$L = T - V = \frac{1}{2}mR\Big[R\dot{\varphi}^2\Big(1 - 2\kappa\cos\varphi + \kappa^2 + \alpha\Big) + 2g\kappa\cos\varphi\Big],$$
$$\frac{\partial L}{\partial\dot{\varphi}} = \frac{1}{2}mR\Big[2R\dot{\varphi}\Big(1 - 2\kappa\cos\varphi + \kappa^2 + \alpha\Big)\Big]\ ,$$
$$\frac{\mathrm{d}}{\mathrm{d}t}\Big(\frac{\partial L}{\partial\dot{\varphi}}\Big) = \frac{1}{2}mR\Big[2R\ddot{\varphi}\Big(1 - 2\kappa\cos\varphi + \kappa^2 + \alpha\Big) + 4\kappa R\dot{\varphi}^2\sin\varphi\Big]\ ,$$
$$\frac{\partial L}{\partial\varphi} = mR\Big[\kappa R\dot{\varphi}^2\sin\varphi - \kappa g\sin\varphi\Big]\ .$$

Substituting these expressions into

$$\frac{\mathrm{d}}{\mathrm{d}t}\Big(\frac{\partial L}{\partial\dot{\varphi}}\Big) - \frac{\partial L}{\partial\varphi} = 0$$

yields the equation of motion

$$\underline{\underline{\ddot{\varphi}(1 - 2\kappa\cos\varphi + \kappa^2 + \alpha) + \kappa\dot{\varphi}^2\sin\varphi + \kappa\frac{g}{R}\sin\varphi = 0\ .}}$$

P9.8 **Problem 9.8** A homogeneous bar (mass m, length l) is attached to a thread of length $l/2$, which is pinned at point A. The mass of the thread is negligible.

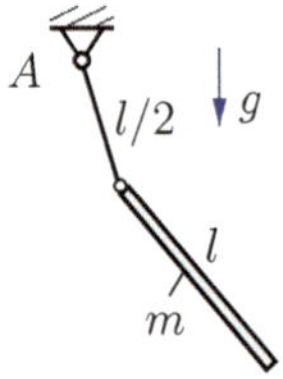

Derive the equations of motion using the Lagrange formalism.

Solution The system is conservative and has two degrees of freedom. As generalized coordinates we choose φ_1 and φ_2 and assume the zero level of the potential at the height of A. With

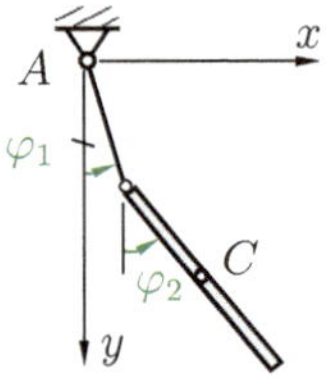

$$x_c = (l/2)(\sin\varphi_1 + \sin\varphi_2)\,,$$
$$y_c = (l/2)(\cos\varphi_1 + \cos\varphi_2)\,,$$
$$\dot x_c = (l/2)(\dot\varphi_1\cos\varphi_1 + \dot\varphi_2\cos\varphi_2), \qquad \dot y_c = (l/2)(\dot\varphi_1\sin\varphi_1 - \dot\varphi_2\sin\varphi_2)$$

and $\Theta_C = ml^2/12$ follow the required energies

$$V = -mg\frac{l}{2}(\cos\varphi_1 + \cos\varphi_2)\,,$$
$$T = \frac{1}{2}m(\dot x_c^2 + \dot y_c^2) + \frac{1}{2}\Theta_C\dot\varphi_2^2$$
$$= \frac{1}{8}ml^2[\dot\varphi_1^2 + \dot\varphi_2^2 + 2\dot\varphi_1\dot\varphi_2\cos(\varphi_1+\varphi_2)] + \frac{1}{24}ml^2\dot\varphi_2^2$$

for the Lagrangean $L = T - V$. Introducing the derivatives

$$\frac{\partial L}{\partial\varphi_1} = -\frac{ml^2}{4}\dot\varphi_1\dot\varphi_2\sin(\varphi_1+\varphi_2) - \frac{1}{2}mgl\sin\varphi_1\,,$$
$$\frac{\partial L}{\partial\varphi_2} = -\frac{ml^2}{4}\dot\varphi_1\dot\varphi_2\sin(\varphi_1+\varphi_2) - \frac{1}{2}mgl\sin\varphi_2\,,$$
$$\frac{\mathrm{d}}{\mathrm{d}t}\left(\frac{\partial L}{\partial\dot\varphi_1}\right) = \frac{ml^2}{4}[\ddot\varphi_1 + \ddot\varphi_2\cos(\varphi_1+\varphi_2) - \dot\varphi_2(\dot\varphi_1+\dot\varphi_2)\sin(\varphi_1+\varphi_2)]\,,$$
$$\frac{\mathrm{d}}{\mathrm{d}t}\left(\frac{\partial L}{\partial\dot\varphi_2}\right) = \frac{ml^2}{12}[4\ddot\varphi_2 + 3\ddot\varphi_1\cos(\varphi_1+\varphi_2) - 3\dot\varphi_1(\dot\varphi_1+\dot\varphi_2)\sin(\varphi_1+\varphi_2)]$$

into the Lagrange equations

$$\frac{\mathrm{d}}{\mathrm{d}t}\left(\frac{\partial L}{\partial\dot\varphi_1}\right) - \frac{\partial L}{\partial\varphi_1} = 0\,, \qquad \frac{\mathrm{d}}{\mathrm{d}t}\left(\frac{\partial L}{\partial\dot\varphi_2}\right) - \frac{\partial L}{\partial\varphi_2} = 0$$

yields the coupled equations of motion

$$\underline{\underline{\ddot\varphi_1 + \ddot\varphi_2\cos(\varphi_1+\varphi_2) - \dot\varphi_2^2\sin(\varphi_1+\varphi_2) + 2(g/l)\sin\varphi_1 = 0}}\,,$$
$$\underline{\underline{4\ddot\varphi_2 + 3\ddot\varphi_1\cos(\varphi_1+\varphi_2) - 3\dot\varphi_1^2\sin(\varphi_1+\varphi_2) + 6(g/l)\sin\varphi_2 = 0}}\,.$$

Problem 9.9 A simple pendulum is attached to point 0 of a disk which rotates with constant angular velocity Ω about the vertical axis. P9.9

Derive the equation of motion using the Lagrange formalism. Disregard the weight of the point mass.

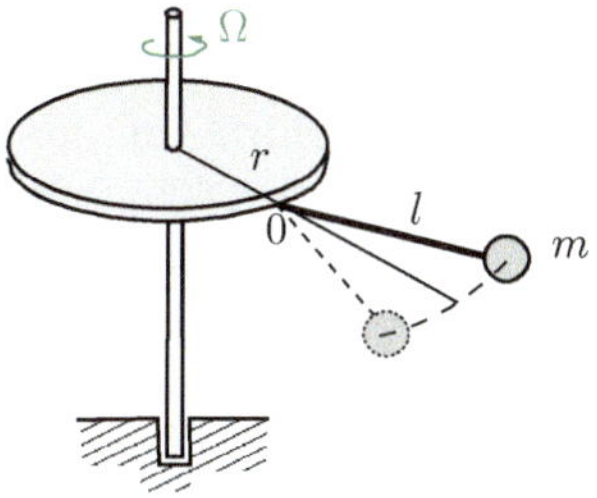

Solution The system is conservative. Since the weight is neglected ($V = 0$), only the kinetic T energy is needed in the Lagrangean. To describe the motion, we introduce the angle $\psi = \Omega t$, prescribed by the rotating disk, and the angle φ relative to the disk. With

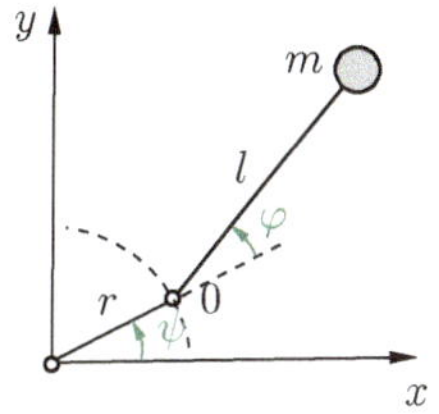

$$
\begin{aligned}
x &= r\cos\psi + l\cos(\psi+\varphi)\,,\\
y &= r\sin\psi + l\sin(\psi+\varphi)\,,\\
\dot{x} &= -r\Omega\sin\psi - l(\Omega+\dot{\varphi})\sin(\psi+\varphi)\,,\\
\dot{y} &= r\Omega\cos\psi + l(\Omega+\dot{\varphi})\cos(\psi+\varphi)\,,\\
\dot{x}^2+\dot{y}^2 &= r^2\Omega^2 + l^2(\Omega+\dot{\varphi})^2 + 2rl\Omega(\Omega+\dot{\varphi})\cos\varphi
\end{aligned}
$$

follows

$$L = T = \frac{1}{2}m(\dot{x}^2+\dot{y}^2) = \frac{1}{2}m\Big[r^2\Omega^2+l^2(\Omega+\dot{\varphi})^2+2rl\Omega(\Omega+\dot{\varphi})\cos\varphi\Big].$$

Introducing the derivatives

$$
\begin{aligned}
&\frac{\partial L}{\partial\dot{\varphi}} = \frac{m}{2}\Big(2l^2\Omega + 2l^2\dot{\varphi} + 2rl\Omega\cos\varphi\Big),\\
&\frac{\mathrm{d}}{\mathrm{d}t}\Big(\frac{\partial L}{\partial\dot{\varphi}}\Big) = \frac{m}{2}\Big(2l^2\ddot{\varphi} - 2rl\Omega\dot{\varphi}\sin\varphi\Big), \qquad \frac{\partial L}{\partial\varphi} = -\frac{m}{2}\,2rl\Omega(\Omega+\dot{\varphi})\sin\varphi
\end{aligned}
$$

into the Lagrange equation

$$\frac{\mathrm{d}}{\mathrm{d}t}\Big(\frac{\partial L}{\partial\dot{\varphi}}\Big) - \frac{\partial L}{\partial\varphi} = 0$$

leads to the equation of motion

$$\underline{\underline{\ddot{\varphi} + \frac{r}{l}\Omega^2\sin\varphi = 0\,.}}$$

Remark: For small angles ($\sin\varphi \approx \varphi$) this equation describes harmonic vibrations with $\omega = \Omega\sqrt{r/l}$ (see also Problem 8.9).

P9.10 **Problem 9.10** A homogeneous bar (length $2a$, weight $W = mg$) is suspended to the support A, which can move horizontally and is held by a spring (spring constant k).

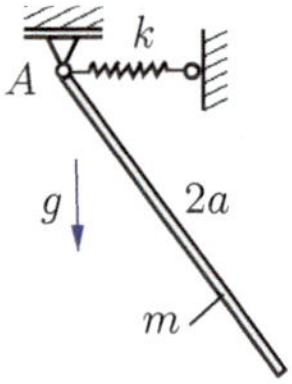

Find the equations of motion using the Langrange formalism.

Solution The system is conservative and has two degrees of freedom. As generalized coordinates we choose the displacement w and angle φ, both measured from the equilibrium position. With

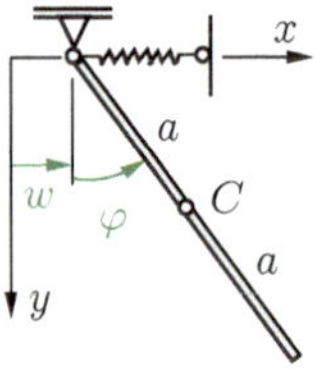

$$x_C = w + a\sin\varphi\,, \quad y_C = a\cos\varphi\,,$$
$$\dot{x}_C = \dot{w} + a\dot{\varphi}\cos\varphi\,, \quad \dot{y}_C = -a\dot{\varphi}\sin\varphi$$

and $\Theta_C = ma^2/3$ follow the energies

$$V = -mga\cos\varphi + kw^2/2\,,$$
$$T = (m/2)\big(x_C^2 + y_C^2\big) + (\Theta_C/2)\dot{\varphi}^2$$
$$= (m/2)\big(\dot{w}^2 + a^2\dot{\varphi}^2 + 2a\dot{w}\dot{\varphi}\cos\varphi\big) + (ma^2/6)\dot{\varphi}^2\,.$$

Herewith, the derivatives of the Lagrangean $L = T - V$ are

$$\frac{\mathrm{d}}{\mathrm{d}t}\left(\frac{\partial L}{\partial \dot{w}}\right) = m\Big(\ddot{w} + a\ddot{\varphi}\cos\varphi - a\dot{\varphi}^2\sin\varphi\Big), \qquad \frac{\partial L}{\partial w} = -k\,w\,,$$
$$\frac{\mathrm{d}}{\mathrm{d}t}\left(\frac{\partial L}{\partial \dot{\varphi}}\right) = m\Big(\frac{4}{3}\,a^2\ddot{\varphi} + a\ddot{w}\cos\varphi - a\dot{w}\dot{\varphi}\sin\varphi\Big),$$
$$\frac{\partial L}{\partial \varphi} = -m\Big(a\dot{w}\dot{\varphi} + ag\Big)\sin\varphi\,,$$

and the Lagrange equations

$$\frac{\mathrm{d}}{\mathrm{d}t}\left(\frac{\partial L}{\partial \dot{w}}\right) - \frac{\partial L}{\partial w} = 0\,, \qquad \frac{\mathrm{d}}{\mathrm{d}t}\left(\frac{\partial L}{\partial \dot{\varphi}}\right) - \frac{\partial L}{\partial \varphi} = 0$$

yield the coupled equations of motion

$$\underline{\underline{\ddot{w} + a\,\ddot{\varphi}\cos\varphi - a\,\dot{\varphi}^2\sin\varphi + \frac{k}{m}\,w = 0}}\,,$$

$$\underline{\underline{\frac{4}{3}\,a\,\ddot{\varphi} + \ddot{w}\cos\varphi + g\,\sin\varphi = 0}}\,.$$

Problem 9.11 The double pendulum consists of 4 bars of equal length l and mass m. The spring is unstrained in the vertical position of the hanging bars. P9.11

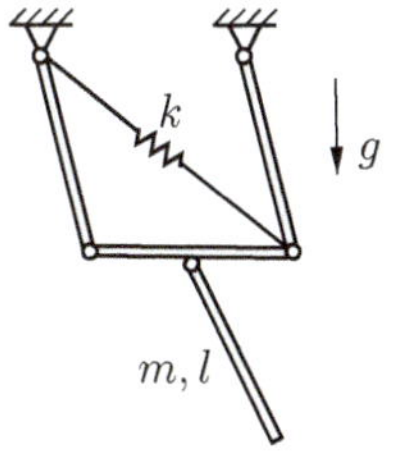

a) Determine the equations of motion by using the LAGRANGE equations.
b) Linearize the equations for small amplitudes.

Solution **a)** The position of the *conservative* system is uniquely described by the angles α and β. Accordingly, it has two degrees of freedom.
To formulate the potential and kinetic energies, we introduce the x, y-coordinate system and determine first the coordinates and velocities of points ①, ② as well as the length change Δ of the spring:

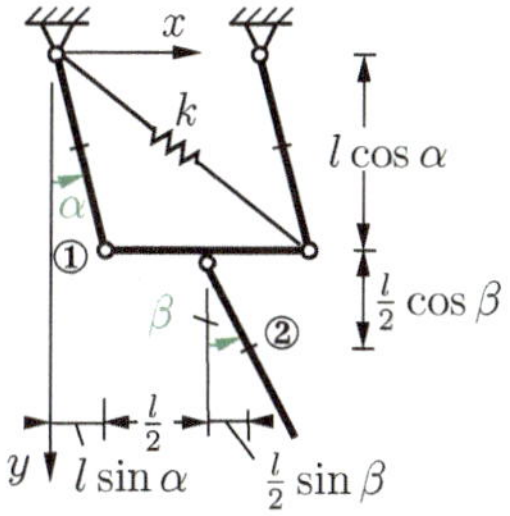

$$x_1 = l\sin\alpha\,, \qquad x_2 = l(\sin\alpha + \tfrac{1}{2} + \tfrac{1}{2}\sin\beta)\,,$$
$$\dot{x}_1 = l\dot{\alpha}\cos\alpha\,, \qquad \dot{x}_2 = l(\dot{\alpha}\cos\alpha + \tfrac{1}{2}\dot{\beta}\cos\beta)\,,$$
$$y_1 = l\cos\alpha\,, \qquad y_2 = l(\cos\alpha + \tfrac{1}{2}\cos\beta)\,,$$
$$\dot{y}_1 = -l\dot{\alpha}\sin\alpha\,, \quad \dot{y}_2 = -l(\dot{\alpha}\sin\alpha + \tfrac{1}{2}\dot{\beta}\sin\beta)\,,$$
$$v_1^2 = \dot{x}_1^2 + \dot{y}_1^2 = l^2\dot{\alpha}^2(\cos^2\alpha + \sin^2\alpha) = l^2\dot{\alpha}^2\,,$$
$$v_2^2 = \dot{x}_2^2 + \dot{y}_2^2 = l^2(\dot{\alpha}^2\cos^2\alpha + \dot{\alpha}\dot{\beta}\cos\alpha\cos\beta + \tfrac{1}{4}\dot{\beta}^2\cos^2\beta$$
$$+ \dot{\alpha}^2\sin^2\alpha + \dot{\alpha}\dot{\beta}\sin\alpha\sin\beta + \tfrac{1}{4}\dot{\beta}^2\sin^2\beta)$$
$$= l^2[\dot{\alpha}^2 + \dot{\alpha}\dot{\beta}\cos(\alpha-\beta) + \tfrac{1}{4}\dot{\beta}^2]\,,$$
$$\Delta = \sqrt{(l\sin\alpha + l)^2 + l^2\cos^2\alpha} - l\sqrt{2} = l\sqrt{2}\left(\sqrt{1+\sin\alpha} - 1\right),$$
$$\Delta^2 = 2l^2\left(2 + \sin\alpha - 2\sqrt{1+\sin\alpha}\right).$$

Herewith, the potential energy is obtained as

$$V = -2mg\frac{y_1}{2} - mgy_1 - mgy_2 + \frac{1}{2}k\Delta^2$$
$$= -mgl\left(3\cos\alpha + \frac{1}{2}\cos\beta\right) + kl^2\left(2 + \sin\alpha - 2\sqrt{1+\sin\alpha}\right).$$

When determining the kinetic energy, we consider that the motion of the upper bars is a pure rotation (angular velocity $\dot{\alpha}$) about the pins, the motion of the horizontal bar is a pure translation (velocity v_1) and the motion of the lower bar is a combination of translation (velocity v_2) plus rotation (angular velocity $\dot\beta$). With the moment of inertia $\Theta_1 = ml^2/3$ of an upper bar with respect to the pin and the moment of inertia $\Theta_2 = ml^2/12$ of the lower bar with respect to its center of mass ②, it follows

$$\begin{aligned} T &= 2\frac{1}{2}\Theta_1\dot\alpha^2 + \frac{1}{2}mv_1^2 + \Big(\frac{1}{2}mv_2^2 + \frac{1}{2}\Theta_2\dot\beta^2\Big) \\ &= ml^2\Big[\frac{4}{3}\dot\alpha^2 + \frac{1}{6}\dot\beta^2 + \frac{1}{2}\dot\alpha\dot\beta\cos(\alpha-\beta)\Big]. \end{aligned}$$

With $L = T - V$ and the derivatives

$$\begin{aligned} \frac{\partial L}{\partial \alpha} &= -\frac{ml^2}{2}\dot\alpha\dot\beta\sin(\alpha-\beta) - 3mgl\sin\alpha - kl^2\cos\alpha\Big(1 - \frac{1}{\sqrt{1+\sin\alpha}}\Big), \\ \frac{\partial L}{\partial \beta} &= \frac{ml^2}{2}\dot\alpha\dot\beta\sin(\alpha-\beta) - \frac{1}{2}mgl\sin\beta\,, \\ \frac{\mathrm{d}}{\mathrm{d}t}\Big(\frac{\partial L}{\partial\dot\alpha}\Big) &= ml^2\Big[\frac{8}{3}\ddot\alpha + \frac{1}{2}\ddot\beta\cos(\alpha-\beta) - \frac{1}{2}\dot\beta(\dot\alpha-\dot\beta)\sin(\alpha-\beta)\Big], \\ \frac{\mathrm{d}}{\mathrm{d}t}\Big(\frac{\partial L}{\partial\dot\beta}\Big) &= ml^2\Big[\frac{1}{6}\ddot\beta + \frac{1}{2}\ddot\alpha\cos(\alpha-\beta) - \frac{1}{2}\dot\alpha(\dot\alpha-\dot\beta)\sin(\alpha-\beta)\Big], \end{aligned}$$

we obtain from the LAGRANGE equations

$$\frac{\mathrm{d}}{\mathrm{d}t}\Big(\frac{\partial L}{\partial\dot\alpha}\Big) - \frac{\partial L}{\partial\alpha} = 0\,, \qquad \frac{\mathrm{d}}{\mathrm{d}t}\Big(\frac{\partial L}{\partial\dot\beta}\Big) - \frac{\partial L}{\partial\beta} = 0$$

the equations of motion:

$$\underline{\underline{\frac{8}{3}\ddot\alpha + \frac{1}{2}\ddot\beta\cos(\alpha-\beta) + \frac{1}{2}\dot\beta^2\sin(\alpha-\beta) + 3\frac{g}{l}\sin\alpha + \frac{k}{m}\cos\alpha\Big(1 - \frac{1}{\sqrt{1+\sin\alpha}}\Big) = 0,}}$$

$$\underline{\underline{\frac{1}{6}\ddot\beta + \frac{1}{2}\ddot\alpha\cos(\alpha-\beta) - \frac{1}{2}\dot\alpha^2\sin(\alpha-\beta) + \frac{1}{2}\frac{g}{l}\sin\beta = 0.}}$$

b) For small amplitudes $\alpha \ll 1$, $\beta \ll 1$ and $\dot\alpha \ll 1$, $\dot\beta \ll 1$ the following linearizations apply:

$$\begin{gathered} \sin\alpha \approx \alpha\,, \quad \cos\alpha \approx 1\,, \quad \sin(\alpha-\beta) \approx (\alpha-\beta)\,, \quad \cos(\alpha-\beta) \approx 1\,, \\ \frac{1}{\sqrt{1+\sin\alpha}} \approx \frac{1}{\sqrt{1+\alpha}} \approx 1-\alpha\,, \quad \dot\beta^2\sin(\alpha-\beta) \approx 0\,, \quad \dot\alpha^2\sin(\alpha-\beta) \approx 0\,. \end{gathered}$$

Herewith, the equations of motion simplify to

$$\underline{\underline{\ddot\alpha + \frac{3}{16}\ddot\beta + \Big(\frac{9\,g}{8\,l} + \frac{3\,k}{8\,m}\Big)\alpha = 0}}\,, \qquad \underline{\underline{\ddot\beta + 3\,\ddot\alpha + 3\,\frac{g}{l}\,\beta = 0\,.}}$$

Chapter 10

Hydrodynamics

10

The **velocity field** $\boldsymbol{v}(\boldsymbol{x}(t), t)$ describes the motion of a fluid. The vector $\boldsymbol{x}$ assignes to each location in the fluid a velocity $\boldsymbol{v}$ at time t.

The velocity field is *stationary* for $\partial \boldsymbol{v}/\partial t = 0$, otherwise *instationary*.

Pathline: trajectory that a material point of the fluid (fluid element) follows over a time period. The pathline $\boldsymbol{x}(t)$ yields from the solution of the differential equation

$$\frac{\mathrm{d}}{\mathrm{d}t}\boldsymbol{x}(t) = \boldsymbol{v}(\boldsymbol{x}(t), t)\,.$$

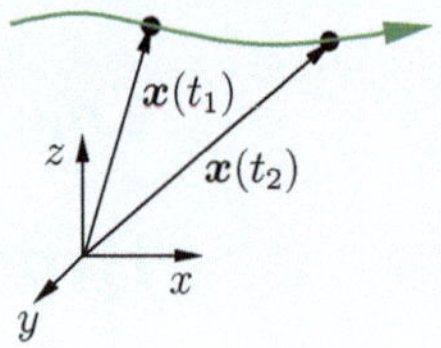

Streamlines: family of curves whose tangents coincide in each point $\boldsymbol{x}$ with the direction of the local velocity vector. The streamline field follows from the differential equation

$$\frac{\mathrm{d}}{\mathrm{d}s}\boldsymbol{x}(s) = \boldsymbol{v}(\boldsymbol{x}(s), t)\,,$$

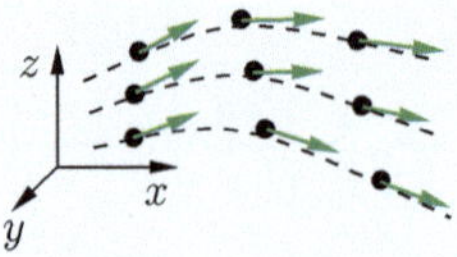

where s is the arclength of the streamline.

Notice: pathlines and streamlines are identical for a *stationary* velocity field.

Stream Filament Theory: In what follows we restrict ourselves to the *stationary* motion of an *incompressible* fluid in a streamtube, where the flow behavior is characterized by the behavior at a median streamline. This one-dimensional theory is described by the following basic equations:

a) Continuity equation

$$A_1 v_1 = A_2 v_2 \quad \text{or} \quad Q = Av = \text{const}$$

where $Q = Av$ is the volume flow.

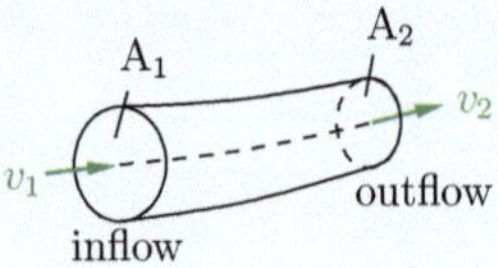

b) BERNOULLI's theorem

For an *inviscid* fluid holds

$$\frac{1}{2}\varrho v^2 + \varrho g z + p = \text{const} \quad \text{or} \quad \frac{v^2}{2g} + z + \frac{p}{\varrho g} = H = \text{const}\,.$$

where

$\varrho v^2/2$ = dynamic pressure (specific kinetic energy),

$\varrho g z$ = geodetic pressure (specific potential energy),

p = static pressure (pressure energy),

H = hydraulic (total) head,

$v^2/2g$ = velocity head,

z = elevation head,

$p/\varrho g$ = pressure head.

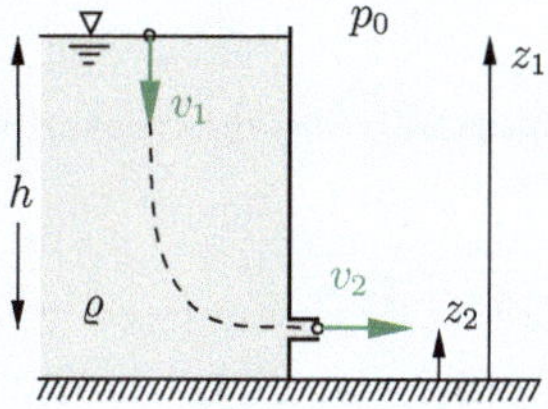

Example: outflow from a reservoir:

$$\frac{1}{2}\varrho v_1^2 + \varrho g z_1 + p_0 = \frac{1}{2}\varrho v_2^2 + \varrho g z_2 + p_0 \,.$$

In the special case $v_1 = 0$ (h =const) follows **TORRICELLI's law** (outflow from big fluid tanks).

$$v_2 = \sqrt{2gh} \,.$$

For viscous fluids (flow with energy losses) the **generalized BERNOULLI's theorem**

$$\frac{1}{2}\varrho v_1^2 + \varrho g z_1 + p_1 = \frac{1}{2}\varrho v_2^2 + \varrho g z_2 + p_2 + \Delta p_v$$

is valid, where

$$\Delta p_v = \zeta \frac{1}{2}\varrho v_1^2 = \text{pressure loss}, \quad \zeta = \text{pressure loss coefficient.}$$

c) Balance of Momentum

$$\boldsymbol{F} = \varrho Q\,(\boldsymbol{v}_2 - \boldsymbol{v}_1)$$

or

$$F_x = \dot{m}\,(v_{2x} - v_{1x}) \,,$$
$$F_y = \dot{m}\,(v_{2y} - v_{1y}) \,,$$
$$F_z = \dot{m}\,(v_{2z} - v_{1z}) \,,$$

where

$\boldsymbol{F}$ = resulting force exerted on the closed fluid volume within the streamtube (control volume),

ϱQ = $\dot{m}$ = mass flow,

$\varrho Q \boldsymbol{v}_1$ = inflowing momentum,

$\varrho Q \boldsymbol{v}_2$ = outflowing momentum.

P10.1

Problem 10.1 A flow is described by the plane velocity field

$$\boldsymbol{v}(\boldsymbol{x},t) = 2ax\,\boldsymbol{e}_x - 2ay\,\boldsymbol{e}_y$$

Determine the equation for the streamlines and sketch the profile for the specific streamline through the point A with coordinates $x = 0.5\,\mathrm{m}$; $y = 4\,\mathrm{m}$.

Solution The differential equation for the streamlines reads in components

$$\frac{\mathrm{d}x}{\mathrm{d}s} = v_x = 2ax\ , \qquad \frac{\mathrm{d}y}{\mathrm{d}s} = v_y = -2ay\ .$$

Dividing the 1st by the 2nd equation yields

$$\frac{\mathrm{d}x}{\mathrm{d}y} = \frac{v_x}{v_y} = -\frac{x}{y}$$

and by separation of variables it follows

$$\frac{\mathrm{d}x}{x} + \frac{\mathrm{d}y}{y} = 0\ .$$

Integration leads to

$$\ln x + \ln y = \ln xy = C =: \ln c \quad \rightsquigarrow \quad \ln xy = \ln c\ .$$

Accordingly, the streamlines are given by the hyperbola

$$\underline{\underline{y = \frac{c}{x}}}\ .$$

For a streamline through point A, the integration constant c is calculated with the given data as

$$c = 0.5\,\mathrm{m}\cdot 4\,\mathrm{m} = 2\,\mathrm{m}^2\ .$$

Having c, the profile of the streamline can be sketched.

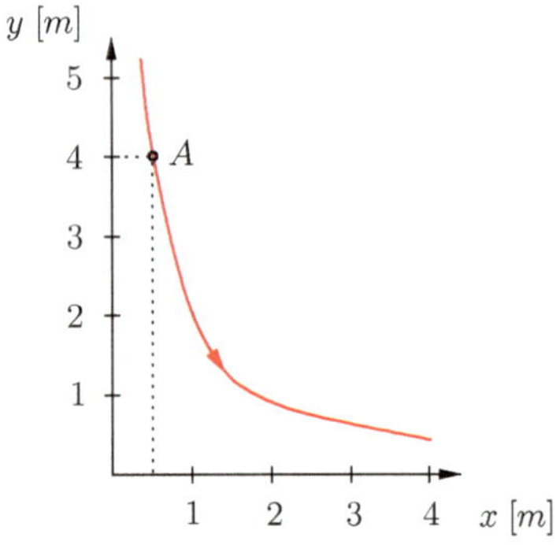

Remarks:

- For $x \to \infty$, the y-component of the velocity vector vanishes: $\boldsymbol{v} \to 2ax\,\boldsymbol{e}_x$.
- Because the flow is stationary ($\partial \boldsymbol{v}/\partial t = 0$), the streamlines and the pathlines coincide.

Problem 10.2 A planar flow is described by the velocity field P10.2

$$\boldsymbol{v}(\boldsymbol{x},t) = ax\,\boldsymbol{e}_x + b\,\mathrm{e}^{-t}\boldsymbol{e}_y\,,$$

where a and b are given constants.

a) Determine the pathline of the particle, which at time $t = 0$ is at point $P = (1,1)$.

b) Determine the streamline, which at time $t = 0$ passes the point $P = (1,1)$.

Solution Because the flow is instationary, the pathlines and streamlines do *not* coincide!

a) The components of the pathline are determined from

$$\frac{\mathrm{d}x}{\mathrm{d}t} = ax \quad \rightsquigarrow \quad \int \frac{\mathrm{d}x}{x} = \int a\,\mathrm{d}t \quad \rightsquigarrow \quad \ln\frac{x}{C_1} = at \quad \rightsquigarrow \quad x = C_1\,\mathrm{e}^{at},$$

$$\frac{\mathrm{d}y}{\mathrm{d}t} = b\,\mathrm{e}^{-t} \quad \rightsquigarrow \quad y = -b\,\mathrm{e}^{-t} + C_2\,.$$

With the initial conditions $x\,(t{=}0) = 1$, $y\,(t{=}0) = 1$, we obtain

$$\underline{\underline{x(t) = \mathrm{e}^{at}}}\,, \qquad \underline{\underline{y(t) = b\,(1 - \mathrm{e}^{-t}) + 1}}\,.$$

b) For $t = 0$, the differential equations of the streamlines are given by

$$\frac{\mathrm{d}x}{\mathrm{d}s} = ax\,, \qquad \frac{\mathrm{d}y}{\mathrm{d}s} = b \qquad \rightsquigarrow \qquad \frac{\mathrm{d}x}{\mathrm{d}y} = \frac{a}{b}\,x\,.$$

Separation of variables and integration lead to

$$\frac{b}{a}\int \frac{\mathrm{d}x}{x} = \int \mathrm{d}y \qquad \rightsquigarrow \qquad y = \frac{b}{a}\ln x + C_3\,.$$

The boundary condition yields

$$y(x = 1) = 1 \qquad \rightsquigarrow \qquad C_3 = 1\,,$$

and thus, it follows

$$\underline{\underline{y(x) = \frac{b}{a}\ln x + 1}}\,.$$

Remark: From the parameter representation of the pathline in a), by eliminating t ($t = \frac{1}{a}\ln x$), we can obtain the representation $y(x) = b\,(1 - x^{-1/a})$.

P10.3 **Problem 10.3** The velocity field of a planar, instationary flow is described by

$$\boldsymbol{v}(x,y,t) = a\,xy\,\boldsymbol{e}_x + b\,t\,\boldsymbol{e}_y$$

where a, b are given constants. As initial conditions, $x = x_0$, $y = y_0$ at $t = 0$ are prescribed.

a) Determine the pathlines and streamlines.

b) Where has a fluid particle been at time $t = 0$, which was detected at time $t_1 = 1\,\mathrm{s}$ at point $(x_1, y_1) = (1, 0)\,\mathrm{m}$?

Solution **a)** The pathlines are determined from

$$\frac{\mathrm{d}x}{\mathrm{d}t} = a\,xy\,, \qquad \frac{\mathrm{d}y}{\mathrm{d}t} = b\,t\,.$$

With $y(t{=}0) = y_0$ the second equation yields

$$\underline{\underline{y(t) = y_0 + \frac{1}{2}\,b\,t^2}}\,.$$

Introducing the result into the first equation, after separation of variables, integration and using $x(t{=}0) = x_0$ we obtain

$$\frac{\mathrm{d}x}{x} = a\left(y_0 + \frac{1}{2}\,b\,t^2\right)\mathrm{d}t \qquad \rightsquigarrow \qquad \underline{\underline{x(t) = x_0\,\mathrm{e}^{a\,(y_0 t + bt^3/6)}}}\,.$$

The streamlines are calculated from

$$\frac{\mathrm{d}x}{\mathrm{d}s} = a\,xy\,, \qquad \frac{\mathrm{d}y}{\mathrm{d}s} = b\,t\,.$$

The second equation in conjunction with $y(s{=}0) = y_0$ yields

$$\underline{\underline{y(s) = b\,t\,s + y_0}}\,.$$

Again, introducing the result into the first equation, applying separation of variables and using $x(s{=}0) = x_0$ leads to

$$\frac{\mathrm{d}x}{x} = a\,(b\,t\,s + y_0)\,\mathrm{d}s \qquad \rightsquigarrow \qquad \underline{\underline{x(s) = x_0\,\mathrm{e}^{a\,(y_0 s + bt^2/2)}}}\,.$$

b) Introducing the conditions $x = x_1 = 1\,\mathrm{m}$, $y = y_1 = 0$ for $t = 1\,\mathrm{s}$ into the pathline yields

$$\underline{\underline{y_0 = -b/2}}\,, \qquad \underline{\underline{x_0 = \mathrm{e}^{ab/3}}}\,.$$

Problem 10.4 From a big reservoir water is taken via a pipe. To increase the mass flow Q, a diffuser is attached at the end of the pipe. Because cavitation danger is to be avoided, the pressure must not drop below p_{min} at any location of the pipe. **P10.4**

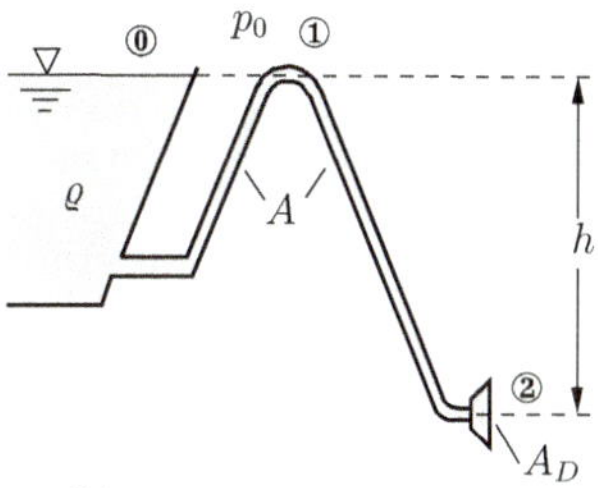

a) Determine the maximum allowable diffuser cross-section $A_{D_{max}}$.

b) Calculate the temporal mass flow Q for this case.

c) Determine the height h^*, the highest point of the pipe can be lifted, if the diffuser is not present.

Solution **a)** Considering $v_0 = 0$ (big reservoir) and applying BERNOULLI's theorem for the points ⓪ and ① as well as for ⓪ and ② of a streamline, the corresponding velocities can be determined:

$$\tfrac{1}{2}\varrho v_0^2 + p_0 + \varrho g h = \tfrac{1}{2}\varrho v_1^2 + p_1 + \varrho g h \quad \leadsto \quad v_1 = \sqrt{\frac{2}{\varrho}(p_0 - p_1)}\ ,$$

$$\tfrac{1}{2}\varrho v_0^2 + p_0 + \varrho g h = \tfrac{1}{2}\varrho v_2^2 + p_0 \quad \leadsto \quad v_2 = \sqrt{2gh}\ .$$

Herewith follows from the continuity equation the cross section of the diffuser:

$$A_D\, v_2 = A\, v_1 \quad \leadsto \quad A_D = A\,\frac{v_1}{v_2} = A\sqrt{\frac{p_0 - p_1}{\varrho g h}}\ .$$

It is a maximum, when we insert for p_1 the minimum allowable pressure p_{min}:

$$\underline{\underline{A_{D_{max}} = A\sqrt{\frac{p_0 - p_{min}}{\varrho g h}}}}\ .$$

b) The temporal mass flow Q in this case is gicen by

$$\underline{\underline{Q}} = v_2\, A_D = \underline{\underline{\sqrt{2gh}\, A_{D_{max}}}}\ .$$

c) In the same way as in a) we obtain for a streamline between ⓪ and ① as well as between ⓪ and ② after lifting point ① to the height h^*

$$p_0 + \varrho g h = \tfrac{1}{2}\varrho v_1^2 + p_{min} + \varrho g h^*\ , \qquad v_2 = \sqrt{2gh}\ .$$

With $A_D = A$, the continuity equation yields $v_1 = v_2$. By introducing this and solving for h^* we obtain

$$\underline{\underline{h^* = \frac{p_0 - p_{min}}{\varrho g}}}\ .$$

P 10.5

Problem 10.5 The lower trapezoidal tank (constant depth f) is filled via a pipe (cross section A_p) from a big reservoir located above.

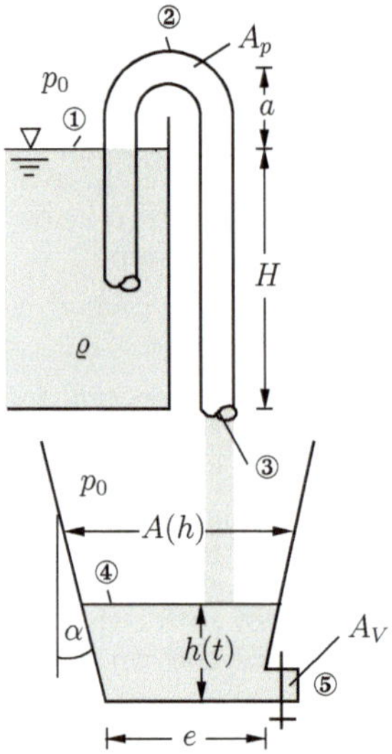

a) Determine the maximum height $a = a_{\max}$, the pipe may protrude the fluid level of the reservoir, such that the pressure in the pipe does not drop below p_D.

b) When is the height $h(t) = H/2$ reached in the lower tank?

c) When $h(t) = H/2$ is reached, the valve of the lower tank is opened. Determine the cross section A_V of the valve, such that the fluid level keeps constant.

Solution a) According to TORRICELLI's law (big reservoir), the velocity at point ③ is given by

$$v_3 = \sqrt{2gH} \; .$$

Thus, from the continuity equation follows at point ②

$$A_p \, v_2 = A_p \, v_3 \qquad \leadsto \qquad v_2 = v_3 = \sqrt{2gH} \; .$$

Using BERNOULLI's theorem for a streamline between the points ① and ②

$$p_0 + 0 + 0 = p_D + \varrho \, g \, a_{\max} + \frac{1}{2} \, \varrho \, v_2^2 ,$$

and introducing v_2, the maximum height a_{max} can be determined:

$$\underline{\underline{a_{\max}}} = \frac{1}{\varrho \, g} \left[p_0 - p_D - \frac{1}{2} \varrho \, (2gH) \right] = \underline{\underline{\frac{p_0 - p_D}{\varrho \, g} - H}} \; .$$

b) The relation between the filling height h and time t follows from the continuity equation between point ③ and point ④

$$A_R \, v_3 = A(h) \, v_4$$

in conjunction with the rise velocity of the fluid

$$v_4 = \frac{\mathrm{d}h}{\mathrm{d}t}$$

and the cross section of the tank

$$A(h) = (e + 2h \tan \alpha) f$$

as

$$\frac{\mathrm{d}h}{\mathrm{d}t} = \frac{A_p}{A(h)} v_3 .$$

Separation of variables and integration leads to

$$\int_{t_0=0}^{t} A_p \sqrt{2gH} \, \mathrm{d}t = \int_{h_0=0}^{h} (e + 2h \tan \alpha) f \, \mathrm{d}h .$$

This yields

$$A_p \sqrt{2gH} \, t = (e\,h + h^2 \tan \alpha) f .$$

From this result the required time t_r, to reach the filling height $H/2$, is found by introducing $h = H/2$:

$$\underline{\underline{t_r = \left(e \frac{H}{2} + \frac{H^2}{4} \tan \alpha \right) \frac{f}{A_R \sqrt{2gH}}}} .$$

c) From BERNOULLI's theorem, applied between the points ④ and ⑤,

$$p_0 + \varrho\, g \frac{H}{2} + \frac{1}{2} \varrho\, v_4^2 = p_0 + 0 + \frac{1}{2} \varrho\, v_5^2 ,$$

the velocity at the valve can be calculated. It leads with the requirement of a constant fluid level, i.e. $v_4 = 0$, to

$$v_5 = \sqrt{gH} .$$

Finally, using the continuity equation

$$A_V \, v_5 = A_p \, v_3 ,$$

the cross section of the valve is obtained as

$$\underline{\underline{A_V = \sqrt{2} \, A_p}} .$$

P10.6 **Problem 10.6** From a big reservoir, an ideal fluid (density ϱ) flows out through a pipe (cross section A_2) with a local smooth contraction (cross section A_1).

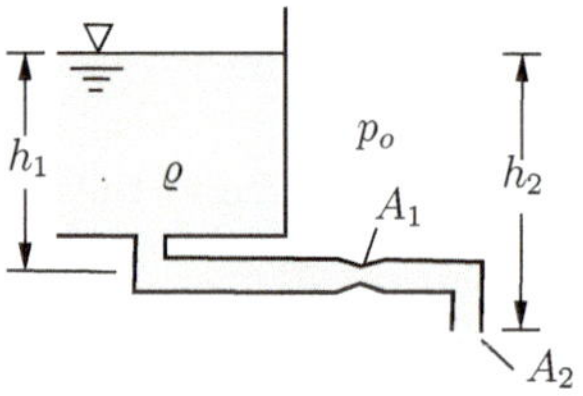

a) Determine the pressure p_1 in the cross section A_1.

b) The pipe will now be spot drilled at A_1. Calculate h_2, such that no fluid leaks from the drilled hole.

c) Now a vertical standpipe is connected to the drilled hole, which dips into a lower fluid tank. Find the necessary cross section ratio A_2/A_1, such that fluid is sucked from the tank.

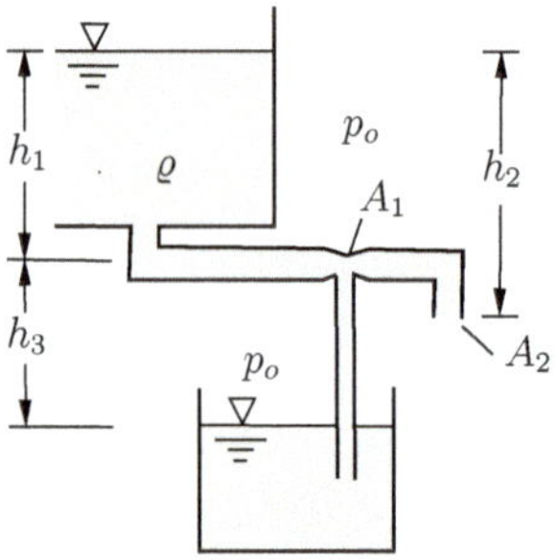

Solution **a)** First, from TORRICELLI's law follows the outflow velocity at the point ②

$$v_2 = \sqrt{2gh_2}$$

and from that, using the continuity equation, the velocity in the cross section A_1:

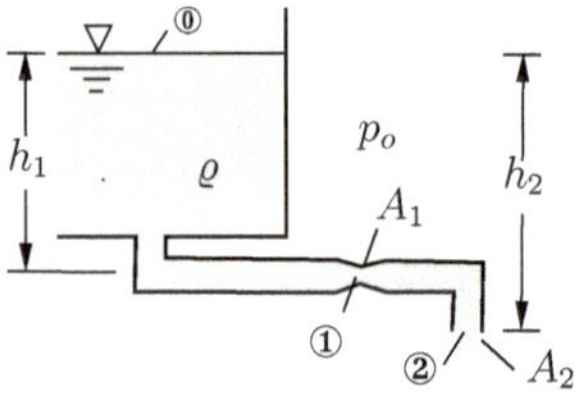

$$A_1\, v_1 = A_2\, v_2$$

$$\rightsquigarrow \quad v_1 = \frac{A_2}{A_1}\, v_2 = \frac{A_2}{A_1}\, \sqrt{2gh_2}\,.$$

Introducing v_1 into BERNOULLI's theorem for a streamline between the points ⓪ and ① yields the pressure p_1:

$$\frac{1}{2}\varrho v_1^2 + 0 + p_1 = 0 + \varrho g h_1 + p_0 \quad \rightsquigarrow \quad \underline{\underline{p_1 = p_0 + \varrho g\Big[h_1 - \Big(\frac{A_2}{A_1}\Big)^2 h_2\Big]}}\,.$$

b) No fluid leaks from the drilled hole, if the pressure p_1 is below the ambient pressure p_0:

$$p_1 < p_0 \quad \rightsquigarrow \quad p_0 + \varrho g\left[h_1 - \left(\frac{A_2}{A_1}\right)^2 h_2\right] < p_0$$
$$\rightsquigarrow \quad h_1 - \left(\frac{A_2}{A_1}\right)^2 h_2 < 0\,.$$

This leads to the condition

$$\underline{\underline{h_2 > \left(\frac{A_1}{A_2}\right)^2 h_1\,.}}$$

c) With the pressures $p_4 = p_1$ and $p_3 = p_0$ at the locations ④ and ③, we obtain from BERNOULLI's theorem for a streamline between the points ④ and ③

$$\frac{1}{2}\varrho v_4^2 + \varrho g h_3 + p_1 = 0 + 0 + p_0\,.$$

Introducing p_1 and using the condition $v_4^2 > 0$ leads to

$$p_0 - p_1 - \varrho g h_3 > 0$$
$$\rightsquigarrow \quad h_1 - \left(\frac{A_2}{A_1}\right)^2 h_2 + h_3 < 0$$

and finally to

$$\underline{\underline{\frac{A_2}{A_1} > \sqrt{\frac{h_1 + h_3}{h_2}}\,.}}$$

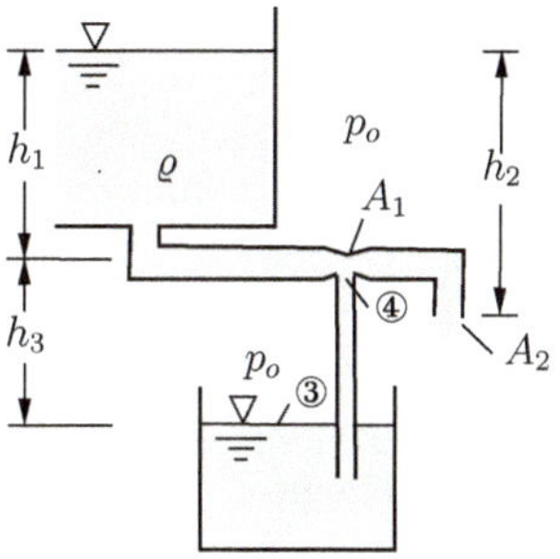

Remark: At the location ④ of the standpipe prevails the pressure p_1 but not the velocity v_1!

P 10.7

Problem 10.7 A tank is filled by a pump through an opening. At the same time, fluid flows out through a leak at the bottom.

a) Which stationary fluid level H will be reached?

b) Determine for this case the loss of volume flow through the leak.

c) Now the pump is shut down and the inflow is closed. How long does it take until the tank is empty?

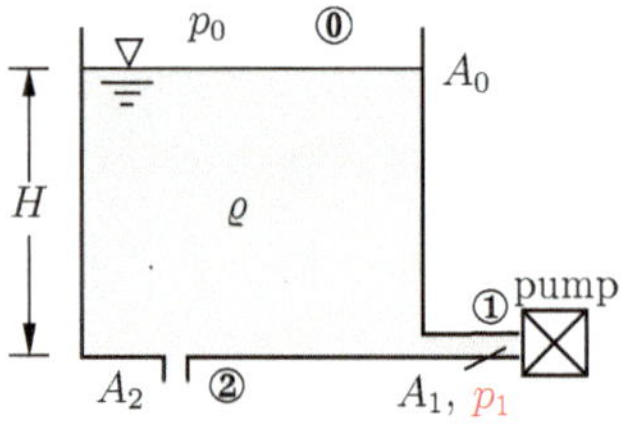

Solution **a)** With the BERNOULLI theorem applied between the points ⓪ and ① as well as ① and ②,

$$\frac{1}{2}\varrho v_0^2 + p_0 + \varrho g H = \frac{1}{2}\varrho v_1^2 + p_1 \,,$$

$$\frac{1}{2}\varrho v_1^2 + p_1 = \frac{1}{2}\varrho v_2^2 + p_0 \,,$$

and the continuity equation

$$A_1 v_1 = A_2 v_2$$

in conjunction with the stationarity condition $v_0 = 0$, we have three equations for the three unknowns v_1, v_2 and H. Solving the equations yields the velocities

$$v_1 = v_2 \frac{A_2}{A_1} = \sqrt{\frac{2(p_1 - p_0)\, A_2^2}{\varrho\,(A_1^2 - A_2^2)}}$$

and the stationary fluid height

$$\underline{\underline{H = \frac{p_1 - p_0}{\varrho g} \frac{A_1^2}{A_1^2 - A_2^2}}} \,.$$

It can be seen that a stationary state is only possible for $A_2 < A_1$.

b) The loss of volume flow Q_V is determined by using the continuity equation

$$Q_V = A_2 v_2 = A_1 v_1 \,.$$

Introducing v_1 and v_2, respectively, yields

$$\underline{\underline{Q_V = A_1 A_2 \sqrt{\frac{2(p_1 - p_0)}{\varrho\,(A_1^2 - A_2^2)}}}}\ .$$

Alternatively, the loss of volume flow can be calculated by using $v_2 = \sqrt{2gH}$ (TORRICELLI) and A_2.

c) Due to the leak in the tank, the fluid level changes continuously. For the velocity of level decrease we have

$$v(z) = -\frac{\mathrm{d}z}{\mathrm{d}t}\ ,$$

where z is the actual fluid level in the tank. Thus, BERNOULLI's theorem for a streamline between a point on the fluid surface and point ② reads

$$\frac{1}{2}\varrho\, v(z)^2 + p_0 + \varrho\, g z = \frac{1}{2}\varrho\, v_2^2 + p_0\ .$$

Using the continuity equation

$$A_0\, v(z) = A_2\, v_2$$

we obtain for the decrease velocity of the fluid surface

$$v(z) = -\frac{\mathrm{d}z}{\mathrm{d}t} = \sqrt{\frac{2gzA_2^2}{A_0^2 - A_2^2}}\ .$$

The time T, required to empty the tank, can be determined by separation of variables and integration:

$$-\int_H^0 \frac{\mathrm{d}z}{\sqrt{z}} = \sqrt{\frac{2gA_2^2}{A_0^2 - A_2^2}} \int_0^T \mathrm{d}t \quad \leadsto \quad \underline{\underline{T = 2\sqrt{\frac{A_0^2 - A_2^2}{2gA_2^2}}\sqrt{H}}}\ .$$

Here, for H, the result from **a)** can finally be introduced:

$$\underline{\underline{T = \frac{\sqrt{2}}{g}\sqrt{\frac{p_1 - p_0}{\rho}\,\frac{A_1^2\,(A_0^2 - A_2^2)}{A_2^2\,(A_1^2 - A_2^2)}}}}\ .$$

P10.8 **Problem 10.8** From a big reservoir, fluid flows out through a pipe with a smoothly changing cross section.

Determine the fluid levels z_1 and z_2 in the standpipes

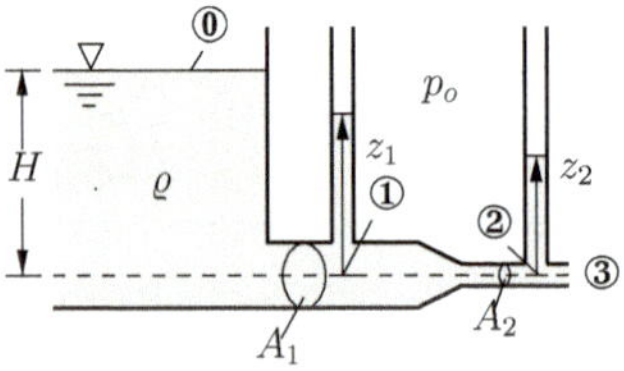

Solution The outflow velocity at point ③ follows from TORRICELLI's law as

$$v_3 = \sqrt{2gH} \, .$$

The pressures at the locations ① and ② with the fluid levels z_1 and z_2 in the standpipes are given by

$$p_1 = p_0 + \varrho g z_1 \, , \qquad p_2 = p_0 + \varrho g z_2 \, .$$

Thus, applying BERNOULLI's theorem for a stremline between the points ① and ③,

$$\tfrac{1}{2}\varrho v_1^2 + 0 + p_1 = \tfrac{1}{2}\varrho v_3^2 + 0 + p_0 \, ,$$

and using the continuity equation

$$A_1 v_1 = A_2 v_3 \quad \rightsquigarrow \quad v_1 = \frac{A_2}{A_1} v_3 \, ,$$

we first obtain

$$p_1 = p_0 + \frac{1}{2}\varrho v_3^2 \Big[1 - \Big(\frac{A_2}{A_1}\Big)^2\Big] \, .$$

Introducing $v_3 = \sqrt{2gH}$ yields the fluid level in the standpipe:

$$\underline{\underline{z_1 =}} \frac{p_1 - p_0}{\rho g} = \underline{\underline{H\Big[1 - \Big(\frac{A_2}{A_1}\Big)^2\Big]}} \, .$$

In the same way, the pressure p_2 is calculated by applying BERNOULLI's theorem for a streamline between ② and ③,

$$\tfrac{1}{2}\varrho v_2^2 + 0 + p_2 = \tfrac{1}{2}\varrho v_3^2 + 0 + p_0 \, ,$$

and using the continuity equation $v_2 = v_3$. Thus, it follows for the pressure $p_2 = p_0$ and for the fluid level

$$\underline{\underline{z_2 = 0}} \, .$$

Problem 10.9 From the drainpipe of a big reservoir, a water-jet hits a pivoted plate of weight W. **P10.9**

Determine
a) the pressure in the pipe ②③ as a function of the coordinate z,
b) the angle φ, the plate is rotated, if the jet flows off in plate direction.

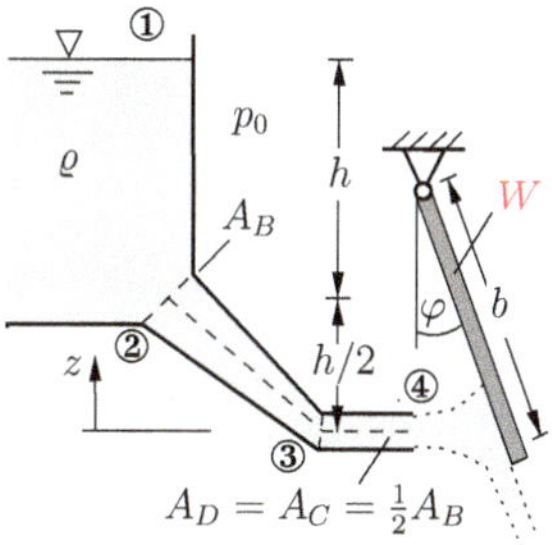

Solution **a)** The cross section of the drainpipe ②③ is given by

$$A(z) = 2(A_2 - A_3)\,z/h + A_3 \; .$$

The continuity equation for an arbitrary point in the range ②③ and the point ③ reads $A(z)\,v(z) = A_3\,v_3$. The velocity at point ③ is $v_3 = v_4 = \sqrt{3gh}$. Thus, the velocity in the range ②③ follows as

$$v(z) = \frac{hA_3}{2(A_2 - A_3)z + hA_3}\,\sqrt{3gh} \; .$$

BERNOULLI's theorem

$$\tfrac{1}{2}\varrho v_A^2 + p_0 + \tfrac{3}{2}\varrho gh = \tfrac{1}{2}\varrho v(z)^2 + p(z) + \varrho gz$$

between point ① and a point in the range ②③ leads with $v_A = 0$ (big reservoir) to the pressure in the drainpipe:

$$\underline{\underline{p(z) = p_0 + \frac{3}{2}\varrho gh\left[1 - \frac{h^2}{(2z+h)^2}\right] - \varrho gz}} \; .$$

b) Using the sketched control volume, the momentum balance in the direction of the normal force N exerted to the plate yields

$$\varrho Q(0 - v_4\cos\varphi) = -N \; .$$

With the volume flow $Q = A_4\,v_4$ follows

$$N = \varrho\,A_4 v_4^2\cos\varphi \; .$$

Finally, from the equilibrium condition

$$\overset{\curvearrowleft}{B}: \quad N\frac{a}{\cos\varphi} - W\frac{b}{2}\sin\varphi = 0$$

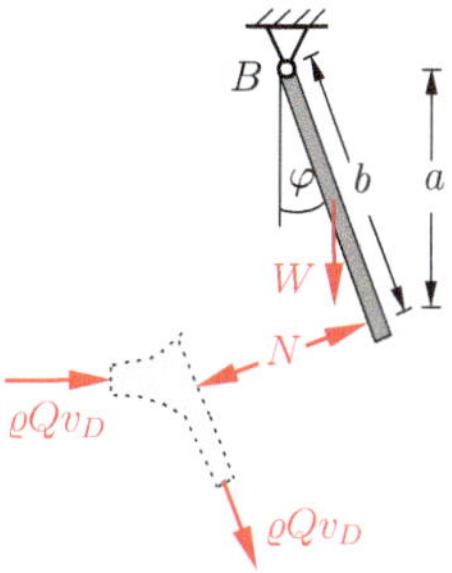

the angle φ is determined: $\quad \underline{\underline{\sin\varphi = \frac{6\varrho g A_4 a h}{Wb}}}$.

P10.10 **Problem 10.10** A trapezoidal tank of rectangular cross section ($b_1 \times b/2$) is filled by a pump via a pipe. The pump produces a constant volume flow Q and the pipe has the cross section $b^2/10$.

a) Determine the rise velocity of the fluid in the tank.

b) Calculate for $z = h/2$ the resulting force exerted on the pipe.

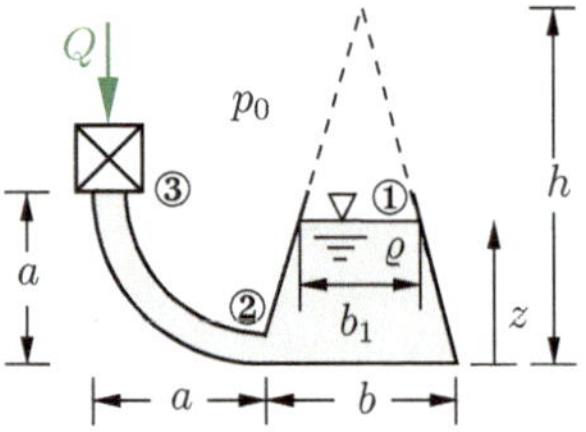

Solution **a)** With the varying width

$$b_1(z) = \frac{b}{h}(h - z)$$

of the trapezoidal tank the fluid surface is given by

$$A_1(z) = b_1(z)\,\frac{b}{2} = \frac{b^2}{2h}(h - z)\ .$$

Thus, due to the constant volume flow Q, the velocity in ① follows as

$$\underline{\underline{v_1(z) = \frac{Q}{A_1(z)} = \frac{2\,Qh}{b^2(h - z)}}}\ .$$

b) The force exerted by the fluid on the pipe can be determined from the momentum balance. For this purpose, we first calculate for $z = h/2$ the pressures and velocities at the points ② and ③ by using BERNOULLI's theorem and the continuity equation.

For point ② follows from BERNOULLI's theorem between ① and ②

$$\frac{1}{2}\varrho\, v_1^2 + p_0 + \varrho\, g\,\frac{h}{2} = \frac{1}{2}\varrho\, v_2^2 + p_2 + 0\,,$$

and from the volume flow

$$Q = v_2\, A_2 = v_2\,\frac{b^2}{10}$$

follows the velocity

$$v_2 = \frac{10\,Q}{b^2}\ .$$

Thus, we obtain for the pressure

$$p_2 = p_0 + \varrho\, g \frac{h}{2} - \varrho\, \frac{42\, Q^2}{b^4}\ .$$

From the continuity equation

$$Q = v_3\, A_3 = v_2\, A_2$$

between points ② and ③ follows

$$v_3 = v_2 = \frac{10\, Q}{b^2}\ .$$

Thus, BERNOULLI's theorem

$$\frac{v_3^2}{2g} + \frac{p_3}{\varrho\, g} + a = \frac{v_2^2}{2g} + \frac{p_2}{\varrho\, g} + 0$$

between points ② and ③ yields the pressure at point ③ as

$$p_3 = p_0 + \varrho\, g \left(\frac{h}{2} - a\right) - \varrho\, \frac{42\, Q^2}{b^4}\ .$$

As control volume for the balance of momentum, we now choose the fluid within the pipe. It is loaded by its weight, by the pressure forces at ② and ③ and by the forces R_x, R_y exerted from the pipe-wall. The opposite forces are exerted from the fluid to the pipe-wall. Thus, the balance of momentum reads in components

$$\rightarrow:\quad \varrho\, Q(v_2 - 0) \qquad = -p_2 A_2 + R_x\ ,$$

$$\uparrow:\quad \varrho\, Q(0 - (-v_3)) = -p_3 A_3 + R_y - \varrho\, g V\ .$$

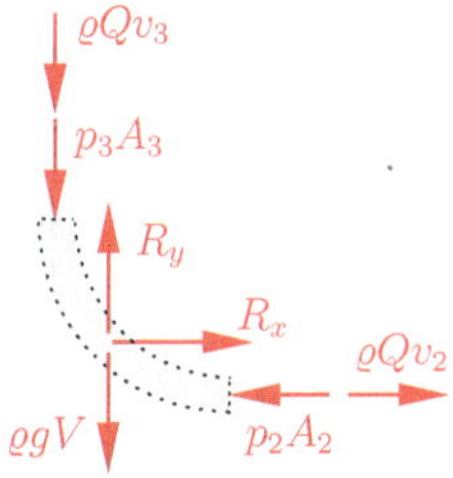

Introducing the pressures and velocities yields with the fluid volume $V = \pi a b^2/20$ in the pipe the force components exerted on the pipe-wall

$$\underline{\underline{R_x = \frac{b^2}{10}\left(p_0 + \varrho\, g \frac{h}{2}\right) + \frac{29}{5}\, \frac{\varrho\, Q^2}{b^2}}}\ ,$$

$$\underline{\underline{R_y = \frac{b^2}{10}\left(p_0 + \varrho\, g \frac{h}{2} - \varrho\, g a\right) + \frac{29}{5}\, \frac{\varrho\, Q^2}{b^2} + \varrho\, g \frac{\pi a b^2}{20}}}\ .$$

P 10.11

Problem 10.11 In a horizontal plane, a fluid jet hits with the velocity v_0 a wall under the angle α. The jet depth in vertical direction is constant: $h = \text{const}$.

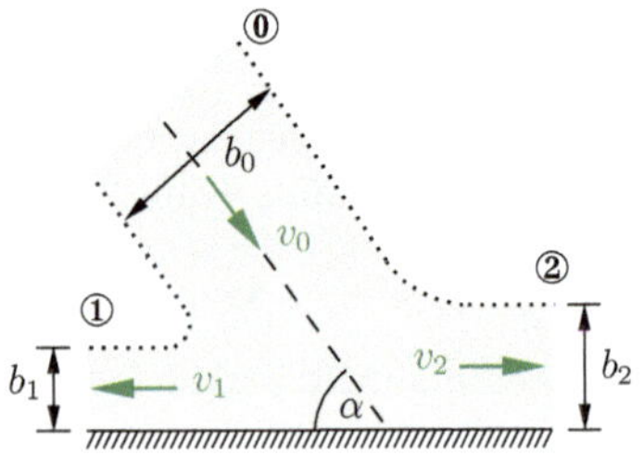

a) Determine the velocities v_1 and v_2 of the two off-flowing jets.

b) Calculate the width b_1 and b_2 of the off-flowing jets.

c) Determine the normal force exerted on the wall.

Solution **a)** The velocities v_1 and v_2 can be determined from BERNOULLI's theorem applied to streamlines between the points ⓪ and ① and between the points ⓪ and ②:

$$\frac{1}{2}\varrho v_0^2 + p_0 = \frac{1}{2}\varrho v_1^2 + p_0 \quad \rightsquigarrow \quad \underline{\underline{v_1 = v_0}} \,,$$

$$\frac{1}{2}\varrho v_0^2 + p_0 = \frac{1}{2}\varrho v_2^2 + p_0 \quad \rightsquigarrow \quad \underline{\underline{v_2 = v_0}} \,.$$

b) From the continuity equation follows with v_1 and v_2 the relation between the jet-widths:

$$v_0 b_0 h = v_1 b_1 h + v_2 b_2 h \quad \rightsquigarrow \quad b_0 = b_1 + b_2 \,.$$

The balance of momentum in wall direction

$$\rightarrow: \quad \varrho Q_2 v_2 - \varrho Q_1 v_1 - \varrho Q_0 v_0 \cos\alpha = 0$$

yields with $Q_0 = b_0 h v_0, \; Q_1 = b_1 h v_1, \; Q_2 = b_2 h v_2$

$$b_0 \cos\alpha = b_2 - b_1 \,.$$

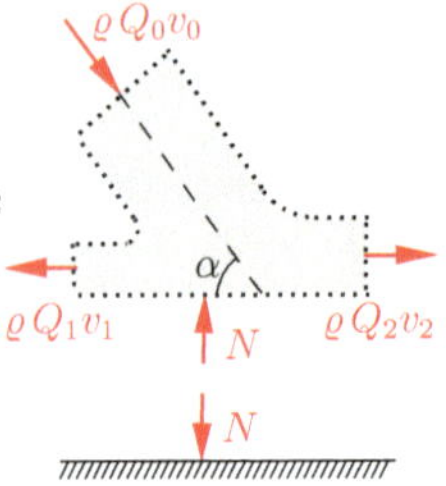

This provides the jet-widths

$$\underline{\underline{b_1 = \frac{1}{2} b_0 (1 - \cos\alpha)}} \,, \qquad \underline{\underline{b_2 = \frac{1}{2} b_0 (1 + \cos\alpha)}} \,.$$

c) The normal force exerted on the wall is directly obtained from the balance of momentum perpendicular to the wall:

$$\uparrow: \quad \varrho Q_0 v_0 \sin\alpha = N \quad \rightsquigarrow \quad \underline{\underline{N = \varrho v_0^2 \, b_0 h \sin\alpha}} \,.$$

Problem 10.12 A horizontally placed bend of a pressure pipe is held by a concrete block B. **P10.12**

Determine the horizontal and vertical force component exerted from the bend to the concrete block.

Given: v_1, p_1, A_1, A_2, α

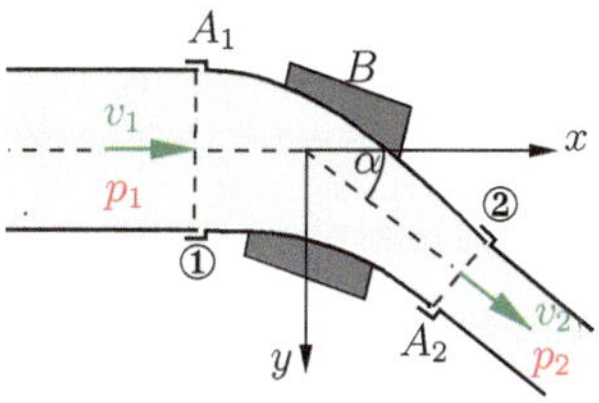

Solution Using the continuity equation, the outflow velocity v_2 can be determined:

$$Q = A_1\, v_1 = A_2\, v_2 \quad \rightsquigarrow \quad v_2 = \frac{A_1}{A_2}\, v_1\,.$$

The pressure at location ② follows from BERNOULLI's theorem applied to a streamline between points ① and ②:

$$\frac{1}{2}\varrho\, v_1^2 + p_1 = \frac{1}{2}\varrho\, v_2^2 + p_2 \qquad \rightsquigarrow \qquad p_2 = p_1 + \frac{\varrho}{2} v_1^2 \left[1 - \left(\frac{A_1}{A_2}\right)^2\right]\,.$$

The fluid within the bend (control volume) is loaded by the forces R_x, R_y (exerted by the bend-wall) and the pressure forces at ① and ②. Thus, the balace of momentum $\boldsymbol{F} = \varrho\, Q(\boldsymbol{v}_2 - \boldsymbol{v}_1)$ reads in components

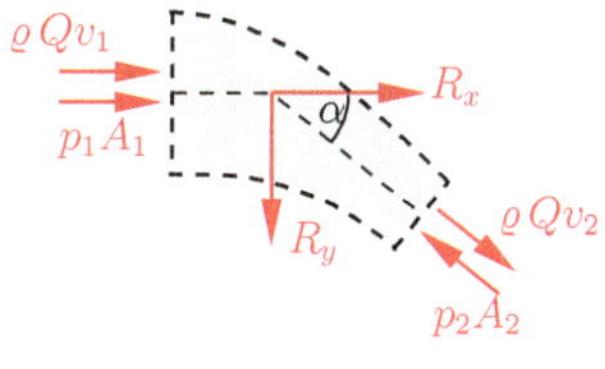

$$\rightarrow: \quad \varrho\, Q(v_2 \cos\alpha - v_1) = p_1 A_1 - p_2 A_2 \cos\alpha + R_x\,,$$

$$\downarrow: \quad \varrho\, Q(v_2 \sin\alpha - 0) = \qquad - p_2 A_2 \sin\alpha + R_y\,.$$

This leads to the forces

$$\underline{\underline{R_x = A_1 \left\{\frac{\varrho}{2} v_1^2 \left[-2 + \left(\frac{A_1}{A_2} + \frac{A_2}{A_1}\right)\cos\alpha\right] - p_1\left(1 - \frac{A_2}{A_1}\cos\alpha\right)\right\}}}\,,$$

$$\underline{\underline{R_y = A_1 \left\{\frac{\varrho}{2} v_A^2 \left(\frac{A_1}{A_2} + \frac{A_2}{A_1}\right) + p_1 \frac{A_2}{A_1}\right\}\sin\alpha}}\,.$$

Because of the equilibrium conditions these forces must be carried by the concrete block.

P 10.13 **Problem 10.13** A pipeline in a plane splits into two line sections which then again merge. In the line ①, a globe valve G (pressure-loss coefficient ζ_1) is built in, while in line ②, a clack valve C (pressure-loss coefficient ζ_2) is present. The total volume flow throug the pipe system is given by Q.

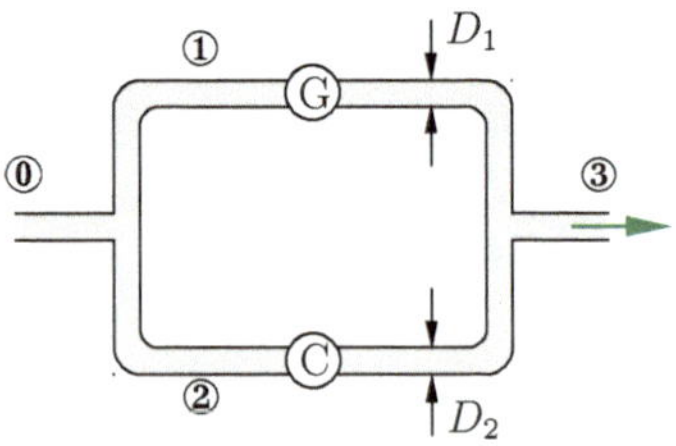

a) Calculate the volume flows in the pipes ① and ②.

b) Determine the pressure loss (related to ρg) between inflow and off-flow.

Given: $D_1 = 1.4\,\text{m}$, $D_2 = 0.8\,\text{m}$, $Q = 5.0\,\text{m}^3/\text{s}$, $\zeta_1 = 1.3$, $\zeta_2 = 0.3$.

Solution **a)** The generalized BERNOULLI's theorem between points ⓪ and ③ (via pipe ① and via pipe ②) yields

$$\left.\begin{aligned} \tfrac{1}{2}\varrho v_0^2 + p_0 &= \tfrac{1}{2}\varrho v_1^2 + p_1 = \tfrac{1}{2}\varrho v_3^2 + p_3 + \Delta p_{v_1} \\ \tfrac{1}{2}\varrho v_0^2 + p_0 &= \tfrac{1}{2}\varrho v_2^2 + p_2 = \tfrac{1}{2}\varrho v_3^2 + p_3 + \Delta p_{v_2} \end{aligned}\right\} \leadsto \quad \begin{aligned} \Delta p_{v_1} &= \Delta p_{v_2} \,, \\ \zeta_1 \frac{v_1^2 \varrho}{2} &= \zeta_2 \frac{v_2^2 \varrho}{2} \,. \end{aligned}$$

With the volume flow $Q = vA = vD^2\pi/4$ follows

$$\left(\frac{\zeta_1}{\zeta_2}\right) = \left(\frac{v_2}{v_1}\right)^2 = \left(\frac{Q_2}{Q_1}\right)^2 \left(\frac{D_1}{D_2}\right)^4 .$$

Using the continuity equation $Q = Q_1 + Q_2$ leads to the volume flow in pipe ①:

$$\underline{\underline{Q_1}} = \frac{Q}{1 + \sqrt{\frac{\zeta_1}{\zeta_2}}\left(\frac{D_2}{D_1}\right)^2} = \frac{5}{1 + \sqrt{\frac{1.3}{0.3}}\left(\frac{0.8}{1.4}\right)^2} = \underline{\underline{2.98\,\text{m}^3/\text{s}}} .$$

Thus, the volume flow in pipe ② is given by

$$\underline{\underline{Q_2}} = Q - Q_1 = 5 - 2.98 = \underline{\underline{2.02\,\text{m}^3/\text{s}}} .$$

b) From the volume flows the flow velocities are determined as

$$v_1 = \frac{4\,Q_1}{D_1^2\,\pi} = \frac{4 \cdot 2.98}{1.4^2\,\pi} = 1.94\,\frac{\text{m}}{s} , \qquad v_2 = \frac{4\,Q_2}{D_1^2\,\pi} = \frac{4 \cdot 2.02}{0.8^2\,\pi} = 4.02\,\frac{\text{m}}{s} .$$

Herewith, the (related) pressure loss can be calculated:

$$\underline{\underline{\frac{\Delta p_{v_1}}{\varrho\, g} = \frac{\zeta_1\, v_1^2}{2g} = 0.25\,\text{m}}} , \qquad \underline{\underline{\frac{\Delta p_{v_2}}{\varrho\, g} = \frac{\zeta_2\, v_2^2}{2g} = 0.25\,\text{m}}} .$$